The Invisible Empire

The Invisible Empire

A History of the
Telecommunications Industry
in Canada, 1846–1956

JEAN-GUY RENS

Translated by Käthe Roth

McGill-Queen's University Press
Montreal & Kingston · London · Ithaca

Legal deposit second quarter 2001
Bibliothèque nationale du Québec

Printed in Canada on acid-free paper

This book is a revised and updated version of *L'Empire
invisible: histoire des télécommunications au Canada de
1846 à 1956*, published by Presses de l'Université du
Québec in 1993.

Translation of this book was made possible by a grant
from the Translation Program of the Canada Council for
the Arts. We also acknowledge the generous support of
the Canada Council for the Press's trade publishing
program.

The author acknowledges the financial support of the
Government of Canada through Heritage Canada and
Industry Canada.

McGill-Queen's University Press acknowledges the
financial support of the Government of Canada through
the Book Publishing Industry Development Program
(BPIDP) for its activities.

Canadian Cataloguing in Publication Data

Rens, Jean-Guy, 1946-
    The invisible empire: a history of the telecommunications
    industry in Canada
    Translation of: L'empire invisible.
    Includes bibliographical references and index.
    Contents: [1]. 1846-1956.
    ISBN 0-7735-2052-X
    1. Telecommunication – Canada – History. I. Roth, Käthe
    II. Title.
    HE7814.R4613 2001   384'.0971   C00-900606-0

This book was typeset by Typo Litho Composition Inc.
in 10/12 Times.

To Huguette Guilhaumon

Contents

Acknowledgments

Every book is the individual expression of a collective effort. Even though *The Invisible Empire* represents only my opinion and is my sole responsibility, it would not have been possible to write this book without the assistance, support, and advice of a small group of patient and devoted friends.

First, I would like to thank Huguette Guilhaumon, former vice-president of Media Relations at Burson-Marsteller, now senior partner at Science Tech for her constant support and her many readings of the manuscript at various stages of its development. Her vision and tenacity are well represented here. She was the first one to offer advice but not the only one.

The following people also read all or part of *The Invisible Empire,* for which I thank them (in alphabetical order): Robert J. Chapuis, former senior advisor to the CCITT at the ITU; author of *100 Years of Telephone Switching* and *Electronics, Computers, and Telephone Switching*; Claire Poitras, researcher at INRS–Urbanization and author of a PhD dissertation on the social impact of telephone networks on urbanization; Ivo Rens, professor of the history of political doctrine, Faculty of Law, University of Geneva; Stanley Swihart, president, Telephone History Institute; author of *Telephone Dials and Push Buttons.*

No book of this sort would be complete without acknowledgment of the support of many researchers who took time to answer my questions. I thank John Barry, former regulations specialist at Bell Canada; Pierre Dion, former librarian at Bell; and Ed Toombs, his successor.

I also thank the staff at Bell Canada's History department for their help, in particular Anna Supino and Lise Noël, who patiently found, photocopied, and sometimes searched a second time for the many documents that were used in the writing of this book.

The following people also did research or facilitated access to sources of information, meriting special mention: Claude Beauregard, former assistant vice-president, Public Relations, Bell Canada; Stephanie Boyd, director, Special Information Centre, Bell Canada; Richard French, former vice-president, Bell Canada; Monique Perrier, former head of the central library, Communications Canada (now Industry Canada); Stephanie Sykes, former director, History department, Bell Canada; A.G. El-Zanati, former head of the central library, documentation, and archives, ITU.

Finally, this book is based on interviews with some of the main actors in the telecommunications world: Tony Cashman, former historian at AGT; author, *Singing Wires*; Alex G. Lester, former executive vice-president, Bell Canada; Vernon O. Marquez, former president, Nortel; Cyril A. Peachey, former executive vice-president, Nortel (then Northern Electric); and Tony Zeitoun, Telecommunications Sector, Canadian International Development Agency.

Particularly valuable was Alex Lester, one of the greatest Canadian engineers who, through his very special combination of high-mindedness and practicality, provided a clear look at many events of the past. Unfortunately, Lester has died and thus cannot judge the result of his advice; I therefore hope his family will accept my gratitude.

Revisions to the text to bring the material in the original French version up to date were made possible by the backing of Heritage Canada and Industry Canada.

Illustrations from Bell Canada were graciously provided by the Bell Canada Historical Telephone Collection; those from AT&T by the AT&T Archives. Illustrations from other firms – BC Tel, Cantel, Québec-Téléphone, Bell-Northern Research, Nortel Networks, MT&T, Rogers Communications, Videotron, Telesat Canada, Spar Aerospace, Unitel, Skywave Electronics, MCI Communications, and Teleglobe Canada – were graciously provided by the Public Relations departments of the respective firms.

Note. The opinions and interpretations in this book are mine and can in no case be attributed to the people mentioned above, nor to the institutions at which they work or worked. Any errors contained herein are my responsibility alone.

Abbreviations

CLC   Canadian Labour Congress
CMS   Call Management Services
CN   Canadian National
CNCP   Canadian National/Canadian Pacific
COTC   Canadian Overseas Telecommunications Corporation
CP   Canadian Pacific
CRBC   Canadian Radio Broadcasting Commission
CRC   Communications Research Council
CRTC   Canadian Radio-Television and Telecommunications Commission
(formerly Canadian Radio-Television Commission)
CTC   Canadian Transport Commission
CTEA   Canadian Telephone Employees' Association
CTUA   Commercial Telegraphers' Union of North America
CTV   Canadian Television Network
CWA   Communications Workers of America
CWC   Communications Workers of Canada
DBS   Direct-broadcast satellite
DDD   Direct Distance Dialing
DEW   Distant Early Warning
DMS   Digital Multiplex System
DRTE   Defence Research Telecommunications Establishment
DTH   Direct-to-home (satellites)
DTS   Dominion Telephone System
EDI   Electronic data interchange
FCC   Federal Communications Commission (United States)
FTW   Federation of Telephone Workers of British Columbia
GFO   Government Finance Office (Saskatchewan)
GTE   General Telephone and Electronics Corporation
HDTV   High-definition television
IBEW   International Brotherhood of Electrical Workers
ICAC   Imperial Communications Advisory Committee
IEEE   Institute of Electrical and Electronic Engineers
IFRB   International Frequency Registration Board
ILO   International Labour Organization
INMARSAT   International Maritime Satellite Organization
INRS   Institut national de la recherche scientifique
INTELSAT   International Telecommunications Satellite Organization
IRCC   International Radio Consultative Committee
ISAT   International Society of Aeronautical Telecommunications
ISDN   Integrated Services Digital Network
ISS   International Switching Symposium
ITT   International Telephone and Telegraph
ITU   International Telegraph Union

ITU   International Telecommunication Union
LMCS   Local Multipoint Communication System
LMS   Local measured service
MCI   Microwave Communications Inc.
MIT   Massachusetts Institute of Technology
MGT   Manitoba Government Telephones
MTS   Manitoba Telephone System
MT&T   Maritime Telegraph and Telephone
NAC   National Archives of Canada
NBEW   National Brotherhood of Electrical Workers
NTI   Northern Telecom Inc.
NTSC   Northern Telecom System Corporation
NTT   Nippon Telegraph and Telephone
NORAD   North American Air Defence Agreement
NTC   National Telephone Company (Britain)
NWS   North Warning System
OECD   Organization for Economic Cooperation and Development
PBX   Private branch exchange
PCS   Personal Communications Systems
PTT   Postal, Telegraph, and Telephone
RCA   Radio Corporation of America
SITA   Société internationale de télécommunications aéronautiques
SONET   Synchronous Optical Network
SGT   Saskatchewan Government Telephone
SONET   Synchronous Optical Standards
SRCI   Stentor Resource Centre Inc.
SWIFT   Society of Worldwide Interbank Financial Telecommunications
TAC   Telephone Association of Canada
TAP   Terminal Attachment Program
TAPAC   Terminal Attachment Program Advisory Committee
TCTS   Trans-Canada Telephone System
TLC   Trades and Labour Congress
TWU   Telecommunications Workers Union
VMS   Voice Message Service
WAP   Western Associated Press
WARC   World Administrative Radio Congress
WLRB   War Labour Relations Board
WTO   World Trade Organization

Introduction

The air we breathe is awash in electromagnetic waves; under the streets we walk flows a torrent of underground copper and fibre-optics cables. The telecommunications network is everywhere, most of it beyond our purview. Yet its technology pervades our workdays, our evening entertainment – in short, our entire lives. It is an invisible empire ruling over our information society.

This "invisible empire" woven into our environment is not new. It was born in 1844, with the invention of the telegraph. Is there any technology more obsolete than telegraphy? Yet the telegraph marked the arrival of "real time" in the world. For the first time, communications were transmitted at the speed of electricity – the speed of light. Today, we have a hard time imagining how shocking this instantaneousness was.

The "global village" came into being in 1866, when North America was linked to Europe; from one day to the next, news, financial transactions, and government dispatches could suddenly be sent around the world in a moment. Colonial empires were built on telegraphy. Corporations extended their influence to the clicking rhythms of the telegraph. Wars were directed long distance by telegraph. The crushing of Louis Riel's Métis revolt in western Canada was conducted live from Ottawa.

Today, it is *trendy* to talk about the acceleration of history, and even of "Internet time." But we forget that Marconi invented the radio in 1896 and founded his first company the following year; by 1903, his monopoly over radiotelegraphy was so powerful that an international conference was convened in Berlin to put "Marconism" on trial. It took three times as long (from 1976 to 1999) for Bill Gates to be accused of monopolistic practices by the United States Department of Justice.

The birth of telecommunications (telegraphy, telephony, radio) was utterly intertwined with the convulsive pace of the electrical world. Fortunes were amassed in record time by young men: Bell was just twenty-nine when he invented the telephone, and Marconi was twenty-two when he invented the radio. They were not exceptions to the rule. A generation of young "techno-freaks" made telephones and radios by hand, improved them, and marketed them in search of a "killer application." In 1897 Dr Demers was twenty-six when he founded a telephone company in the Lower St. Lawrence that for a time competed with, and even rivalled, Bell. His secret was to involve the operators in profit sharing. The company founded by Dr Demers, now called Québec Tel, still exists. The Lorimer brothers were barely twenty in 1901 when they developed the switching system that bears their name, and they set up their plant in Peterborough. This system, which memorizes phone numbers to accelerate circuit selection, was the basis for all modern telecommunications. Fessenden was thirty-five when he made radio talk in 1901. Instead of simply routing telegraph signals, radio became the broadcast vector for news, entertainment, and music. Thus, the modern world was born in the minds of a generation of passionate young men, many of them self-taught, who were ready to defy the wisdom of established science and risk mockery by their elders.

This tumultuous explosion of creativity is largely forgotten today. People living at the end of the twentieth century considered telecommunications a doddering industry in the hands of obsolescent monopolies. We always see photographs of Bell as a grandfather with a flowing white beard, so it is difficult to remember that he and his assistant, Watson, were young men when they developed the telephone. It is these adventurous pioneers whom we shall rediscover in this book.

When I began to write *The Invisible Empire,* my goal was to return to the source of the second industrial wave, which many authors feel bears a great resemblance to the Internet explosion of the 1990s. More specifically, this book was born of an exploration of what propelled Canada into the forefront of telecommunications. Today, as a new millennium dawns, symbolized by the Internet, the issue seems even more important. Will telecommunications remain a significant industry in the information economy? Will Canada be a major player in the new power structure that is being formed?

There is another, more personal reason why I decided to write this book. I worked for about ten years at Bell Canada, and for another ten years as a consultant in telecommunications and new media, and I have come to know the telephone business from the inside. In North America, a telephone industry has existed at least since the turn of the twentieth century, with its own language, common values, meaningful silences, heroes, and exploits – in short, its own culture.

Beyond the image of the bureaucratic monopoly that has often been justly criticized, it is important to understand what enabled this industry to shape the modern economy as it has. Three values are central to telephone people: network reliability, public service, and a fair return on shareholders' investment – in no particular order. I know that critics of telecommunications companies would say that profit is the engine for these institutions. It is true. Telephone service was, in fact, one of the few public services in North America that for over a century didn't lose money. Every decision was scrutinized for its profitability.

Network reliability, public service, and a fair profit – all of these concepts are collective, overarching, and evolutionary, and this has made telephone people "conservative." Introducing a new technology into the telecommunications network makes no sense unless it can be integrated into an overall plan, so that it doesn't interrupt the service of entire groups of subscribers or require all older equipment to be replaced. Innovation must always be submitted to the imperatives of continuity.

The introduction of competition in telecommunications and the eruption of the Internet have called this carefully planned evolution into question. The structural upheaval that has resulted seems to have returned the industry to the very sources from which it sprang. And that is perhaps why this book has an unexpected currency aside from its historical aspect. To understand what will replace the monopoly held up so recently by telecommunications carriers, we must understand the nature of their monopolistic organization, of course, but also have a look at the tumultuous origins of the industry.

To evaluate whether telecommunications companies will succeed in their cultural revolution, we must understand the nature of these firms. Little has been written on the subject in Canada. The few books that have been published deal with sectoral aspects, generally legal and regulatory. There are two historiographies, both funded by telecommunications companies, but their focus is corporate, and they are more than twenty years old – an eternity in this effervescent sector.[1] A more recent book, written by someone who is systematically contemptuous of telecommunications companies, reflects his prejudices and not reality; after all, the fact is that Canadian telephony has been a success, not a failure.[2] More interesting are journalist Lawrence Surtees's books on Jean de Grandpré, president of Bell and founder of BCE,[3] and on the advent of competition in the long-distance market.[4]

I should warn readers that *The Invisible Empire* is no friend to ideology. Although private initiative is one of my explicit themes, its shortcomings will be clearly pointed out; on the other hand, when governments accomplish something positive, I recognize it fully. (We shall see considerable government achievement in the Prairie provinces.) Any ideology interprets reality through a particular analytic grid, reducing to a dull abstraction the iridescent world that shimmers, bright and furtive, before our eyes. What I have tried to do is uncover the facts and hold them up so that they illuminate the evolution of the telecommunications industry.

In *The Invisible Empire*, facts are judged according to three objective parameters: the penetration rate of telephones in society, the cost of telephone service, and the quality of life of telephone-company employees. Canada has always had one of the highest telephone-penetration rates in the world. Residential service rates in Canada have been (and still are) among the lowest in the world; in particular, they were lower on average than those in the United States. On the social side, working conditions for telephone employees in Canada were generally among the best in the world.

There is a fourth, very different criterion: national identity. Canada was built around east–west infrastructures, countering the natural north–south tendency. *The Invisible Empire* judges Canadian telecommunications as a function of this parameter, which, though subjective, is always invoked by the builders of the industry. It must be stated that Canada has existed as an independent nation in North America since 1867, in spite of its injustices, fragility, and unsettled issues.

This book is not an academic work. I am a fieldworker, and I want to provoke thought about an industry of growing importance. It is my hope to encourage dialogue between the corporate and academic worlds and help, in my fashion, to forge links between these two great sectors of human activity.

Because of the extreme compartmentalization of Canadian telecommunications, I have highlighed certain aspects of the industry. While I am aware that such an undertaking may be incomplete, I wanted to offer an overview, looking at the service aspect as well as manufacturing; reviewing developments province by province; examining regulatory, technological, and power issues; and integrating the oft-neglected subject of labour relations. Any part of this book could give rise to an in-depth study: my only aim is to provide a survey of the industry's main points and encourage other researchers to take this work further.

The Bell group is at the centre of the entire Canadian telecommunications industry; along with its associated telecommunications carriers, Bell accounts for 60 percent of Canadian telecommunications, not including manufacturing, and its dominance is only reflected in this book. The company evolved in three distinct phases, so I have divided the history of telecommunications in Canada into three periods, with their own values and the few links between them. This book deals with the first two periods; a second volume will deal with the contemporary period.

First, from 1880 to 1915, under the long reign of Charles Fleetford Sise, a dynamic but stubborn autocrat, Bell acted as a predator, trying to construct a monopoly as cheaply as possible with moves typical of the unfettered capitalism of the nineteenth century. Bell failed to reach this goal and barely avoided nationalization. The technological basis of the company was the manual exchange. There was no network as such but a constellation of exchanges.

The second period in Canadian telecommunications was marked by the triumph of the ideas of Theodore Vail, the president of AT&T during the preceding period. Vail's great vision – universality of telephone service – was attained in fifty years. The telephone companies' monopoly was counterbalanced by a policy of transparency regarding the public and the state, and by social benefits for employees that were clearly ahead of their time. Vail's ideas were introduced to Canada by Bell, which imposed technological and managerial standards throughout the industry through an association of companies. This wide-ranging policy, however, kept Canada within the American sphere of influence.

The book ends in 1956: why this year in particular, and not, for instance, the end of the Second World War? Because 1956 was a crucial year for Canadian telecommunications. First, it was in this year that the Consent Decree between AT&T and the American Department of Justice, which terminated the privileged links that kept Canadian main manufacturer Northern Electric dependent on the American manufacturer Western Electric. Just as important, 1956 was when the first transcontinental telephone cable between North America and Europe was put into service; Newfoundland, where the cable landed on this side of the Atlantic, became the gateway to overseas telecommunications. Telephony moved from regional to global. Finally, by this period (the late 1950s), telephones were to be found in all households, thus becoming truly universal in Canada.

This book asks questions that resonate in today's world. Far from reducing history to a cyclical recitation of events, it seeks to bring to light what is significant about the past. Often, the most complex situation started with a simple idea. By returning to the source of what has become an unending flood of complexity, we may find this lost simplicity.

PART ONE

The Pioneering Era: Inventions and Impediments, 1846–1915

In Canada, the first age of telecommunications began in 1846, when telegraphy linked Toronto to the United States, and ended in 1915, when the founder of the Canadian telephone industry, Charles Sise, retired. It was also in 1915 that telephone technology conquered distance: AT&T linked New York to San Francisco through its long-distance network and to Paris by radio-telephone. It was one of those extraordinary moments in history when the end of one era and the vibrant beginnings of another converged.

During this period, a new industry was built. Firms sprang up with little capital but much enthusiasm and many good ideas. A spider's web of telegraph cables, then telephone wires, spread throughout the continent. In this epoch of unbridled capitalism, rates were ratcheted up if the potential customers' ability to pay permitted, and lowered when competition had to be nipped in the bud – right down to free telephone service, when necessary. Phone company employees, especially the women, were treated like slaves: they were poorly paid and fired at the slightest threat of a strike.

It was a tough period but not without its triumphs. Telegraphy turned the traditional view of world geography on its head with the transatlantic and trans-Canada cables. Telephony progressed more slowly, but its effects were felt in the privacy of the home. Telegraph machines and telephones were the first widespread electrical apparatuses, and their advent permanently changed the industrial order.

SECTION ONE

Telegraphy

Telegraphy did not materialize out of thin air: it came into being as a function of the increasing speed of various modes of transport. When a train rolled on steel rails at forty-five kilometres per hour, it seemed the height of speed, but if it arrived at its destination at the same time as the news of its departure, how could train traffic be regulated?

Newspapers were also searching for instantaneous means of communications. An independent United States and a post-revolutionary France had spread the idea of democracy. The press had played a motivating role in promulgating the exciting new ideology, and now it needed a supply of timely news.

Finally, reinforcement of social organization was needed to establish standardization of behaviours. Companies were growing into giant corporations, armies now comprised hundreds of thousands – even millions – of men, and government was beginning to grow. These expanding institutions had a voracious appetite for information.

Telegraph networks were the arteries of this new, modern world. In fact, modernity for Canada was ushered in with, first, the transatlantic cable in 1866, which provided access to Europe, and then the trans-Canada cable in 1886, which provided access to all of North America. Thus, telegraphy went hand in hand with Canada's emancipation from colonialism.

1   The Birth of Telegraphy

The concept of telegraphy, along with the word, was born at the end of the eighteenth century to designate the transmission of text by semaphore. There were no technical innovations involved: towers were constructed within visual range of each other so that the signalmen could read the transmitted messages. From a technological point of view, this was a pre-industrial invention.

For many years, particularly in the navy, optical signals were used to transmit messages. The novelty of this telegraphy was in its "software": ideographic symbols (for example, white-and-blue checkerboard flag for distress calls, black flag for shipwrecks) were replaced by alphabet letters, allowing for transmission of more complex messages.

A number of semaphore experiments were being conducted at the end of the eighteenth century, but the country that developed the most sophisticated system was France, where Claude Chappe convinced the revolutionary government to build a 230-kilometre line between Paris and Lille. The purpose of the line, inaugurated in August 1794, was to keep the government informed of military operations in northern France. It worked, but it was expensive. Operating and maintaining the semaphore lines required large numbers of people scattered along their considerable length. Nevertheless, the national network eventually comprised up to five thousand kilometres of line, and it was in operation until 1859.[1]

Meanwhile, in what would later be Canada, an original but short-lived experiment was being conducted. Two semaphore lines were erected, one

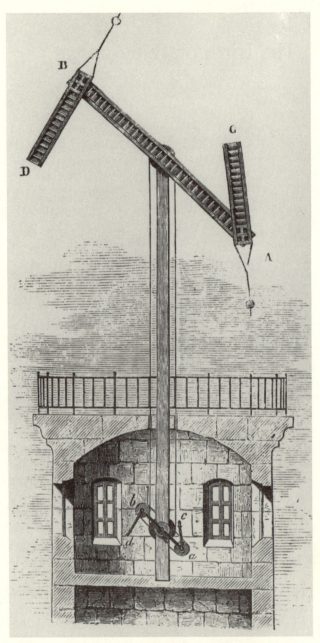

Chappe's optical telegraph was one of the key instruments in revolu-
tionary France's victory over the European coalition in the 1790s.
Napoleon then used the apparatuses to run his empire. Courtesy Parks
Canada.

between Halifax and Annapolis, Nova Scotia, a distance of about 210 kilometres, and the other between Saint John and Fredericton, New Brunswick, a distance of almost 130 kilometres, and ships ferried between them. These lines also had a military vocation: when they were inaugurated in 1798, the corsairs of the new French republic were presenting some threat to British North American settlements.

The commander-in-chief of the Nova Scotia army, Prince Edward,[2] was responsible for the project. A driven, meticulous man, Prince Edward quickly realized the potential of the semaphore system, and he used it to control with an iron fist everything that went on in the colony. A letter from one of his officers tells about this strange expansion of royal power: "The duke returned on Saturday, what he has been about so much longer than he had proposed I have not heard, but I am told they have established telegraphs all the way to Annapolis, so that there was a continual communication kept up of ordering and counter orders while he was away even to the approval of courts martial and ordering the men to be flogged."[3] Prince Edward was not content, however, with conducting floggings by semaphore. He had a real sense of government and he was planning to extend the semaphore line to Quebec City via Baie des Chaleurs. His project called for construction of a line of wooden towers, each topped with a mast and spaced ten or twelve kilometres apart, and for the forest to be cleared between them so that visibility would not be hindered. Six men would be needed per semaphore tower to ensure round-the-clock operation. Unfortunately, British North America lacked the funds and the men to bring such a project to fruition. When Prince Edward left Halifax in 1800, Canadian telegraphy reverted to a purely military vocation, and the Peace of Amiens, concluded in 1802 between France and Great Britain, removed any justification for its existence. Both lines were abandoned, and even a renewal of hostilities wasn't enough to revive interest in a project that had been woven out of whole cloth in an extravagant princely mind.

The details of the code used by the Canadian semaphore system have been lost. It seems that the system involved balls hung from a pole with two arms in the shape of the cross of Lorraine. The letters of the alphabet were assigned numbers, and the numbers were represented by the position of the balls on the arms of the cross. Thus, it was a little more complex than Chappe's system, but it was clearly related to telegraphy and not to a simple exchange of ideographic signals.

Semaphores continued to find limited use in Canadian ports after Prince Edward left. Halifax, Saint John, and Quebec City used semaphore systems to signal the arrival of ships until the 1840s. These systems were used for both civil and military purposes, and conflicts arose over the cost of maintaining them. Halifax ended up with three parallel semaphore systems: one each for the army, the navy, and commercial sea traffic.[4]

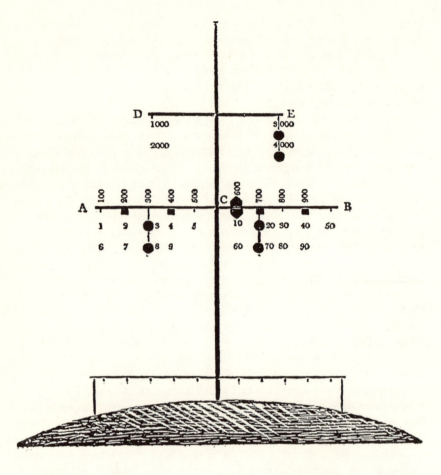

The Canadian optical telegraph was more complex than Chappe's: the alphabet was digitized, and numbers were represented by the position of balls on a structure shaped like a cross of Lorraine. Courtesy Parks Canada.

Morse's apparatus introduced simplicity to telegraphy. A single wire carried binary information in Morse code. Courtesy Canadian National.

Once electricity was harnessed, researchers very quickly came up with the idea of using it to transmit information. Unfortunately, most of the systems invented were terribly unwieldy; the first telegraphic apparatus, created by the Frenchman André Marie Ampère, needed as many wires as there were letters in the alphabet. Much more efficient was a five-wire apparatus developed in 1837 by two Englishmen, William Cooke and Charles Wheatstone, which was widely used in Great Britain until the twentieth century.

The real breakthrough was made from outside the scientific and industrial communities by a second-rate but iron-willed American painter, Samuel Finley Breese Morse. Born in 1791 in Charleston, Massachusetts, Morse eked out a living from his art until he was forty-one. In October 1832, on a ship that was taking him home from art studies in Europe, his interest was piqued by electromagnetism, and he invented a telegraph machine in 1835. It was difficult to use because it worked according to Chinese ideograms (one word, one sign) rather than alphabetically (dividing a word into letters), and it was very crudely made.

Morse's great contribution to telegraphy, of course, was the code that bears his name. Ironically, it was developed by his associate, Alfred Vail, a talented technician. Vail reduced Morse's code to sequences of dashes and dots that corresponded to letters of the alphabet, and the Morse code was born. This binary code has the luminous simplicity of computer code and might be said to be its distant ancestor.

In 1837, when it was ready for testing, Morse asked the American government for funds to build an experimental line. After many setbacks and difficulties, a line was built between Baltimore and Washington, D.C. The first transmission was made on May 28, 1844. It was a short quotation from the Bible: "What hath God wrought!"

In fact, Morse didn't "invent" anything in the scientific sense of the word, but he put the pieces of the scientific and technological puzzle together, thus bringing to telegraphy the technological simplicity that enabled it to be developed in a largely pre-industrial world. Telegraphy, the first practical and commercial application of electricity, came into use well before the electric light bulb, the phonograph, and the telephone, which appeared almost simultaneously in 1876–78.[5]

TELEGRAPHY IN CANADA

On 19 December 1846, the mayor of Toronto sent a telegram of best wishes to his counterpart in Hamilton. The very first words exchanged on this momentous occasion bore no great symbolic significance. The official ceremony was preceded by a laborious dialogue between the two telegraph operators comparing the time in Toronto and in Hamilton:
"What time is it?"
"Eleven twenty-five."
"Don't you mean noon?"
"No, eleven-thirty exactly."
"That's the city time?"
"Yes."
Time zones had not yet been standardized, and Canada went according to solar time. In fact, Toronto and Hamilton were on rather whimsical schedules; the time difference between the two cities was seventeen minutes,[6] and coordination of exact times was necessary so that the two mayors could be on the line in their respective telegraph offices at the same moment. By creating an instantaneous link between two geographic points, the first Canadian telegram fundamentally changed the traditional notion of time.[7]

This first line, 143 kilometres long, belonged to a firm with the portentous name of Toronto, Hamilton, Niagara, and St Catharines Electrical Magnetic Telegraph Company, founded by a Toronto hardware wholesaler, Thomas Denne Harris, after a press campaign orchestrated by local newspapers.[8] The line was constructed by an American entrepreneur, Samuel Porter. At first, traffic amounted to only ten or twelve dispatches per day; the telegraph was not an immediate success with the public, for the price was relatively high and the technology seemed quite mysterious. The slogan gracing the telegraphic forms of this first company conveyed a rather obscure marketing policy: "He directeth it under the whole heaven, and His lightnings unto the ends of the earth."[9]

But Canadian telegraphy was not conceived with the general public in mind. The goal of this line was to supply Toronto newspapers with the latest news from the United States and prices from the American grain market. One month after its inauguration, the Toronto–Hamilton line crossed the Niagara River to link up in Lewiston with the Buffalo–New York line. As of January 1847, Toronto newspapers could receive dispatches from the New York press, which was receiving dispatches from Europe. This situation of cultural dependence was characteristic of the nineteenth century, which was dominated by Victorian Great Britain. Soon, however, telegraphy found a new vocation: transmitting Canadian news items.[10]

Toronto could not maintain its monopoly on the new technology in Canada. The financial and industrial metropolis of British North America was Montreal, and it didn't take long for circles in that city to realize that telegraphy was a serious affair. Annoyed that they had been beaten to the punch by the Torontonians, businessmen at the Montreal Board of Trade brought to Canada the best expert they could find: Morse's first student, Orrin S. Wood, was named president of the newly created Montreal Telegraph Company. Within a few months, two lines were constructed, again by American companies: one to Toronto and one to Quebec City. In August 1847, the first transmissions from New York were received in Montreal.

In contrast to the improvised nature of the Toronto–Hamilton effort, the Montreal Telegraph Company was remarkable for the degree of its industrial organization. By the end of 1847, it had installed 870 kilometres of wire and transmitted 33,000 telegrams. In 1851 Hugh Allan, a very high profile Montreal financier, took over as president.

Allan had headed most of the leading firms in British North America, and he had quickly decided to concentrate his efforts in three sectors: steamships (under his impetus, Canada became the fourth largest maritime power in the world), the railway, and telegraphy. In 1852, one year after his appointment, Montreal Telegraph acquired the pioneering Toronto, Hamilton, Niagara, and St Catharines Electrical Magnetic Telegraph Company. Montreal's "honour" was restored.

Montreal Telegraph quickly imposed its quality criteria in the provinces of British North America, and even in the United States. The telegraph equipment was of Morse's design, while the materials for the transmission lines were imported from Great Britain. Montreal Telegraph had decided to use galvanized steel wire, even though copper was a better conductor, for the copper used in the first American lines broke easily; it wasn't until the late 1880s that cable manufacturers developed copper wire strong enough to meet the needs of telegraph companies.

At first, everything was an experiment, including stringing the telegraph lines. When the wires were strung too tight, they didn't last the winter; contracting with the cold, they snapped like overtuned guitar strings. When they were strung too loose, the wind tangled them. And they had to be insulated from the poles so that the rain didn't transform the poles into grounds, which interrupted communications. After many tests, glass was found to be the best insulator. The cedar poles had to be driven into the ground to a depth of at least 1.5 metres.[11]

The resulting system contrasted with the "quick and dirty" networks, which, in the purest quick-profit tradition, rapidly overran North America. But the cheap systems didn't last: the higher-quality model ended up being the standard throughout the continent. The symbol of Montreal Telegraph, a gloved hand holding a lightning bolt, very quickly became synonymous with telegraphy in the public eye.

Other small enterprises sprang up during this time, all the more easily because Morse's invention was not patented in the provinces of British North America. The most important one, for the future stature of its founder if not for its size, was the British North American Electric Association. Frederick Newton Gisborne, a young English immigrant who had begun his career as a telegraph operator at Montreal Telegraph, set up this company in 1847. His stated goal was to connect Quebec City and Halifax, but in his mind this was just the first step towards the more ambitious goal of a transatlantic cable. (Later, as director of the federal policy on telegraphy, he breathed new life into the dreams of his youth by instigating the transpacific cable project.) However, the line stopped at Rivière-du-Loup, blocked by opposition from the New Brunswick legislature.[12]

A second company, also called the British North American Electric Association, built a line from Quebec City to Montreal and was then amalgamated with its eastern namesake. Combined, however, the companies proved so unprofitable that Montreal Telegraph acquired the Quebec City-Montreal line at one-third of the cost price, with the Quebec City-Rivière-du-Loup line thrown in for free.

From the beginning of Canadian telecommunications, two trends opposed each other: east-west expansion and north-south expansion. The stakes were high, since telegraphy opened the doors wide to news and financial information that would make a mark on society as a whole. The opposition between the north-south continental trend – American expansionism – and the east-west British imperial trend – Canada – was evident throughout the nineteenth century. The early efforts of the Toronto, Hamilton, Niagara, and St Catharines Electrical Magnetic Telegraph Company are an example of the first trend, while Montreal Telegraph and its upstart Quebec City rival illustrate the second. In 1849 Montreal was linked to Troy, New York, by the Montreal and Troy Telegraph Company (a north-south connection), to Rouse's Point and Ogdensburg, both in New York; by the Vermont and Boston Telegraph Company (again, north-south), and to Bytown, now known as Ottawa, by Joseph Aumond and Associates (a point for the east-west trend). All of these early companies were bought out by Montreal Telegraph after a few years, ensuring the predominance of the east-west axis.

In November 1852, the Legislative Assembly of United Canada adopted a law on telegraphy that set out the rules for creating a company in the industry. One of the great problems for telegraph companies, and later for telephone companies, concerned where the poles could be erected. The 1852 law gave telegraph companies a right of way on all Canadian routes. This first Canadian telecommunications legislation was so favourable to the new technology that it gave rise to a second wave of companies.

A firm called the Grand Trunk Telegraph Company[13] built a line between Quebec City and Buffalo, New York; it, too, was bought out by Montreal Telegraph. Immediately after this acquisition, a second line was built on the same route by the Provincial Telegraph Company, which was long able to resist Hugh Allan's empire because it was a subsidiary of the rapidly expanding United States Telegraph Company. Montreal Telegraph had to wait until Western Union acquired the parent company in 1866 to absorb Provincial Telegraph's network.[14] Soon after, People Telegraph put up a line between Montreal and Quebec City, but this isolated company quickly went bankrupt without having the good fortune to be purchased by Montreal Telegraph.

Montreal Telegraph was one of the signatories to a general agreement on cooperation and mutual assistance concluded by all large North American telegraph companies on 29 May 1858. This document defined, among other things, the common standards that would allow cooperation in offices that were points of contact for the Western Union and Montreal Telegraph networks (Whitehall, Oswego, Buffalo, and Detroit). This first Canada-United States telecommunications convention, which paved the way for the intense international consolidation that was to characterize the industry throughout the world, confirmed the uncontested dominance of Montreal Telegraph in central Canada. Hugh Allan even obtained a seat on Western Union's board of directors, thus sealing the amicable sharing of the continent. The proponents of the east-west axis had won the first round.[15]

In the Maritimes, telegraphy had been preceded by a highly colourful system that has remained in the popular imagery: the pony express. Associated Press, a New York press agency, wanted the scoop on the latest news from Europe, whatever the cost, and Cunard's new steamships on the North Atlantic crossing stopped at Halifax before heading for New York. Associated Press came up with an ingenious system based on relays of racehorses (not ponies), which picked up the European dispatches from the postal steamships in Halifax and took them to Digby, in southern Nova Scotia, where a steamship, alerted by a cannon shot, was waiting to weigh anchor. The ship immediately left for Portland, Maine, where the terminus of the telegraph line to New York was located. The passage of the horsemen at a full gallop was a much anticipated event in the towns they went through. Usually, the news arrived in New York a day or two before the postal steamship arrived.

Another system involved steamships sailing up the St Lawrence River to Montreal. A telegraph line had been extended to Father Point, on the shore of the estuary. As the postal ship passed the village, the telegraph operator on duty, a young man named Robert F. Easson, jumped into a boat and went to meet it. According to an arrangement between Cunard and Associated Press, the postal ship would throw overboard a watertight tin box containing the dispatches from Europe. More often than not, the box fell into the water and Easson had to dive in to retrieve it. He then sent the news by telegraph directly to New York.[16]

This organization of the flow of information explains Frederick Gisborne's setback at Rivière-du-Loup. The New Brunswick legislature's opposition to his plan for a line from Quebec City to Halifax signified a rejection not of telegraphy in general, but of the east-west axis. Associated Press, backed by all the New York media, had exercised persistent pressure on behalf of the priority of linking Halifax to the American border.

A firm called New Brunswick Electric Telegraph Company was founded in March 1848 with the support of Associated Press, and the small American town of Calais, Maine, was linked to Amherst, Nova Scotia, via the largest town in New Brunswick, Saint John. In return for financial guarantees granted to New Brunswick Telegraph, Associated Press dispatches had exclusivity on this line.

After New Brunswick's rejection, Gisborne tried his luck in Nova Scotia, again running head-on into opposing American interests. Finally, the provincial government decided to construct the Amherst-Halifax line. Gisborne's consolation prize was the position of manager of the line. On 9 November 1849, the first dispatches were transmitted through the Maritimes by telegram, putting an end to the pony express.

Still, things did not run smoothly. The telegraph companies in the Maritimes and Maine proved incapable of cooperating and adopting common standards, so the dispatches had to be retranscribed when they passed from one network to another, which considerably slowed traffic and multiplied the risk of errors. Finally, Samuel Cunard himself purchased the government lines from Nova Scotia in 1851, and they then became the property of the Nova Scotia Electric Telegraph Company.[17] The American Telegraph Company obtained control of New Brunswick's network through a lease in 1856, then of Nova Scotia's in 1860. The entire Maritimes network was now under American control.

The continentalist trend triumphed across the board, though Montreal Telegraph partially restored the balance by linking Rivière-du-Loup to Halifax in 1851, and just in time. The Maritimes were strategically essential to the plan for a transatlantic cable and thus constituted an extremely important territory for Europe, the United States, and the still embryonic country of Canada.[18]

THE TRANSATLANTIC CABLE

Two men were responsible for laying the first transatlantic cable between Newfoundland and Ireland: Frederick Gisborne and Cyrus Field. Gisborne contributed the idea and the initial impetus, Field, the funding and enormous perseverance. After his disappointments in New Brunswick and mitigated success in Nova Scotia, Gisborne had left his position with the Nova Scotia government enterprise to devote himself to his dream. But before crossing the Atlantic, he had to start by linking New York to Newfoundland.

In 1851, when Gisborne was twenty-eight, he founded the Newfoundland Electric Telegraph Company, and the Newfoundland government gave him a thirty-year telegraphy concession in the colony. Gisborne decided, first, to build a line from St John's, on the eastern tip of the island, to Cap Ray, in the southwest. The second step would be to lay an undersea cable between Cap Ray and Cape Breton Island in Nova Scotia, where the line could be connected with the North American network.

Leading a team of eight men, Gisborne started by surveying the wilderness area on the south coast of Newfoundland. Soon, however, two men on his team deserted, one died, and the others fell sick. With the energy of one possessed, Gisborne didn't give up, and he completed his survey report in three months. On the heels of this first accomplishment, he went to Great Britain to consult with the world's only expert on undersea cables, John Watkins Brett.[19]

In 1850 Brett had laid the first undersea cable under the English Channel, followed by one between Great Britain and Ireland. He was eager to repeat the operation under the North Atlantic, and to this end he had developed a cable that would withstand the rigours of the ocean: four copper wires covered with gutta-percha[20] and protected by two sheaths of tarred hemp wrapped in galvanized iron. In exchange for shares in Newfoundland Electric, Brett gave Gisborne a first lot of cables. Prince Edward Island was to serve as a testing ground. The work was done post-haste, and by November 1852, a cable linked Carleton Head, Prince Edward Island, to Cap Tormentine, New Brunswick.

Encouraged by this success, Gisborne returned to Newfoundland and began to lay the 650 kilometres of cable needed for the Newfoundland part of the transatlantic cable. His first plan was to construct an underground line, but after a few kilometres of exhausting work in the rocky soil, he realized his error and decided to build an open (above-ground) line. It proved just as difficult, however, to drive the poles into the inhospitable land, and the operation ran out of money after just eighty kilometres.

In August 1853 all seemed lost: Gisborne's assets were seized, and he was briefly imprisoned. Undaunted, he went to New York to seek the funding that Newfoundland and Canada had refused him. There, he met Cyrus West Field, a thirty-three-year-old American millionaire who lived off his fortune.

A little-known pioneer in laying telegraphic cable under the Atlantic, Frederick
Gisborne was removed from the project by Cyrus Field. Gisborne became
director-general of federal telegraph services. Courtesy National Archives
(C 59292).

Field liked Gisborne's idea. He supplied the capital and, more impor-
tantly, made his financial network available. They set up a new firm, the New
York, Newfoundland, and London Telegraph Company, which took over
Gisborne's assets and debts. Samuel Morse was appointed chief electrician.
Now, Gisborne could return to Newfoundland with his head high.

Field had to hire six hundred men and invest more than a million dollars to
conquer the natural obstacles that had defeated Gisborne's first attempt.
Even so, running a line across Newfoundland took almost three years. During that time, Field sent Gisborne to lay the 130 kilometres of undersea cable
needed to link Newfoundland to Cape Breton Island; this was accomplished
in 1855, and it was to be Gisborne's last contribution to the transatlantic ven-

ture. Little by little, Field edged the Canadian pioneer out, and Gisborne resigned from his position of chief engineer in February 1857.[21]

When the overland cable was completed, at the end of 1856, Field finally had his uninterrupted Newfoundland–New York telegraph link, and it was time to start construction of the transatlantic portion of the cable. He was now short of money, however, and had to appeal for new capital. It was a bad time; the American market was bearish and attention was turned towards the New York–San Francisco transcontinental link, which was drawing all available investment funds.

It was John Brett who provided the solution. In October 1856, Field and Brett formed the Atlantic Telegraph Company. Great Britain provided most of the capital, along with the necessary expertise and equipment. Field was the only American involved.

American and British fleets had taken ocean soundings, and they soon announced the discovery of an undersea shelf that was quite smooth and shallow. This was good news – the last for some time. Finding the right technique for laying the cable under the Atlantic was a matter of trial and error, and it wasn't until 5 August 1858 that the cable was hooked up to the telegraph office that had been built in Trinity Bay, Newfoundland. When the news was announced, there were gun salutes in New York and celebrations all over the United States and Canada.

Meanwhile, the electricians in charge of cable operations had trouble on their hands: the 3,240 kilometres of transatlantic cable were so resistant to the electric current that no signals were getting through. The current was increased to two thousand volts before it was realized that raising the sensitivity of the receiving galvanometers would be more effective. A total of four hundred messages were sent from Great Britain to its Canadian colony before the insulation of the cables gave way under the considerable tension inflicted on the cable conductors in the first days by increased voltage. On 3 September 1858, the transatlantic cable was dead, electrically speaking.

On the technical level, however, valuable experience had been acquired. Many scientists had predicted that it would be impossible to send an electrical current over such distances, or that the propagation delay would be much too high (a delay of two minutes had been suggested). But the worst prognostications did not materialize; the theoretical work on telegraphy of British physicist William Thomson, the future Lord Kelvin, and his development of the galvanometer provided the final proof of the feasibility of the cable.[22]

After the American Civil War, Cyrus Field tried twice more to lay a transatlantic cable. Even with a brand-new and specially built ship, the *Great Eastern*, and an improved cable design, it wasn't until his fifth attempt, on 27 July 1866, that he was successful. After an authoritative "Without fail," a first message was sent from Newfoundland to Ireland. The outport of Heart's Content at the mouth of Trinity Bay suddenly became the telegraphic gateway to

The *Great Eastern*, launched in 1858, was the biggest ship in the world for forty years. Cyrus Field converted it into a cable ship from stem to stern. In July, 1866, the *Great Eastern* laid the first cable under the Atlantic. Courtesy National Archives (C 4484).

North America. The inaugural rate was one hundred dollars US (equivalent to $750 in today's dollars) for the first ten words. At such a price, it is easy to see why the first transmissions over the transatlantic cable were limited to business communications (governments, the press, business).

THE OVERLAND PROJECT: COMPETITION FOR THE TRANSATLANTIC CABLE

Field's triumph sounded the death knell for another project that had been just as ambitious. An American who had been a trade agent in Russia, Perry McDonough Collins, wanted to connect North America to Europe via British Columbia, Alaska, the Bering Strait, then Siberia and Russia. The Overland project was the competition for Field's undersea plan, which at the time was stalled after three unsuccessful attempts. Collins, forty-eight, had led an adventurous life; he was also very persuasive and, of course, he had in hand the inevitable letter from Samuel Morse confirming that the project was feasible.

First, Collins obtained a Canadian concession, which enabled him to found the Transmundane Telegraph Company in 1859, and the first part of his plan entailed construction of a line from Montreal to British Columbia and Alaska. He approached the powerful Hudson's Bay Company, which owned the vast hinterland between Ontario and British Columbia, but he was quickly shown the door: the fossilized top management at the Hudson's Bay Company was ready to obstruct any attempt to bring the project to reality.

During this time, Western Union had linked New York to San Francisco. Collins went to see Hiram Sibley, the president of Western Union, and proposed launching the Overland project from San Francisco instead of Montreal. Sibley gave his approval in October 1861. Collins then went to Tsar Alexander II of Russia, who committed himself to building more than 11,000 kilometres of telegraph line in Siberia. A line already ran from Moscow to Irkutsk, which represented three-quarters of the route.

Great Britain was more reticent, taking over a year before agreeing to have the Overland lines, even in scaled-down form, cross its possessions. British business circles, in revolt against the Hudson's Bay Company's obstructionism, urged the Colonial Office to create a transcontinental line, which the Colonial Office felt was incompatible with Overland. Thus, the eternal confrontation between north-south and east-west arose again. A number of transcontinental projects were advanced, none of them realized, and Collins finally got the green light from the Colonial Office. He had had the backing of the small British Columbia legislature from the beginning.

In 1864 Western Union, through a subsidiary, California State Telegraph Company, extended its network to New Westminster, now a suburb of Vancouver, but at the time the capital of British Columbia. The colony, established in 1858, was experiencing its first gold rush, and thousands of immigrants were moving north from California to this distant outpost of the British Empire. At stake was neither more nor less than British sovereignty on the shores of the Pacific.

Work on the line began in mid-1865. A team of five hundred American workers made its way northward, led by ex-officers of the Yankee army and assisted by an inexhaustible supply of Chinese labourers who were little more than slaves. No fewer than twenty-four ships supplied provisions. Perry Collins, however, was not part of the expedition. He was pushed out by financiers, as Frederick Gisborne had been in the transatlantic project, just as construction was getting under way. It seems, though, that he was compensated for his efforts, since the work in British Columbia took place under the name Collins Overland Telegraph Company.

The work was executed like a military operation. Four teams worked simultaneously in British Columbia, Yukon, Alaska, and along the coast of eastern Siberia. The routine was the same everywhere: surveyor-geometers surveyed the land, lumberjacks cleared a wide swath so that falling trees wouldn't damage the line, navvies prepared the land, woodsmen transformed thuja trunks into poles, riggers attached the crosspieces and blue-glass insulators and raised the poles, and linemen strung up the copper wires.

In July 1866 the British Columbia team had passed what is now Hazelton, more than 1,300 kilometres north of New Westminster, when it received by telegram an order to stop work immediately. The transatlantic cable had just been completed. Transportation was so difficult that the teams in Yukon, Alaska, and eastern Siberia didn't find out until the following year. The Yukon workers raised black flags on the useless poles. When he heard, the head of the Siberian team burst into tears. The Overland project was dead.

Western Union absorbed the losses of the Collins Overland Telegraph Company, but the operation wasn't a total write-off. When it read the Overland reports, the American government realized the importance of Alaska, and it purchased the colony from Russia in 1867. In British Columbia, the lines erected by Western Union in the southern part of the province were leased and then purchased by the colonial government; the federal government took them over in 1871, when the province joined Confederation. British Columbia was now linked to the rest of Canada via the American network. North of Quesnel the line was abandoned, and for years native people used the poles for firewood or to build houses.

The failure of the Overland project marked the end of the most serious north-south project in British North America. Had it succeeded, British Columbia no doubt would have met the same fate as Russia's American territory and been integrated into the United States. Overland also represented the end of an era, that of pioneers who dreamed of changing the world with a single asset: an empirical knowledge of telegraphy. From then on, large organizations came to the fore.[23]

2 The Telegraph Industry Gets Organized

At this point, the Canadian telegraph industry had reached maturity. The "one company, one line" concept had given way to that of the integrated network, thanks to Montreal Telegraph, which had gained telegraphy the status of public utility in Canada.

Montreal Telegraph started with the Montreal-Toronto line and grew with the rapid acquisition of a series of north-south links. Under its illustrious president, Hugh Allan, the company laid lines between most population centres, large and small, in Quebec and Ontario, with some bridgeheads in the Maritimes and the northeastern United States. A total of thirty-five new lines for railway usage accompanied the expansion of the railway in Canada.

Not content with imposing quality criteria on telegraphy in North America, Montreal Telegraph was determined to set the prices as well. In 1871, after a series of rate drops, the company set a uniform rate of twenty-five cents for ten words and one cent per word thereafter throughout its territory, from Sarnia, Ontario, to Sackville, New Brunswick. This tidy arrangement was soon challenged by a new company, Dominion Telegraph, which had been born under a dark cloud. A man named Selah Reeves founded Dominion Telegraph in 1868, thanks to an organized subscription among businessmen in the Niagara Peninsula. He then awarded generous contracts for the construction of more than three thousand kilometres of line. Several hundred kilometres had already been built when rumours of misappropriation of funds caused the shareholders to revolt and take control of the business. Buffalo, Detroit, and Quebec City were then connected up in a rather haphazard way.

But Dominion Telegraph really got rolling in 1871, when the company was restructured on an industrial basis. A profitable agreement was made with Direct US Cable Company, whose transatlantic cable ended in Nova Scotia. In 1874 Dominion Telegraph obtained the rights from Direct US for its Europe-Canada traffic, for which it extended its lines to Nova Scotia.

This prosperity was compromised, however when Dominion Telegraph began a price war with Montreal Telegraph by adopting a fixed rate of twenty cents for ten words, undercutting the competition by five cents. The losses of revenue were so severe that it quickly became impossible to expand, or even modernize, the network. In 1878 American Union Telegraph leased Dominion Telegraph's network. When Western Union purchased American Union Telegraph, Dominion Telegraph naturally fell under the control of the American giant, which had long had a cordial relationship with Montreal Telegraph, Dominion Telegraph's arch-enemy.

The battle between Montreal Telegraph and Dominion Telegraph carried over into the telephone sector. In February 1879 Dominion Telegraph acquired Bell's patents and opened telephone exchanges in Montreal, Quebec City, Ottawa, Saint John, Halifax, and several small Ontario towns. That same year, not wanting to be left behind, Montreal Telegraph took over Edison's patents and opened its own exchanges, with the result that cities such as Montreal, Quebec City, Ottawa, and Halifax soon had two parallel telephone systems that couldn't intercommunicate. This wild competition was going nowhere; when, in 1880, Bell offered to repurchase its telephone patents, both Montreal Telegraph and Dominion Telegraph sold out with relief.

Also in 1880, a group of Winnipeg politicians created the Great North Western Company to provide telegraph service to Manitoba and the Northwest Territories. The following year, when operations had just begun, the company was bought by Erastus Wiman. Wiman was born in Churchville, a small town outside of Toronto, but he had made his fortune in New York as director of R.G. Duns, a company of exchange bureaus. More significantly, he sat on the board of Western Union. He had started out as a journalist, and he continued to publish numerous short articles on various subjects, among them the merits of free trade between the United States and Canada. He also justified his financial and business operations with an acerbic pen. A pamphlet entitled *Chances of Success* illustrates the virtuosity with which he wielded his cynicism: "The past 50 years have been so full of Chances that the wonder is that the rich man is a rarity and that the poor are so plentiful."[1] Given his propensities, Wiman would not have purchased Great North Western for philanthropic reasons.

Western Union wanted to divest itself of its lease on Dominion Telegraph because of the disastrous price war it had instigated and offered the lease to Montreal Telegraph. But Hugh Allan had long since retired, and his company was only a shadow of its former self. Montreal Telegraph committed a cardinal error: it declined to purchase Dominion Telegraph's network on the grounds that Dominion's assets merely duplicated its own main lines.

Wiman jumped at the opportunity. He took over Western Union's lease on Dominion Telegraph and offered to lease Montreal Telegraph's lines at a preferential rate, with Western Union's financial guarantee – and with thinly veiled threats if Montreal Telegraph rejected the offer. The balance of power had shifted, and in July of 1881 top management of Montreal Telegraph succumbed to the onslaught.

Montreal Telegraph brought 28,000 kilometres of line to the merger, Dominion Telegraph, 8,000 kilometres. With this acquisition, Great North Western re-established a quasi monopoly over Canadian telegraphy. Wiman, who had guided the takeover from his New York office, named Orrin Wood, the first president of Montreal Telegraph, as vice-president and *de facto* manager of the company. Federal approval was required, so Wiman organized a powerful lobbying effort, travelling to Ottawa to circumvent Prime Minister John A. Macdonald when this proved necessary. The bill sailed smoothly through Parliament.

Soon after Wiman's financial *tour de force*, telegraph rates began to rise again. The merger also took a terrible toll on personnel: hundreds of employees were summarily fired with a brutality that was shocking even at a time of unbridled capitalism. Such excessiveness was one of Wiman's hallmarks.

Wiman, the man who had sold out the telegraph business to American interests, further provoked the wrath of Montreal business circles. One of the first measures adopted by Great North Western was to supply Associated Press's "wire" to Canadian newspapers through two lines, Buffalo–Quebec City and Buffalo–London, Ontario. The Quebec and Ontario press thus had a single American source for all news of the outside world (the service covered neither the Maritimes, which were served by Western Union, nor Manitoba, which wasn't directly linked with eastern Canada). Would the new information industry be organized on the north-south axis?

The Press and Business News Division of Great North Western fell under the aegis of Robert Easson, the man who had helped to pioneer timely news in the 1850s. The intrepid telegraph operator from Father Point innovated yet again: he transformed all Great North Western agents into reporters, who filed stories on murders, robberies, deaths of local notables, fires, and so on. Canadian dispatches were added to Associated Press's international dispatches. This was when the telegraphic code marking the end of a message, "–30– ," entered the typographic argot of newspapers to indicate the end of articles.

Journalists called this local supplement the "Easson Service." It was in fact the first Canadian press agency, and it was welcomed by the newspapers, which had not yet come up with the idea of news correspondents. But Great

Nineteenth-century telegraph office. One of the best signs of the new technology's acceptance was the almost immediate popularity of the telegraph operators. Their craft retained many aspects of pre-industrial craftsmanship, including a love of work well done and a spirit of independence. Courtesy National Archives (C 7632).

North Western did not keep its monopoly over news in Canada for long. Canadian Pacific Railway's Telegraph Division launched a similar service in 1886, and the opening of the transcontinental line gave a decisive advantage to CP's dispatches: Great North Western had no line west of Winnipeg. CP's intervention ensured the ultimate triumph of the east-west axis in Canadian telegraphy.[2]

As the industry was reshuffling itself, how did the public view telegraphy? After an initial mistrustful reaction, telegrams became a common means of communication, even though it was indirect, requiring the intervention of a qualified operator. By the mid-1850s, people regularly exchanged greetings and news by telegram. Some wealthy people even had a telegraph machine installed at home so that they could play long-distance chess.

It was the news media that popularized the telegraph. The telegraph office became a meeting-place during such important events as elections, the posting of wheat prices, and declarations of war. In Quebec City, historic milestones like the taking of Sebastopol in 1855 and the fall of Napoleon III in 1870 drew joyful or solemn crowds to the telegraph offices to hear the latest news.

One of the most obvious signs of success of the new technology was the popularity of telegraph operators. The best operators, who could tap out forty to fifty words a minute (the average was twenty-five to thirty-five words a minute), could earn up to seventy-five dollars a month, which was a more-than-respectable salary in the mid-nineteenth century. Lewis McFarlane, who became the third president of Bell Telephone, William Van Horne, who built CP, and Thomas Ahearn, who launched one of the first electric tram systems in Canada, all started as telegraph operators. Linemen, for their part, went from site to site anywhere new networks were being built. They were called "boomers," and the entire continent was their home.[3]

The trades of telegraph operator and boomer belonged to what might be called the aristocracy of the working class, drawing on attributes of the pre-industrial craftsman, particularly a love for work well done and a spirit of independence. Throughout the nineteenth century, the lack of a clear distinction between the telegraph company and the press agency provided many telegraph operators with an opportunity to double as reporters. These telegraph operator-reporters were also the advance guard of a nascent unionism. Starting in the 1870s, they joined the Knights of Labour, in the ranks of which they constituted the powerful Brotherhood of Telegraph Operators.

GETTING ORGANIZED: TELEGRAPHY UNIONS

It is estimated that the Brotherhood of Telegraph Operators, with branches in Great Britain, Ireland, Belgium, Australia, and New Zealand, had up to 2,500 members in Canada. Women were admitted, a rarity in nineteenth-century unions, as were workers without professsional qualifications. This humanistic Christian union encouraged awareness of workers' rights, but it was poorly armed for action and preferred forming cooperatives and conducting political training to organizing strikes.

Nevertheless, it was the Brotherhood of Telegraph Operators that, in 1883, held the first national strike in Canada. In fact, it was a continentwide strike, organized in Pittsburgh, for salary increases and shorter working hours. It affected only "business" telegraph companies, sparing the railways, and thus about half of all telegraph operators. The increase demanded was substantial: thirty-five dollars more for all who made less than forty-five dollars a month. At the time, a male telegraph operator earned an average of thirty-seven dollars a month in Canada; a female telegraph operator earned twenty dollars. Equal pay for men and women was one of the demands; another was a reduction in working hours from nine to eight on the dayshift and from eight to seven hours on the nightshift. Finally, Canadian telegraph operators wanted parity with their American counterparts.

None of the telegraph companies in Canada would negotiate (except for one small independent company). On 19 July, a little before noon, the strike committee circulated a coded message to all telegraph offices in Canada and the United States: "General Grant is dead." The strike was instantaneous. In

Montreal, fifty-four telegraph operators out of sixty walked out, and it seems that this high rate of participation held throughout the country. The owners' reaction came from the irrascible head of Great North Western, Erastus Wiman. From his New York office, the apostle of free trade between Canada and the United States railed against the American character of the strike: "[It is] a most extravagant demand, made in an offensive manner; which, if acceded to, would have the effect of placing the entire telegraphic communications of the Dominion in the hands of an organized mob, governed by regulations emanating from a band of agitators in a foreign country, enforced by intimidation and coercion in some cases almost bordering on brutality".[4]

Thanks to strikebreakers, service was resumed after a few days of confusion. Rather than setting up picket lines, the strikers wasted much energy trying to create an alternative telegraphic cooperative, which never became operational.

After one month, on 18 August, the union heads had to ask their troops to go back to work. The telegraph companies directed by Wiman, not satisfied with this victory, required employees to sign renunciations to their right of association. Beaten and humiliated, the Knights of Labour lost much of its prestige. By October, there was not one union local active in the telegraph industry. Workers had to content themselves with creating secret associations, which had no future because of the great mobility of labour.[5]

Unionization in business telegraphy really took hold in 1902, with the creation of the Commercial Telegraphers' Union of North America in Chicago. This union comprised up to 80 percent of Canadian telegraph operators outside of the railways, and it lasted more than sixty years. With the adoption of the Industrial Disputes Investigation Act (the "Lemieux Law") in 1907, the CTUA came to be recognized by telegraph companies and provided a basis for more stable labour relations.[6]

In the railway companies, workers belonged to the Big Four, a group of purely Canadian unions covering the major trades in the railway industry, which was joined by organizations representing connected trades, including telegraph operators. The first local of the Order of Railroad Telegraphists was created in Montreal in 1886. Independent of the major workers' organizations, the Big Four specialized in negotiations with the Department of Railways and in the financial protection of its members. Insurance rates for railway employees were out of sight and one of the major benefits of the Big Four was its life- and disability-insurance co-op for its members.[7]

Overall, this union grouping was very jealous of its autonomy, and it refused any link with the other unions. This explains the persistent division betweeen "commercial" and "railway" telegraph operators. However, telegraph operator-reporters had an influence out of proportion to their position within their unions. They very often had the support of newspapers and even of politicians who had begun their career as telegraph operators. At the beginning of the twentieth century, a senator and minister of railways, J.D. Reid, served as a natural conduit, coming to the defence of his former colleagues whenever they came under attack.

Finally, it should be noted that most of the linemen, who moved from site to site around the country in gangs, were Quebecers. If the telegraph promoters were uniformly Anglo-Saxon, the actual builders of the lines were French Canadian. The villages and forests through which these pioneers of telegraphy passed echoed with French songs.[8]

CONQUERING THE WEST

Although private enterprise managed to serve central and Atlantic Canada quite well, it was a different story for the vast hinterland that stretched between Ontario and British Columbia. This huge expanse remained for some time under the fluid reign of the First Nations and the Métis; British sovereignty was confined to the Hudson's Bay Company.

The idea of a telegraph line to link British Columbia to Ontario had been in the air since the late 1850s. Following Perry Collins's dealings with the Colonial Office, the North-West Transportation Navigation and Railway Company was founded to build a railway and a telegraph line. Soon after, the president of the Grand Trunk Railroad Company, British businessman Edward William Watkin, founded the Atlantic and Pacific Transit and Telegraph Company for the same purpose. Neither of these privately financed projects came to fruition, due to insufficient capital.

The only company that could have amassed the necessary funds, the Hudson's Bay Company, was opposed to colonization of the West. The historian Harold Innis came up with a now-famous table showing the conflict between the economies of the fur trade and of agriculture. Since the Hudson's Bay Company was based on the fur trade, it naturally opposed the railway–telegraph pairing that would encourage groups of people to settle and, as a consequence, put its primary business activity at risk. No one, not even the Colonial Office, dared to mount a frontal attack on the Hudson's Bay Company's operating monopoly on the Prairies.[9]

Considering a change in policy, the company sent its most famous explorer, Dr John Rae, to the Canadian West to study, among other things, construction of a transcontinental telegraph line. When he returned to Great Britain in 1864, Rae produced a favourable report. Nevertheless, nothing changed until after Confederation in 1867, when the *status quo* was challenged by two new facts. First, in 1870 the new Canadian nation acquired the property titles to the Hudson's Bay Company in Rupert's Land and the Northwest Territories. Second, British Columbia joined Confederation in

1871 on condition that a transcontinental railway be constructed. Of course, telegraph lines went hand in hand with the rail links. Sometimes, as eventually happened in the West, the telegraph preceded the railway.

With the nationalization of the Hudson's Bay Company, Canada more than doubled in size. When the transaction was announced, the Métis of the Red River region, who had not been consulted, rebelled, under the leadership of Louis Riel. At the same time, the Fenian Society, an Irish Catholic association based in the United States, launched raids on the new Canadian territories. The authority of the new federal government was threatened throughout the West. John Alexander Macdonald's government had to assert itself. In 1870 it granted provincial status to Manitoba. This was a partial victory for the Red River Métis. But the new province was completely isolated, without modern means of communication. The first telegraph line was established in 1871 between Minnesota and Manitoba by an American company, the Northwestern Telegraph Company. Far from consolidating Canadian sovereignty, telegraphy was undermining it.

One man who changed the course of events was Sandford Fleming, an engineer from Scotland, a fervent British imperialist, and one of the protagonists in the establishment of the transcontinental railway and the creation of the first Canadian postage stamp (the threepence beaver). He had just standardized the Canadian time system (Eastern Standard Time) and was planning to do so worldwide.[10]

Fleming saw a strategic purpose in telegraphy: to confirm British sovereignty in North America – a purpose much too important to be left to the vagaries of private enterprise. To have military value, the transcontinental link had to follow a northern route as far as possible from the United States border. But the few populated areas were along the border, to the south. This was even more reason not to leave things to private enterprise.

When the Liberals, led by Alexander Mackenzie, came to power in Ottawa in 1873, Parliament finally rallied to Fleming's cause and pronounced itself in favour of a government telegraph line through the Prairies. In the face of myriad difficulties, five hundred pioneers laid a line of beacons through the Canadian West from Port Arthur (now Thunder Bay) on the shore of Lake Superior to Fort Edmonton, following a northern route, with no roads or support from local populations. But this feat was instigated by a series of entrepreneurs whose scruples were as weak as their ties to the Liberal party were strong, and who were chosen through the farcical charade of a call for tenders.

At first glance, the result was impressive: an immense territory between the Great Lakes and the Rockies was spanned by two thousand kilometres of telegraph line by the end of 1876. But this terminus fell far short of the objective of the original project, which was Vancouver. Edmonton was only a military outpost lost in the heart of the Northwest Territories. Furthermore, the poles were made of poplar, which quickly rotted at the base; in some loca-

Engineer Sanford Fleming, Scottish by birth and a fervent British imperialist, was the motive force behind Western telegraphy in 1876 and the transpacific cable in 1902. Courtesy National Archives (C 14128).

tions, they had been driven into ice and disaster struck when spring came; elsewhere, the lines had been attached directly to roughly pruned trees. Bison rubbed against the poles, overturning any that were not well anchored. To top it all off, devastating forest fires annihilated large sections of the network on the Prairies.

A commission of inquiry threw a troubling light on irregularities that had dogged construction of the western telegraph line. Meanwhile, business circles were unhappy with the federal government's intervention in what they believed should have been a private enterprise plum. With some complacency contemporary media sources pointed out the deficiencies of the publicly built line. For whatever reason, Fleming was cleared of the accusations of corruption made against him. It seems that all the problems could be traced back to the contractors.[11]

For the first Prairie settlers, the telegraph line was a cause of great discontent; rates were high and the service was irregular. Sending ten words from Port Arthur to Winnipeg cost an exorbitant two dollars – the best incentive to write in "telegraphic" style.[12]

In 1882 the federal government finally reacted. Rather than adopt an over-all telecommunications policy, it tried to salvage what it could. All of the lines built with government money were grouped into a single entity, the Government Telegraph and Signal Service, under the authority of the Department of Public Works. The two principal axes were the Great Lakes-to-Edmonton line (sixteen hundred kilometres) and the Vancouver-to-Quesnel line (six hundred kilometres). Appointed to manage this public administration was Frederick Gisborne, and one of his first moves was to purchase Western Union's assets in British Columbia.[13]

The erstwhile promoter of the transatlantic cable spent the last years of his life trying to put some order into the chaotic situation in the West. He carefully avoided competing with private companies, not hesitating to hand over sections of the network that might be doing double duty and reconstructing line, sometimes at great cost, in areas where he felt the state must intervene (the Battleford-Edmonton line was rebuilt in part with metal poles). He also linked the northern population centres to the new CP telegraph line with north-south feeder lines, such as the one from Prince Albert to Humboldt, which played an important role in the Métis rebellion.

Gisborne went out on the land himself, not afraid to confront often rowdy crowds, such as the colonists of Prince Albert, who had burned the telegraph poles because the new line wasn't placed where they wanted it. Gisborne's mission reports read like adventure stories and give a good picture of the difficulties of technological penetration of the West. His linemen resembled cowboys more than skilled electricians.

In spite of such efforts, the western lines remained underutilized because of the low population density and the cost of sending telegrams. Unexpected uses sprang up to fill the gap. Telegraph operators provided long-distance first aid in the region, which was terribly short on physicians, and when time weighed too heavily, they organized intertown telegraphic chess tournaments.[14]

Aside from the two major western lines (Prairies and Overland), the public network included lines to the lower St Lawrence and the Gaspé Peninsula in Quebec, linking isolated communities on the Lower North Shore as far as the Strait of Belle-Isle. In Manitoba, it connected Winnipeg to the American network; in British Columbia, it linked the islands and mining towns in the north of the province; in addition, it served all isolated government installations (lighthouses, native reservations, and Northwest Mounted Police outposts).[15]

TRANS-CANADA TELEGRAPHY
AND THE MÉTIS REBELLION

In the end, the transcontinental telegraph line was completed by a private company (with massive financial aid from the government). Canadian Pacific was created in February 1881 with a mandate of build an east-west railway, but the charter granted to it by the federal government authorized it to offer telegraphy to the public as well as telephone services, which were just becoming available in Canada. Contrary to Fleming's dream, the railway followed the American border. This solution was dictated by economic imperatives – the majority of the population was there – while Fleming, as we have seen, invoked political and military reasons to justify his northern route.

As a birthday gift, the new company received the few sections of the government's western telegraph network that might be useful to it (most of the government installations were too far north). Construction of the transcontinental telegraph line progressed in parallel with that of the railway. The line was finished without problem in December 1886, one year after the track, and the service was opened to the public a month later. This time, it was a true public utility – in January 1886 CP had created an independent communications service in Montreal called CP Telegraphs – not a line jerry-built by pioneers and out of order every other day. It is to be noted, however, that the first American transcontinental line had been inaugurated in 1861, twenty-five years earlier.

Even before it was completed, the new line proved its strategic importance. As it advanced westward, the paired railway-telegraph line gave rise to the same sort of dislocations that had led in 1870 to the Red River rebellion in Manitoba. The compromise that had ended the rebellion was not powerful enough to protect the rights of francophone Métis and native peoples, who retreated farther west to the Saskatchewan Valley, where they abandoned their traditional semi-nomadic way of life and became sedentary. The pace of Anglo-Saxon colonization accelerated with the advance of the railway. In 1885 it provoked a new rebellion.

The Métis and native peoples appealed to the hero of 1870, Louis Riel. This time, the central government decided to suppress the revolt and dispatched the army to the future province of Saskatchewan. From beginning to end, the military operations were influenced by telecommunications; indeed, the minister of Militia and Defence, Joseph-Philippe Caron, had a telegraph office installed right in the Parliament building. Telegraphy was used to receive information from the battlefield and to orchestrate federal propaganda in the field: dispatches from Ottawa dictated the distribution to native people of flour, lard, tea, tobacco, and even livestock. General Middleton's troops, progressing in three separate columns, remained in constant telegraphic contact.[16]

The other side had no idea that this system existed. The rebels used "moccasin telegraphy," spreading the news incredibly fast by the traditional word-of-mouth method. But information was distorted according to the mood of the messenger, and the diffusion zone was limited to the Prairies. No one realized how inadequate word of mouth was compared to telegraphic technology.[17]

An outpost of telegraph service in the west, the Humboldt station played a key role in helping the Canadian army crush the Métis revolt. Courtesy National Archives (C 753).

Of course, one of Louis Riel's first moves, on 18 March 1885, was to cut the telegraph line from Batoche to Prince Albert (the next day he formed the provisional government of Saskatchewan at Batoche). Throughout the hostilities, Riel's partisans would occasionally repeat this act of sabotage, and each time, the telegraph operators of the government and CP networks worked miracles in re-establishing service.

If the Métis had been fully aware of the importance of the telegraph, they would have ripped out the line for several kilometres to make patched-together repairs an impossibility. In particular, they should have cut CP's east-west line and not contented themselves with cutting the government network's north-south feeder line between Prince Albert and Batoche. Nor had the rebels thought of using the telegraph themselves, as the conservative French Canadian daily *La Minerve* noted: "It is worth remarking, in this regard, that the dispatches in question come from only one source, only one side, the insurgents being absolutely deprived of communications with the outside world."[18] The members of the provisional government saw the telegraph as a symbol of Ottawa's power and not as a strategic weapon.

During this time, the Canadian troops remained in permanent contact with Ottawa. They were inventing modern war communications, in which the military power on the ground is in constant contact with the political powers – and with public opinion. Indeed, a new type of journalist came into being at this time: the war correspondent. Four English Canadian dailies, the *Montreal Star*, *Montreal Witness*, *Toronto Globe*, and *Toronto Mail*, sent their star reporters to the front. On the French-language side, *La Minerve* followed suit.

This new form of journalism introduced "human interest" into the reporting of the war. An article that began with the words "Riel is 40 miles northwest of us" (*La Minerve*) gave readers in Montreal a sense of immediacy.

Every day, the names of those who had died the day before were published and military leaders – and sometimes ordinary soldiers – were interviewed. Thus, the war went beyond a strictly military framework to become an event that emotionally involved each reader.[19] In comparison, the grandiloquent editorials written in Montreal seemed irrelevant. "It's the ancient struggle between civilization and barbarism," wrote *La Presse*. If the tone struck a false note, it was because telegraphy had moved the paper's centre of gravity from the editor's desk to the scene of the action.[20]

Such press coverage explains the great hold that the Métis rebellion had on public opinion in both Quebec and Ontario. The large Anglo-Saxon companies of Montreal and Toronto granted leave (on full salary) to all of their employees who wanted to volunteer to fight the Métis; in the telecommunications sector, this included Montreal Telegraph, Great North Western, and Bell Telephone, which provided the Canadian troops with a large number of skilled personnel at the front. Mobilization crossed linguistic lines, and francophones participated fully in the spirit of repression, as *La Presse* testified: "Military enthusiasm is spreading throughout Canada."[21]

Throughout the military operations, the newspapers, fed through the telegraph, rallied all Canadians around an economically based – and victorious – nationalism on the American model of the conquest of the western frontier. But the sentencing to death of Riel in August 1885 caused an uproar. Suddenly, the French-language press forgot all political alliances and took up the cause of Riel's pardon, while the English-language press continued to demand his blood. *La Minerve*, which had called for repression throughout the war, began its 16 November lead article, "Riel was executed this morning at 8: 23. He died a courageous soldier."[22]

In Quebec, the mobilizing capacity of the telegraph intensified with the trial of Riel. Fanfares of pan-Canadian nationalism immediately gave way to the hotter rhythms of tribal drums. In Montreal a huge demonstration brought together fifty thousand people at the Champ-de-Mars, signalling the beginning of the decline of the Conservative party in the province, while the Liberal party shot out roots that turned out to be immensely durable (Wilfrid Laurier began his political rise at this event). How can such a spectacular turn in public opinion be explained?

Telegraphy encouraged people to respond viscerally to the news they received and identify with one or the other "side." During the military part of the Métis rebellion, English Canadians had identified with Riel's victims, who were also English speaking. French Canadians followed along, since they had no communication with the rebels in the Saskatchewan provisional government. The federal government used the telegraph skilfully to distribute the news that suited it best.

From the moment when French-speaking reporters established contact with Riel – that is, when the trial began – French Canadians began to identify with Riel. The federal government's arguments, which had not changed throughout the military and judicial parts of the Riel affair, suddenly seemed cold and threatening, according to francophone public opinion. Telegraphy was not able to transmit the juridical nuances, only the passions. From the moment when the federal government dragged Riel before a jury, it put him in contact, through the telegraph, with the French Canadian people of Quebec, allowing them to identify with him.

The most obvious sign of the importance of communications in the Saskatchewan Métis rebellion was the sudden jump in CP's telegraphy revenues, from $70,000 in 1884 to $145,000 in 1885, mainly thanks to military use and press coverage. The death of Riel coincided, almost to the week, with the inauguration ceremony for the transcontinental railway, promoting an increase in the value of CP shares. Cornelius William Van Horne, at the time vice-president of CP, jokingly suggested that his company erect a monument to the memory of Louis Riel.[23]

THE MEDIA REBEL AGAINST CP

With the conquest of the West, Canadian Pacific Telegraphs slowly acquired a monopoly on the market for distribution of international news. Its main competitor, Great North Western, lost the contract for routing Associated Press dispatches in 1894, since its network did not extend beyond Winnipeg.

It will be remembered that both telegraph companies already functioned as press agencies for Canadian news. In what was no doubt an extreme case, operators in a telecommunications network intervened in the definition and creation of information content. The limited nature of the Canadian market partly explains this confusion of genres; the domination of a foreign agency, Associated Press, accounts for the rest.

It was an error by CP Telegraphs itself that ended this incestuous marriage between telegraphy and the press. In July 1907 the Montreal company quietly modified distribution of the international press service in the West, which raised rates by two to three hundred percent. Even more than the money, it was the arrogance of the company's move that displeased the western media.

Three competing Winnipeg dailies, the *Telegram*, *Tribune*, and *Free Press*, formed their own press agency, Western Associated Press (WAP), in September 1907. They pooled their network of correspondents across Canada and replaced their Associated Press feed with a combination of three small American agencies. Although none of these agencies had as extensive a network as Associated Press's, WAP made do, and it quickly became a success in the West, where most newspapers subscribed to the service.

CP Telegraphs reacted by invoking its own rules limiting the press rate to newspapers only. Because WAP was a press agency, not a newspaper, CP charged it an extra 50 percent. The Winnipeg mutineers got around this by having dispatches addressed to one newspaper, which distributed them anonymously to the others. The CP Telegraphs agents weren't fooled; they picked through the newspapers to track WAP's news items, even when they weren't identified, and charged them for their retransmission, thus reinventing control of the press in all good conscience.

WAP migrated to other telegraph companies: Great North Western east of Winnipeg and Canadian Northern to the west. However, there were still cities and towns served only by CP Telegraphs. The wily WAP journalists talked of requisitioning carrier pigeons ... Meanwhile, CP Telegraphs raised the press rate again, this time across the board. The press revolt spread to British Columbia, where the Nelson *News* was cut off from the Associated Press wire for misconduct.

The rate decision, made by a lower-down at CP Telegraphs, reflected the company's arrogance. The Toronto press followed in the footsteps of the western press, and the spectre of nationalization once again reared its head. The threat was even more credible because, two years before, Bell Telephone had barely escaped nationalization at the federal level, and Manitoba was in the process of nationalizing Bell's provincial installations (see chapter six). In addition, the Winnipeg *Telegram* belonged to interests close to the Manitoba premier, Rodmond P. Roblin. The affair finally came to the attention of Prime Minister Laurier, who put pressure on CP. The company capitulated and, in mid-October 1907, proposed a return to the *status quo ante*.

Although the crisis had passed, the issue of rates for WAP remained unresolved; the agency still wasn't benefiting from the press rate. When the authority of the Railway Commission was extended in 1909 to telegraphy WAP naturally used that forum to raise the issue. Why should it pay a higher rate than its competitor, CP Telegraphs? The telegraph company's lawyer answered with a metaphor. Let us compare news to charcoal and its transmission to the transportation of charcoal from a mine owned by the company, he said: "A railway company may sell its surplus coal at so much per ton to residents at Winnipeg and what it costs to haul that surplus coal to Winnipeg need not enter into the price. The sale price of coal sold under these circumstances need not be considered on a complaint as to the rate charged by the company for hauling coal to Winnipeg. The two matters are absolutely separate and distinct and bear no relation to each other."[24]

In its judgment of January 1910, the Railway Commission continued this unfortunate metaphor, adding some lumps of coal for CP: "Suppose a railway company has a coal (news gathering) mine at Montreal and the applicants have a like mine at Winnipeg, and Saskatoon is an important point of consumption. Can the company deliver its commodity to the Saskatoon consumer at $4 a ton, including both value of the commodity and cost of haul, and charge the Winnipeg producer $5 per ton for hauling alone? If this were permissible, railway companies owning coal mines could close up every mine except their own."[25]

The judgment satisfied the western newspapers but not those in central Canada. Since CP telegraphs claimed to be supplying the service at a loss, it now had to raise its rates in Montreal and Toronto. In fact the new CP rates pleased no one, and in March 1910 the Railway Commission was once again faced with a united front by the press. After sessions in Ottawa and Winnipeg, the commission pronounced a historic decision in June, forbidding telegraph companies from acting as press agencies.

From then on, an information "carrier" was absolutely barred from creating, processing, or distributing the content of information. This decision was the cornerstone of all Canadian rights in communications. The telegraph and, later, telephone companies were defined as public services with an exclusively technical vocation.

The press thus emerged victorious from its confrontation with the telegraph companies. But while the victory brought the western companies what they wanted, Ontario and Quebec newspapers were saddled with a reponsibility they had not sought. Making the best of their bad luck, they joined with their western colleagues to renegotiate the distribution contract with Associated Press that CP Telegraphs had had to abandon, and Canadian Press was created in December 1910. Seven years later, WAP merged with Canadian Press, which it had done so much to create.[26]

A BRIEF RETURN TO PIONEERING TIMES:
THE KLONDIKE AND THE PACIFIC

At the turn of the twentieth century, two ventures breathed new life into Canadian telegraphy. The discovery of rich veins of gold in the Klondike Valley in 1896 gave rise to the biggest gold rush in the history of North America. Once again, the United States contested the placement of the Canadian border, hoping to claim the little Yukon valley for Alaska, and the Canadian government hastily erected a telegraph link between the gold mines and Canadian frontier outposts. For reasons of national security, Prime Minister Laurier did not leave it to private enterprise to provide this critical service.

At the same time, the British Columbian network was expanding northward. Following in the path of the Overland project, a line was built from Quesnel to Atlin, in the far north of British Columbia but at a respectable distance south of Dawson, where most of the Klondike gold mines were concentrated. These two lines, built from 1899 to 1901 through the Rockies, one of the most inaccessible regions of Canada, brought the myth of the Wild West back to life for a few years.

The other great venture of the early twentieth century was the cable under the Pacific, a project instigated by Sandford Fleming, the promoter of the transcontinental telegraph and railway. His imperial dream had not ended with the creation of a line from the Atlantic to the Pacific; his goal was the "All Red" line, which would circle the globe entirely within the British Empire (red was the colour of British possessions on maps). This fervent believer in public intervention in telecommunications and transportation liked to say that all civilized nations had government telegraph networks, and Canada should quickly join the ranks of such nations by nationalizing its installations.[27]

Since 1879, when he was still engineer-in-chief for the government railway, Fleming had been interested in extending the transcontinental telegraph line by means of a transpacific cable. The abandonment of the public project and its privatization when CP was created in 1881 hadn't discouraged him. He enlisted the support of Frederick Gisborne, who was then the director of federal telegraphy services. Unfortunately, Gisborne died in 1892. Fleming himself was getting old, but he never gave up. For twenty years, he battled in Great Britain, Canada, Australia, and New Zealand for the construction of the cable: "If we resort to the agencies of steam and electricity, the people of Australasia and the people of Canada may, for all practical purposes, become neighbours. And why, it may be asked, should they not be neighbours, as far as it is possible for art and science to make them? Are they not one in language, in laws, and in loyalty?"[28]

This fiery speech, delivered to the governmental delegates of the various British colonies at the Colonial Conference in 1887, ran headlong into the inertia of the parent country's bureaucrats. Fleming's missionary zeal in favour of state intervention also met with hot opposition from the British company Eastern Telegraph, which had a monopoly on communications between London and British possessions in Asia and Oceania. Nothing discouraged Fleming, whose principal argument was military: what would happen to communications between Great Britain and its empire if Turkey, which the Eastern Telegraph line crossed, interrupted traffic? (That is precisely what did happen during the First World War, when Turkey allied itself with Germany against Great Britain.)

The project should have failed many times, because of Fleming's refusal to let the cable run through an unallied country; he wanted it to be one hundred percent on British territory. However, the technology of the time was not advanced enough to cross the Pacific with a single continuous cable, so surface links had to be planned. Japan was asked to cede its sovereignty over one of the Kuril Islands, but it refused. Fleming went so far as to finance an expedition to a rocky outcropping called Necker, in the middle of the Pacific, in order to take possession of it in the name of the British Crown (it had been claimed by Hawaii, which was an independent country at the time).

In the end, Fanning (now Tabuaeran), an island on the equator, was chosen. But the distance from Fanning to Vancouver was greater than the length of any existing undersea cable. Great Britain agreed to construct a gigantic cable-ship, the *Colonia*, which was capable of transporting ten thousand kilometres of cable, to link Vancouver to Fanning, then to Fiji and Norfolk, with one branch to Australia and another to New Zealand. On 31 October 1902, Fleming sent from Ottawa two messages around the world, one to the east and one to the west. The old imperial battler had won his wager.

The date is important, for it marks the advent of the world telecommunications network. The time of adventure was giving way to a period of technological planning and financial rationalization (even if, in Canada, the latter would often be revealed to be deficient).

This globalization of information exchange arose from the notion of imperial Great Britain and not from any international ideology. The bringing together of peoples mentioned in the official discourse of the era concerned exclusively the various branches of the British Empire. In the creation of this world network, Canada did not follow a strictly national plan but played a leading role.

MOVING TOWARDS DUOPOLY

Even within Canada, the situation evaded any attempt at order. With the erection of Canadian Pacific's line, the telegraphic monopoly established by the magnate of Great North Western, Erastus Wiman, was broken. As in the time of the duel between Montreal Telegraph and Dominion Telegraph, the Canadian telegraph network was drawn to an unusual structure: duopoly.

Wiman had, in fact, tried to purchase CP's Telegraph Division in 1883, when the company found itself short of funds in the middle of constructing the transcontinental railway. He was stopped short by the categorical veto of the general manager of the company, William Cornelius Van Horne. An intense hatred grew between the two men, and in spite of the financial support of Western Union, Wiman completely failed in his initiative. Ironically, it was a transplanted American, Van Horne, who defended the interests of Canada against a Canadian, Wiman, who had made his home in the United States.[29]

Great North Western and CP did not constitute a perfect duopoly since only CP's network covered all of Canada. Great North Western, which offered only telegraph services, was concentrated in the most populous areas of Canada (Ontario, Quebec) and the Maritimes.

At the turn of the century two new transcontinental railways were started by Canadian Northern Railways (1899–1915) and Grand Trunk Railway of Canada (1902–14). Naturally, each of these rail lines was doubled by a telegraph line. Canadian Northern Telegraph was founded in 1902, and Grand Trunk Pacific Telegraph in 1906.

The creation of these companies corresponded for more to political speculation than to economic necessity. The wave of populism[30] that was breaking over Canada at the beginning of the century, especially in the West, had raised public opinion against CP's monopoly. The federal and Manitoba governments bowed to this pressure and largely encouraged and financed the new companies. Since business circles could not let an opportunity to enrich themselves go by, companies sprang up over a few years outside of any economic justification.

Was this the definitive breaking up of the duopoly structure of Canadian telegraphy? There was no room in Canada for three parallel railway networks from sea to sea, nor for a proliferation of telegraph networks. The first victim of this artificial and vicious competition was Great North Western, which was having increasing difficulty surviving the combined assaults of CP and the newcomers. On 1 January 1915, Canadian Northern purchased Great North Western from Western Union and merged the two networks.

Canadian Northern and Grand Trunk Railway were in desperate straits, however, having both gone into debt and become insolvent. An inevitable second restructuring movement a few years later led Canadian telecommunications to its "manifest destiny," which seemed, yet again, to be duopoly.

The Maritimes, it should be noted, constituted a special case. The New Brunswick and Nova Scotia systems remained the property of Western Union, so the duopoly consisted of CP and Western Union. Prince Edward Island and Newfoundland remained the exclusive domain of Anglo-American Telegraph.[31]

Canada embarked on the twentieth century in a unique position. While almost everywhere else telegraphy was a monopoly – government monopolies in Europe, a private monopoly in the United States – Canada seemed to be in a state of complete anarchy. The public sector had yielded to the private sector, then created artificial competition by a policy of arbitrary grants to railway companies.

It should be remembered that there was a small government sector limited to unprofitable lines. The Government Telegraph and Signal Service responded to social (services to isolated populations) and political (presence in the Far North) needs. Never did the federal government consider endowing itself with a tool of direct intervention in the telecommunications sector, as was the case with the European PTTs.[32] Imperialists like Sandford Fleming had planned a coherent strategy around nationalization, but their influence, real though it may have been, was never dominant. On the contrary, Canada seemed incapable of resolving to practise a policy of *laisser-faire* analogous to that in the United States, perhaps precisely because of the economic weight of her large neighbour. The result was repeated but contradictory interventions, without an overall plan and often with counterproductive results.

Finally, telegraphy developed in Canada as if the environment were purely Anglo-Saxon. The main reason for this was no doubt historic: telegraphy arrived in Canada right at the time of the failed rebellions of 1837–38. The French Canadian landed gentry had been defeated and were incapable of catching on the fly business opportunities that should have been theirs in the dawning industrial age. Since it concerned the field of telecommunications, which meant opening to the outside world, their failure had major consequences. In Canada, telecommunications ended up one hundred percent Anglo-Saxon, and the situation was no different, as we shall see, when it came to the telephone.

TELEGRAPHY IN THE REST OF THE WORLD

Before moving on to the telephone, here is a brief comparison of Canadian telegraphy with the industry in the rest of the world. One of the best parameters for measuring telegraphic activities in a country is to calculate the number of telegrams sent per year per thousand inhabitants (see Table 1). I chose to compare Canada with a group of industrialized countries, including two other Commonwealth countries, Australia and New Zealand. I will use these countries as a basis of comparison throughout the work.

Telegraphy was the technology *par excellence* of Victorian Great Britain. The British Empire was founded on the basis of written communications transmitted over long distances by electricity. With regard to telegraph service, North America lagged well behing Great Britain. This is not because the importance of communication was not well understood but because in the New World telegraphy was very quickly overtaken by rapid development of the telephone.

On 15 May 1865 representatives from twenty European countries met in Paris and signed the International Telegraph Convention standardizing telegraphic traffic rules. This conference also founded the International Telegraph Union, the first international organization, predating even the Universal Postal Union. Russia and Turkey were part of the initial group, but the United States, Canada, and even Great Britain were not, since by unwritten rule membership was limited to countries whose telegraph services were operated by a government. Great Britain did not become a member until 1871, when her telegraph system was rationalized. The Paris conference adopted the Morse alphabet as the international code and French, the perfect diplomatic language, as the working language.

The ITU's convention was completely rewritten at St Petersburg in 1875, and the result was deemed so satisfactory that there were no more conferences of plenipotentiaries until 1932, when one was held in Madrid. (After St Petersburg, Great Britain obtained two votes, one for her own telegraph system and one for India's telegraph system. Canada was not a beneficiary of

Table 1
World Telegraph Density (dispatches per 1,000 habitants)

	1865	1875	1885	1895	1905	1910
Australia	n/a	n/a	2,580	2,026	2,835	3,690
Canada[a]	n/a	n/a	905	1,198	1,183	1,256
France	83	298	858	1,170	1,436	1,648
Germany	90	326	408	713	838	907
Great Britain	n/a	877	1,484	2,680	2,752	2,540
Japan	n/a	n/a	70	216	506	n/a
New Zealand	n/a	n/a	2,950	3,109	6,160	8,090
Sweden	222	268	254	442	627	772
Switzerland	236	1,110	1,057	1,353	1,572	1,544
United States[b]	n/a	477	745	832	887	815

Notes: a. The figures for 1885, 1895, and 1905 were not available, so figures shown are for
1887, 1897, and 1904. See Fortner, *Messiahs*, 9. Includes telegrams for Great North
Western, CP Telegraphs, Western Union, and the federal government.

b. Western Union Annual Reports, US Department of Commerce, Bureau of the
Census.

Source: Statistical yearbooks of the International Telegraph Union.

this voting system until 1907 since its telegraphic lines were mainly private.
Canada was never part of this first telegraphic organization, and it wasn't until Madrid and the birth of the current ITU that it joined the ranks of member
countries.[33]

The Telephone

The nineteenth century marked the apogee of the printed word, which triumphed thanks to the marriage of the press and the steam engine. Not everyone knew how to read and write, but writing was considered the noblest form of communication.

When the telephone came into being, there were doubts about how long the technology would last. Who would dream of making orders on an instrument that left no written record? Business management and government administration – in short, serious matters – were expressed in writing and, if possible, printed.

Nevertheless, the telephone made a remarkably favourable impression on public opinion in the nineteenth century. The hitch was that it was treated like a luxury toy. It wasn't the poor relative of the telegraph but – worse – its spoiled child. On top of that, aside from scientific curiosity, of what use was the telephone?

3 Invention of the Telephone

Like telegraphy, telephony resulted from the confluence of many factors. In this case, they all converged on one man, Alexander Graham Bell, Scottish humanist, and on one city, Boston, and the effervescent scientific atmosphere at the Massachusetts Institute of Technology. The ties the Bell family established with Canada provided this country with a special place in the geography of the new technology.

As it had been for the telegraph, research on the telephone was linked to advances in research on electricity, which had begun in the early nineteenth century. In 1837 two American physicists, Joseph Henry and Charles Grafton Page, discovered that a metal rod subjected to rapid fluctuations in a magnetic field emitted sounds. In 1860 a German professor of physics and music, Philipp Reis, transmitted a melody from a cone covered with a membrane (the transmitter) through an electrical circuit to a knitting needle (the receiver). Variations in current in the knitting needle reproduced certain sounds. Three years later, Reis found that words could be reproduced in this way.[1] Thus, when Alexander Graham Bell conducted his decisive experiments, research on telephony was far less advanced than telegraphic research had been when Samuel Morse proposed his simple solution, and the process of invention was clearly more sophisticated.

Graham Bell[2] was born in Edinburgh on 3 March 1847 into a family of professors of diction. His grandfather had inaugurated the tradition; his father, Alexander Melville Bell, had continued it and added a scientific dimension. Melville developed a universal sign alphabet for deaf-mutes, and in 1867 he published a book called *Speech Made Visible.*

The Bell family's house in Brantford, Ontario (in one of the most temperate climates in Canada), between 1870 and 1881. Alexander Graham Bell regained his health there in the winter of 1870–71, and returned regularly for vacations. Courtesy National Library of Congress.

In the space of three years, Graham's mother went deaf, his two brothers were killed by tuberculosis, and he himself seemed to fall ill with the disease. A Canadian Baptist minister, Thomas Henderson, convinced the Bell family that the climate in southern Ontario would be beneficial to their surviving son's health. So in August 1870 the family emigrated to Canada and settled in Brantford, Ontario, a small town a hundred kilometres southeast of Toronto.

Brantford was a tonic for the Bell family; Graham quickly regained his health and spent the autumn of 1870 studying the work of German physicist and physiologist Herman von Helmholtz on the decomposition of speech sounds into harmonics.[3] In April 1871 he went to live in Boston, where he conducted the experiments described by Helmholtz. Since he lacked sufficient knowledge of electricity to build an artificial-speech apparatus, he went to the Massachusetts Institute of Technology to study electricity and learn about the state of electrical research in Europe. He remained an acoustical specialist with a good knowledge of electricity, and never became the reverse. Later he admitted, "if I had known more about electricity and less about sound, I never would have invented the telephone."

Bell's first experiment involved putting a tuning fork in contact with a bath of mercury. With each vibration of the tuning fork, an electrical circuit was closed; by induction, another tuning fork began to vibrate at the other end of the circuit. This ingenious apparatus was far from a telephone but close enough to be called "harmonic telegraphy," or multiplexing. Bell's timing was good since at that moment telegraph companies were engaged in a race to develop multiplexing technology; he was eager to sell them his improved apparatus in order to finance his research on speech.

In the spring of 1874 Bell gave a series of lectures on speech at MIT and received permission to use the institution's scientific instruments. In the MIT lab, he discovered a phonautograph composed of a diaphragm and a bristle. When a voice made the diaphragm vibrate, the bristle made an undulating curve on a plate of smoked glass. Though he didn't yet know it, Bell now had all the theoretical material that he needed to invent the telephone. What remained was to put together the few pieces of the puzzle. That summer, during a stay at the family home in Brantford, his mind made the final leap. He explained to his father, in a conversation that has become famous, the new orientation of his research: "If I could vary the intensity of the electrical current in exact proportion to the variation of the air density in the production of words, I would be able to transmit speech by telegraph."

THE TELEPHONE: WELL-PLANNED R&D

The one thing Bell lacked to conduct his final experiments was money. The fathers of two of his students, Thomas Sanders and Gardiner Greene Hubbard, provided him with venture capital, and he rented a workshop in the attic of an electrical shop belonging Charles Williams.

Williams also lent Bell the services of Thomas A. Watson, a twenty-one-year-old mechanic. The son of a stable worker, Watson had quit school to try out different trades, and he was noted for his initiative. He was Bell's indispensable Sherpa and shared in the honour of the invention.

With Watson's help, Bell's apparatuses acquired a technical precision that they had heretofore lacked, and the tuning forks were replaced by reeds similar to those in clarinets or saxophones. In June 1875 a routine verification of one of the reeds in the system resulted in the first transmission of musical sounds. Bell thus discovered that a single reed could transmit compound sounds, and he took a giant step on the road to simplicity: no need to plan for a reed for every pitch. Immediately, he abandoned telegraphic multiplexing for telephony.

Bell built a new apparatus in the shape of a vertical stand with a crosspiece in which a parchment membrane moved a single reed. The apparatus could both transmit and receive sounds; it was completely "reversible." In July 1875 the first human sounds were transmitted from one room to another in the attic of the Williams workshop. However, the communication was too jumbled for the words to be distinguishable, and Watson heard Bell's voice more clearly through the walls than through the prototype telephone.

Bell and Watson then launched themselves into a race to the finish. The American scientific community was not large, and they knew that Thomas Edison and Elisha Gray were conducting similar research. The thirty-nine-year-old Gray was a self-made man: the son of a small-holding farmer, he had had to leave school when his father died and had become a jack of all trades, studying electricity at night, and he had patented some improvements to the telegraph. In 1869 he and some associates had acquired a small manu-facturer of telegraphic equipment that would later become a major player: Western Electric.

In September 1875 Bell went to Brantford and wrote a first draft of his patent application, which he called "Updating of the Telegraph," in order to justify the aid contributed by Sanders and Hubbard. In fact, their financial support was running out. Equipment was more and more expensive. Bell ceded half the rights for the British Empire (including Canada) to brothers George and Gordon Brown of Toronto for the modest sum of twenty-five dollars per month for six months. George Brown was one of the most influ-ential Liberal MPs in Parliament (he had briefly been prime minister) and the publisher of the *Toronto Globe*. To be valid in Great Britain, a patent application had to be exclusive, so the telephone had to be patented in Great Britain before the United States. George Brown was on his way to England in any case, and Bell asked him to take the necessary steps. In the end, however, Brown did not pay for this first telephone "subscription" and did not deposit the patent application in London for fear of looking ridiculous.[4]

Meanwhile, Hubbard, who was awaiting word from England before de-positing a patent application in the United States, thought he might lose on both fronts. Fortunately, he gave in to his own impatience and deposited the application at the Patents Office in Washington at two o'clock in the after-noon on 14 February 1876. The time of day was crucial, since Elisha Gray deposited his application at four o'clock on the same day. The two tech-niques described in the applications were almost identical; only the time of deposit distinguished them.

BORN IN BOSTON

Bell may have won the patent race, but his telephone still didn't work; neither did Gray's. The Patents Office officially registered Bell's application on 7 March 1876. Neither the word "telephone" nor the word "speech" ap-peared anywhere; the term "vocal or other sounds" was used. A few days later, Bell and Watson finally found the winning combination: they replaced the metal reed with a wire submerged in a bath of water and acid. The varia-tions in resistance of the mixture in the mouthpiece created a modulation of

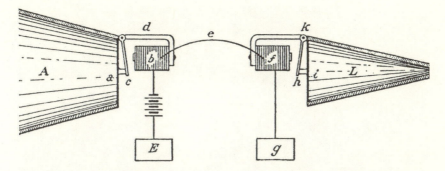

Diagram from patent No. 174456 granted to Bell on March 7, 1876, accompanied by the following explanation: "The armature, c ... is fastened loosely by one extremity to leg d of the electromagnet b, and its other extremity is attached to the centre of membrane a. A cone, A, is used to converge sound-vibrations upon the membrane. When a sound is uttered in the cone, the membrane a is set in vibration, the armature c is forced to partake of the motion, and thus electrical unulations are created upon the circuit Ebefg. These undulations are similar ... to the air vibrations caused by the sound."

current intensity in the wire in the same way that sound waves oscillated in the atmosphere. On 10 March 1876, Watson heard in his instrument Bell's famous words, "Mr Watson, come here, I want you," followed by the singing of "God Save the Queen." The telephone was born.

Strangely, the transmitter into which Bell pronounced the first telephonically transmitted words was different from the apparatus described in his patent application; in fact, it was absolutely identical to Gray's. Bell most certainly went to Washington after the simultaneous deposit of applications, and he would have known about Gray's concept of the liquid-filled transmitter. Of course, he always maintained that the examiner at the Patents Office had not shown him Gray's application but had simply described the nature of his transmitter. When Gray learned of this years later, he was convinced that Bell had had access to confidential information.[5]

In any case, it was Graham Bell who built the first functioning telephone, while Gray was still resolving theoretical problems. In 1876 only Bell had an overall vision of the telephone and was able to integrate the various practical and theoretical elements in one instrument.

Bell officially launched the telephone in June 1876 at the Philadelphia World's Fair, which attracted a large number of scientific and political celebrities. Bell exhibited his still-primitive instrument, not expecting to draw much attention. But then Pedro II, emperor of Brazil, stopped in the room where Bell was and picked up the telephone. "My God," he exclaimed, "this thing speaks!"

Alexander Graham Bell was twenty-nine when he invented the telephone.

THE FIRST LONG-DISTANCE CALLS

When he got back from Philadelphia, Bell went to Brantford, as was his custom, to spend the summer at the family home. That year, he used his visit to make further advances to the telephone. Talking between two rooms was fine, but it wasn't the stunning demonstration Bell wanted; he wanted to prove that the telephone could link cities. Since it would be too expensive to lay dedicated telephone lines for a single experiment, he wanted to use the telegraph wires of the local company, Dominion Telegraph.

Bell wrote to the general manager of Dominion Telegraph in Toronto, Thomas Swinyard, to ask if he could lease the telegraph line between Brantford and Paris, a village thirteen kilometres away, for one hour. "Another crackpot!" Swinyard exclaimed when he read the letter explaining the reason for this unusual request, and he made a note to the attention of the Toronto office manager, Lewis McFarlane: "To file in the wastepaper basket." McFarlane understood at once the significance of the new technology and supported Dominion Telegraph's participation in the experiment. He convinced Swinyard to change his mind. (Forty years later, McFarlane became the third president of the Bell Telephone Company of Canada.)

On 3 and 4 August 1876, two preliminary experiments took place between Brantford and Mount Pleasant, a neighbouring town. Bell's uncle, David, recited the Hamlet soliloquy "To Be or Not to Be" and sang songs, while five kilometres away a few Mount Pleasant residents, gathered in the shoemaker's shop, which was also the telegraph office, picked out a few words among the sonic snow.

On 10 August 1876 the experiment was repeated between Brantford and Paris, Ontario (where the telegraph office was also in a shoemaker's shop). This time, the mayor and the entire village listened for an hour while David Bell recited speeches from *Macbeth*; they refused to leave the telegraph office at the designated time. Since the line went only one way, from Brantford to Paris, they had to telegraph to ask for an extension. A dialogue – telegraphic in one direction, telephonic in the other – took place, during which the listeners requested their favourite song or poem. The small crowd finally dispersed at eleven o'clock at night, terribly late in prudish Ontario.[6]

When Bell rejoined Watson in Boston, he succeeded in developing a bidirectional system, and in October 1876 the two men exchanged the first telephone conversation, between the Boston workshop and its branch in the neighbouring city of Cambridge. After a new round of fruitful experiments, Bell deposited a new patent in January 1877 to cover the most recent improvements: replacement of the electromagnet and battery with a permanent magnet, and of the parchment membrane with a metal membrane. This two-way telephone turned out to be easy enough to use to launch commercially.

A CANADIAN OR AN AMERICAN INVENTION?

From the beginning, Canada and the United States have had a tug-of-war over where exactly the telephone was invented. On a personal level, Alexander Graham Bell was British by birth, which automatically made him a "citizen" of the young dominion of Canada. He applied for American citizenship in October 1874, to simplify the steps in obtaining patents on the telephone but did not obtain his naturalization until November 1882. Thus, at the time the telephone was invented, Bell could not legally be considered an

Above: In August, 1876, Alexander Graham Bell conducted a unidirectional test of long-distance telephony from Brantford, Ontario. When he heard the voice of his father, Melville, he asked by telegraph, "Was the last speaker my father?"
Below: At the other end of the line, thirteen kilometres away in Paris, Ontario, Melville Bell responded by telephone, "Yes, Alec, it was I." Courtesy Bell Canada.

Alexander Graham Bell at his workbench in Washington. After he left National Bell, he divided his time between Washington, where he became president of the National Geographic Society, and Nova Scotia, where he built a summerhouse. Courtesy National Library of Congress.

American. On the other hand, he had lived in Canada for only a few months, from August 1870 to April 1871. Although he spent every summer until 1876 at the family home in Brantford, he did not develop much sense of being Canadian; rather, he saw himself as a Scotsman living in the United States *en route* to American naturalization.

Nor did Bell invent the telephone by himself. He worked with Watson, who was to Bell what Sherpa Tenzing Norgay was to Sir Edmund Percival Hillary in the conquest of Mount Everest. Watson was an active participant in the invention of the telephone and Bell was the first to recognize his contribution, to the point of ceding to him a portion of his rights on exploitation of the patents.

Those who support the "Canadian connection" have a fall-back position: it was during his 1874 summer vacation in Brantford, on Canadian soil, that Bell had the "revelations" linking his research on the harmonic telegraph and his MIT experiments on the phonautograph. But can this "revelation" be considered the invention of the telephone? Bell himself, asked to choose between the two countries, gave a Solomonesque answer: "The telephone was conceived in Brantford and born in Boston."[7] It is true that Bell mentioned the "Brantford invention" in certain documents, but these had been modified on the advice of his lawyers to demonstrate that his work preceded that of his competitors. By 1874, in fact, Bell's research had not yet given rise to anything concrete.

More importantly, an invention cannot be isolated from the fertile ground on which it comes to life. It is obvious that the telephone was born in 1876 in the burgeoning technological culture of MIT and supported by Boston capital. The primary motivation of Bell's investors and associates was to sell their research to Western Union with its unlimited material resources. Looking beyond anecdotal evidence to the conditions of the invention, and especially its commercial development, it is undeniable that the telephone is an American invention. This fact was to weigh very heavily on the beginnings of Canadian telephony by stamping the imprint of the north-south axis upon it.

THEODORE VAIL AND THE TELECOMMUNICATIONS EMPIRE

A number of companies quickly sprang up to exploit the new invention. The most important was, of course, the telegraph giant Western Union. After long neglecting the telephone (Hubbard had offered it the rights to the technology for $100,000 in the winter of 1876–77), Western Union purchased the Edison patent and launched a major marketing effort. The Edison telephone was better than Bell's but its patent used elements that were part of the overall concept of telephony, which fell under Bell's patent. Western Union thought that it could intimidate Hubbard and Sanders with its huge assets: not only did it instigate a price war against companies that used the Bell telephone, but it went so far as to chop down their telephone poles. In June 1878 Graham Bell's original backers, Hubbard and Sanders, enlisted two men to help them to counter attack: Colonel William H. Forbes, a rich merchant of French extraction who headed the American Bell Company, and Theodore Newton Vail, the American postmaster general. The new management's first move was to bring lawsuits against Western Union. Its second move was to change the statutes of Bell Telephone Company so that its shares could be sold on the stock market. Finally, the Vail-Forbes tandem pushed Hubbard and Sanders aside. Bell voluntarily left the board of directors, where he felt like a stranger.

On Forbe's request, Bell worked one more year for the company and applied for six patents, only one of which was used in a commercial application (the twisted-wire metallic circuit), but his mind was elsewhere. He expressed his lack of interest later: "I have not kept up with the literature of telephonic research."[10]

Watson continued to work for National Bell until 1881 as chief engineer. Throughout these decisive years, he fulfilled the role in technology that Vail fulfilled in management, tirelessly travelling the United States to make sure that exchanges were working properly and inspecting plants that made telephone equipment under Bell's licence. He then went on to a life free of material worries: he studied geology at MIT, owned land, led the ephemeral Nationalist (read: populist) Party, and founded one of the largest shipyards in the United States (which he lost in 1903). His last career was as a Shakespearean actor and memoirist.[11]

Even after his departure, Bell maintained ties with National Bell, then with American Bell and AT&T, and he was a witness in the huge number of trials (six hundred of them) in the battle of the patents. He divided his time between Washington, DC, where he became editor of the magazine *National Geographic*, founded by Hubbard, and, Baddeck, Nova Scotia, where he built a summerhouse. He used his fortune to help people who were hard of hearing and worked on various inventions, none of which reached the commercial stage. His most notable success was again linked to the telephone, which he transformed into an audiometer to measure degrees of deafness. The scientific community adopted his name as a standard of international measurement of relative differences in the intensity of sounds, the decibel. When Bell died, in August, 1922, all telephone communications in the United States, Canada, and Mexico were stopped for one minute.[12] At the time, there were thirteen million telephones in the world.

4 The Telephone Comes to Canada

The telephone came to Canada thanks to Alexander Graham Bell's father, Melville. This rather unusual introduction gave the Canadian industry a special connection to the United States, a relationship that persisted in various forms until the gentle separation of Bell Canada from AT&T in June 1975.

On 10 July 1877 Melville Bell received 75 percent of the rights to the Canadian patent on the telephone. As a professor, he was not predisposed to a career as a corporate leader, and he immediately turned to Thomas Henderson, the Baptist minister, now retired, who had advised him to come to Canada a number of years before. The fifty-eight-year-old professor and the sixty-one-year-old pastor organized a speaking tour throughout eastern Canada to explain and sell the telephone: "We have the feeling of entering a supernatural world and hearing the very essence of sounds. It is up to the interlocutor to make sure that these airborne shapes are imprecise or have clear contours. The telephone transmits them weakly, but with perfect definition, if they are articulated in such a manner by the lips of the interlocutor."[1]

This naïve formulation by the founders of the Canadian telephone industry may seem whimsical, but it should not be underestimated. Bell and Henderson had a very clear vision of the telephone's potential, unlike most businessmen of the day: "Experiments are still going on to render the instrument more sensitive and more resonant; and there seems no reason to doubt that it will ultimately be made to convey the voice over our longest submarine cables, and to bring England and America, Australia and China all within mutual speaking distance. At all events we know that the telephone in its present condition will speak to the remotest of our cities – to the extremities of our deepest mines – and to the most distant of our lighthouses and lightships."[2]

When they started, Henderson and Bell didn't think of forming a company; they simply wanted to rent out telephones at forty dollars a pair – Church ministers, teachers, and physicians were offered a reduced rate or even an exemption. Subscribers were invited to connect their equipment by means of the iron wire of metal fences, a common practice in the infancy of North American telephony.

On 18 October 1877 three Hamilton businessmen received telephones. They were the first subscribers in Canada. Fifteen days later, Bell and Henderson managed to rent a pair of telephones to Prime Minister Alexander Mackenzie, linking his office to the residence of the governor general, Lord Dufferin. Salesmen thereafter cited the latter, more prestigious subscription as the first in Canada, evidence that they had quickly caught on to the tricks of marketing. However, things did not go that well at first: Mackenzie complained about the poor quality of his apparatus and threatened to cancel his subscription. Fortunately, Lady Dufferin liked to play the piano and sing on the telephone; she pleaded in favour of the instrument, and the prime minister kept his phone.

Among those first three Hamilton subscribers was Hugh Cossart Baker, the president of a local railway company. The following year, he acquired the small Hamilton District Telegraph Company, which had laid telegraph lines for subscribers who wanted to be linked to the fire station, police station, or other subscribers. A rudimentary central exchange had been set up for this purpose. Some subscribers, including Baker himself, used the telegraph to play long-distance chess. He quickly understood the advantages of the telephone over the telegraph, and in February 1878 he sought permission from the town of Hamilton to erect telephone poles.

In July Baker established the first Canadian telephone exchange. The number of his subscribers nearly quadrupled in less than half a year, from forty in December 1878 to 150 in April 1879. He also obtained from Melville Bell the exclusive right to operate a telephone service between Georgian Bay and Lake Erie, a territory that included Hamilton. Baker had realized that the future of the telephone depended on the creation of a network: subscribers had to be offered access to as large a pool of users as possible.

All of the first telephones marketed in Canada were manufactured in the United States by Charles Williams, Melville Bell's minority shareholder. Very quickly this situation grew untenable. First, Canadian law stipulated that items protected by a patent had to be made in Canada after one year. Second, import duties on the imported apparatuses were prohibitive, practically doubling the subscription price (from ten to nineteen dollars per year). This protectionist policy alone, however, would not have sufficed to protect the Canadian market against American telephones; it was the Americans themselves, completely overwhelmed by their own domestic demand, who could not satisfy Canadian demand. Williams found himself incapable of delivering the one thousand apparatuses due in exchange for his share in the Canadian rights. He delivered half, and those with difficulty.

Bell and Henderson very quickly had to come up with another solution. Just a few months after they began to exploit the patents, they sent a twenty-eight-year-old metal worker, James H. Cowherd, to take a course at the Williams factory in Boston. When he returned, Cowherd built a shop behind the family store in Brantford and in December 1878 the first "Canadian" telephones saw the light of day. In this regard, there are some corrections to be made to the official historiography, according to which the instruments were made entirely in Canada.[3] In fact, the parts were made in Boston and then sent to Brantford for assembly. This distinction is important, as it would be used as the basis for cancellation of the Canadian patents several years later.[4] Nor was Cowherd the young entrepreneur that he is often portrayed as, but a salaried employee of the Williams workshop. A total of 2,398 instruments rolled out of Cowherd's shop before his unexpected death in January 1881 put an end to this unhappy attempt.

Meanwhile, Bell and Henderson were touring Ontario in a small open carriage with a trunk full of wooden telephones, spending considerable energy without really getting beyond the door-to-door stage. They had sales representatives but little accounting expertise, and the maintenance and repair problems were growing along with the number of subscribers. Quebec was neglected, not for any political reason but because the two men didn't have connections there: only two cities in Quebec were served (Montreal and Quebec City), as opposed to twenty in Ontario.

This haphazard marketing strategy caused considerable discontent. Among the first generation of telephone companies created in this era in Canada, only two aside from Baker's obtained an operating permit from Bell: Toronto Telephone Despatch and York Telephone Despatch, both founded by Hugh Neilson. In Winnipeg, a Bell agent sold some telephones but didn't set up an exchange. The others simply ignored patent rights – and who could blame them? A total of fifteen towns in Ontario, Quebec, Nova Scotia, New Brunswick, Manitoba, and British Columbia wanted to set up telephone companies.

Once plans for making a telephone were published in *Scientific American*, large numbers of amateur electricians made their own instruments, and some sold them, although they lacked the sound quality of the Bell apparatuses. The most famous of these entrepreneurs was Thomas Ahearn, a telegraph operator in the Ottawa region who made phones with cigar boxes. This anarchic activity reinforced the idea that the telephone was simply a scientific toy.

One Quebec jeweller, however, was an exception. Cyrille Duquet developed what was probably the first telephone apparantly in which the receiver and transmitter were integrated, as they are now, into a single unit and pat-

Above: The pioneer of Quebec telephony, Cyrille Duquet, founded the first French-speaking telephone company and unsuccessfully competed with Bell. Courtesy Ministère des Communications du Québec.

Below: The first telephone handset in the world was probably made in Quebec. A jeweller in Quebec City, Cyrille Duquet, patented his invention in February 1878. Courtesy National Archives (PA 95 417).

ented it in February 1878. He then started up the Québec and Lévis Telephone Company and began to market his apparatus. A year later he installed
a first line between his downtown store and the suburb of Sillery. The first
telephone line in Montreal, between the Saint-Sulpice seminary and the
Côte-des-Neiges cemetery, used Duquet telephones. Duquet also experi-
mented with long-distance calls between Quebec City and Montreal using
telegraph lines.[5]

TELEGRAPH COMPANIES EYE THE TELEPHONE

The time of amateurs came to an end once the major telegraph companies
became involved. Montreal Telegraph and Dominion Telegraph began systematically to market the telephone in Canada, lending it a "goldrush mentality."[6]

As shown in chapter 2, these two competitors had waged a merciless war
in the telegraph market. Montreal Telegraph, which had always had excellent
relations with Western Union, decided to market Edison and Gray's apparatus, which had better sound quality but was more awkward to use (the crank
that generated electricity had to be turned as the user talked, while the Bell
instrument used permanent magnets that did not require current).

Dominion Telegraph had been collaborating with Bell since the world's
first long-distance call was made from Brantford to Paris, Ontario. In February 1879 Dominion Telegraph became Melville Bell's official representative,
marketing his instruments throughout Canada except in the Hamilton, Toronto, and York regions, where Bell had already sold his rights. Lewis
McFarlane, the enterprising manager of the Toronto office, was named director of Dominion Telegraph's new telephone division.

From 1878 to 1880, full competition characterized the Canadian telephone
industry for the only time in its history until 1992. The price war between
Montreal Telegraph and Dominion Telegraph in telegraphy found its natural
extension in telephony. Doctors and religious leaders often received free service, and during promotional periods new subscribers received three months
of free service. During 1879, the first full year of operation, each company
invested $75,000 for an insignificant return; the public was still divided be-
tween enthusiasm for the novelty of the telephone and scepticism regarding
its practical applications. Competition also meant that Montreal Telegraph
subscribers could not communicate with Dominion Telegraph subscribers,
and city streets were soon invaded by a profusion of poles and wires.

Meanwhile, Melville Bell decided to join his son in the United States and
return to his studies on the education of deaf-mutes. After consulting with
Charles Williams, he set the price for the Canadian rights at $100,000. Dur-
ing the summer of 1879, he contacted Dominion Telegraph, which deemed
his proposition, but not his price, interesting. McFarlane knew that the tele-

phone would be an issue in the war with Montreal Telegraph, but he estimated its value at between $5,000 and $12,000. In Quebec City, Duquet was also contacted, but he could raise no more than $3,000. No one in Canada wanted or was able to purchase the rights to the telephone at the asking price.

At this point, Graham Bell gave things a nudge in the right direction: he persuaded Hugh Baker to come to the aid of his father. Baker rapidly drew his conclusions from the failure of the negotiations with Dominion Telegraph. He turned to the United States, where AT&T (called National Bell at the time) had concluded a triumphant agreement with Western Union in September 1879. Here a curious phenomenon arose, one that recurred a number of times: instead of leaping at the chance to expand their operating territory into Canada, the Americans hesitated. While telegraphy had been marked by a constant pressure by US interests to invade a Canadian market that was already strongly structured, nothing similar occurred in the telephone industry. The Americans seemed to intervene only when called upon to do so, and even then they limited their intervention and withdrew at the first opportunity.

When the new president of National Bell, Colonel William Forbes, agreed to acquire the Canadian rights, he did so for reasons that were strategic rather than financial. Viewed from Boston, the Canadian situation was ominous: two telegraph companies were competing so fiercely that the future of the telephone was at risk in public opinion. Because one of the two parties was Montreal Telegraph, it was not difficult to see the hand of Western Union, National Bell's main rival, in the action.

Thus, two years after Graham Bell ceded the Canadian rights to his father, they returned to the United States. The main beneficiary of this move seemed to be Baker, who was already acting as president of Bell companies in Canada. One question nagged at him: would he establish his head office at Hamilton with the rest of his businesses, or in Toronto, as the national scope of the new company would seem to require? With support from Boston, over the winter of 1879–80 Baker wrote the charter for a new company, Bell Telephone Company of Canada,[7] and set the incorporation procedures in motion. Melville Bell resigned from the board of directors in June 1880 at the first annual general meeting of shareholders and moved to Washington DC, to be with his son, Graham.

CHARLES SISE:
THE FOUNDING FATHER OF BELL IN CANADA

The mess in Canada had to be cleaned up. Hugh Baker did not have the necessary skills to run a large company. It was Forbes who came up with the solution: he recruited Charles Fleetford Sise to represent National Bell's interests in Canada. Sise, who arrived in Montreal on 9 March 1880, was an

Charles Sise ruled Bell with an iron fist between 1880 and 1915. An American exiled to Canada, his only homeland was "his" company. Courtesy Bell Canada.

American renegade because of his Southern sympathies during the Civil War; a "man without a country," in the words of one of his biographers, who would identify body and soul with the Bell enterprise in Canada and not act as a temporary ambassador. Sise had no fall-back position. Nor did he adopt Canada as his country; he retained his American citizenship all his life. His only loyalty was to "his" company.[8]

Born in New Hampshire, Sise was raised in a home dominated by trade and the sea. His marriage to the daughter of a rich Alabama merchant brought the New England seaman into sympathy with the Southern cause.

When the Civil War broke out in 1861, this typical Northener sided with the Confederates, briefly fighting in the secessionist army and probably playing a role in setting up the Southern states' intelligence network. After two years of exile in Great Britain at the end of the war, he returned to Boston, where he was treated as a pariah because of his Confederate sympathies.

A Canadian firm with offices in the United States, Royal Canadian Insurance of Montreal, gave Sise a break by offering him a position as head of its American activities, but the recession of the 1870s was catastrophic for the insurance sector, and Sise had to close Royal Canadian's American office. His personal skills, however, were never in question, as attested to by the firm's president, William Robertson, who congratulated him for the efficiency and self-sacrifice with which he had seen this difficult mission through.

Forbes needed a man he could trust to conduct a commando operation in Canada: first, he had to create a Canadian telephone company to market Bell telephones; second, he had to encourage telegraph companies to cooperate among themselves and with the new company. In effect, Forbes wanted Bell in Canada to gather as much Canadian capital as possible but on the condition that ultimate control would remain in the hands of the American parent company. Charles Sise had the ideal profile. He had good business connections in Canada, he was available on the spot for a mission that Forbes thought would be of short duration, and his political and military involvement in the South would not be a handicap in a foreign country.

Sise was forty-five when he took charge of Bell's interests in Canada and he achieved Forbes's objectives masterfully, but in his own way. His actions were aimed at making himself indispensable to the point where he created a permanent job for himself in Canada. When he arrived in Montreal, he conducted simultaneous negotiations with the two competing telegraph companies and with holders of Bell's operating rights in Canada (Hamilton, York, and Toronto). At the same time, he completed the incorporation of Bell Telephone that Hugh Baker had started. Very soon, things were up and running and Baker was out of power: first, the charter was federal and not Ontarian, as Baker had initially planned; second, the head office was set up in Montreal and not Toronto. On 29 April 1880 the new company received a charter that gave it almost free rein; no mention was made of rates or regulations.

The president named on 1 June 1880, during the first general meeting in Toronto, was William Robertson, not Baker or any of the other telephone pioneers. The advantage of choosing Robertson was that he was completely new to the telecommunications sector and thus neutral – in other words, malleable. Sise became vice-president and general manager, and all real power was in his hands. The board of directors had eight members, three of them, Forbes, Vail, and Sise, American.

One might wonder why Sise left the presidency of Bell Telephone to someone else. The answer is as much due to Sise's fluid status with regard to National Bell as to his concern with conferring on the young Canadian enterprise a façade of respectability that only Robertson, as a known insurer, importer and wholesaler of fabrics, and president of the Port Commission of Montreal, could bring. Robertson played his role discreetly, and there was never any question who held the real power.

Robertson's legacy to Bell was his key decision to locate the head office in Montreal rather than Toronto. When he died, in 1890, he was succeeded as president by Sise, who remained in the position until 1915. In all, Sise was effectively at the head of Bell in Canada for thirty-five years.

Bell's American roots were a controversial issue for a long time. At first glance, AT&T (then called American Bell) held only 24.9 percent of voting shares in Bell Telephone, which was far from the majority and didn't even comprise the largest block of shares. Why did it have this holding, since, after all, it was a principal investor?

A second company had been created less than a month after the general meeting: Canadian Telephone. Following the American model, this company held the patents – and thus the real power – while Bell Telephone's only role was to market telephones under licence. The president of Canadian Telephone was none other than Theodore N. Vail.

This complex structure clearly meant that American Bell controlled the patents directly and their exploitation indirectly. The guardianship was flexible and discreet but real. The creation of a telephone industry in Canada was truly an American act.[9] American Bell owned seventy-five percent of the shares of Canadian Telephone, which in turn owned 44.2 percent of Bell Telephone, for a combined grand total of 69.1 percent – an indication of the extent of American Bell's power over the Canadian telephone industry. In practice, Bell Telephone's inferior position was translated into royalties paid to Canadian Telephone.

The extent of this initial foreign stranglehold makes the events that followed all the more paradoxical. In October 1882, Bell Telephone purchased all Canadian Telephone patents except one,[10] paying for the purchase with a two-for-one stock swap (two Canadian Telephone shares versus one of Bell Telephone's). The main result of this transaction was that American Bell lost its majority share in the Canadian telephone industry. It is true that it remained the largest single stockholder in Bell Telephone by far, with 46.4 percent of shares. But by turning Canadian Telephone into an empty shell, American Bell did away with its instrument of control over Bell Telephone. Canadian Telephone continued to exist on paper until 1892, when it was liquidated.

No satisfactory explanation of what, in hindsight, seemed an extremely important change has thus for emerged from the shadows of the boardrooms.

It is not known how Sise managed to convince American Bell to cede its majority to Canadian Telephone and become a minority shareholder in Bell Telephone. One gathers from his correspondence only that, with precision and flair, he began right away to concentrate all power in Bell Telephone, and he questioned the utility of maintaining a dual structure in Canada. In the United States, where it permitted one central company to keep control, with a minimum of capital, over a number of regional companies, a dual structure made sense. In Canada, however, there was only one operating company, the flagrant lack of capital outside of Montreal and Toronto having forestalled the creation of regional companies. Moreover, a dual structure was costly, especially for Bell Telephone, which was already having difficulty making its scheduled royalty payments to Canadian Telephone.[11]

The direct American hold on Canadian telephony lasted only about two years. No particular political vision should be read into this rapid emancipation, for Sise was completely unaware of the Canadian "national problem." On the other hand, he wanted to be the only master of his enterprise, as he had been the only captain of the ships he had commanded in his youth. He succeeded marvellously with his plan: American Bell's shares never reached the 50 percent threshold; they topped off at 48.8 percent in 1885 and declined continually thereafter, reaching zero in 1975. American Bell contented itself with controlling from afar what was already no longer really its Canadian subsidiary. Notably, it did not integrate it into the Bell System established in 1899.[12]

UNIFICATION OF
THE TELEPHONE INDUSTRY IN CANADA

In 1880, when Charles Sise was doing his utmost to unify the Canadian telephone industry, a major snag cropped up during negotiations. Just like Melville Bell, Sise failed to convince Canadian business circles to invest in the telephone. The directors of the Bell group in Boston had come up with a plan according to which Montreal Telegraph and Dominion Telegraph would form a sort of consortium with the small pioneering telephone companies, which American Bell would manage at arm's length thanks to its rights to the Canadian patents. By this plan, one-third of Bell Telephone's capital would be sold to the two telegraph companies, one-third to Baker and other telephone promoters, and one-third to the public.

The telegraph companies agreed to give up their telephone installations, but they insisted on being paid in cash, not in Bell Telephone shares. In fact, they simply wanted to get out of the telephone business; they were torn with internal strife and had been burned by Western Union's retreat to the United States in the face of American Bell. In Canada, Western Union had a lease to operate Dominion Telegraph and an agreement with Montreal Telegraph, which marketed its telephones. The situation was fluid and different from the analysis that had been made in Boston; however, it was not entirely hostile.

Sise was able to manoeuvre in the narrow window of opportunity that was opening, using the carrot and the stick by turns. He started by allying himself with the telegraph companies and agreeing to purchase their telephone systems. Making contact with Dominion Telegraph was easy because of the connections forged by Melville Bell, but the company still insisted on being paid in cash. American Bell finally purchased Dominion Telegraph's telephone system in July 1880 for $75,000. These installations were quickly traded to Bell Telephone for shares (hence American Bell's 24.9 percent direct ownership of the operating company).

Hugh Allan, the president of Montreal Telegraph, had no reason to be more accommodating: his price was $150,000 in cash for a smaller system than Dominion Telegraph's. Sise got the price down to $75,000, including $25,000 in stock. The old lion of Montreal finance threatened several times to break off discussions, to the point where he was sending ultimatums by employee go-betweens.

In all the deal making, the hardest person to convince was William Forbes, who wanted Bell Telephone to form a company with Canadian capital: "Although we feel that the telephone industry could become very important in Canada and that it has already attained a degree of development that, with good management, could make it very profitable in the near future, nevertheless, our policy there, as in the United States, consists of involving local capital, interests, and managers. The sector is much too large for us to envisage occupying all of it by ourselves."[13]

Sise proposed a leap forward: "If we occupy the entire territory, you will at least recover your investment." Already, he was saying "we" when he talked of Bell in Canada and "you" for Bell interests in the United States. This was in early summer 1880, and he was establishing a distance between Boston and Montreal. Evoking bright promises of the future, he then wrote, "With the monopoly, the difference that we could end up paying to attain a compromise would, I think, quickly be compensated for by the increase in value of the capital shares."[14]

Never were relations between Sise and Forbes so tense. It took all of Sise's diplomacy – and all the reluctance of Canadian business circles to invest in the telephone – for the Americans, and American Bell in particular, finally to take up the slack. In October an agreement was signed with Montreal Telegraph, and the acquisition was concluded the following month.

Sise could not stop when things were going so well, and one by one small local companies in Windsor, Hamilton, London, York, and Toronto, to which Melville Bell had ceded operating rights, also sold him their telephone contracts or traded them for Bell shares. The Bell agent in Winnipeg had not set up an exchange because it wasn't worth it. Sise thus reacquired the rights to what was, at the time, a group of clay-brick cabins and saloons around a fort.[15]

Western Union itself ceded the exchanges it owned in the Maritimes for Bell stock. In 1882 it asked to have a seat on Bell Telephone's board of directors. Sise refused to appoint its representative in Canada, Erastus Wiman, because he had a "bad reputation." Western Union got the hint, and soon after it sold its shares.[16]

By the spring of 1881, Sise had unified all of the Canadian telephone networks under the aegis of Bell Telephone, representing an outlay of about $400,000, of which the major part had been paid in the form of shares to often-reticent owners. Sise had managed to get hold of installations for which he paid with shares in a company that possessed nothing – except patents. He had just risen to the rank of captain of industry.

The last pocket of resistance came from Quebec City, where Cyrille Duquet was threatened with a suit for patent infringement. Duquet coolly replied, "Would you please study the act concerning invention patents for Canada, and article twenty-eight will show you (if you did not already know) that the patent over which you have made such a fuss has lapsed and is null and void because the instruments you have in use throughout Canada have been manufactured in a foreign country such that, according to article twenty-nine, I would have the right to cancel your patent."[17]

The argument was solid; it would be taken up by other litigants and would finally triumph over Bell Telephone in 1885. But Duquet himself lacked the funds and, especially, the support from business circles to have a hope of winning. After a series of bittersweet exchanges, the suit was resolved in 1882 when he sold his patent and the installations of Québec and Lévis Telephone to Canadian Telephone for $2,100.

Charles Sise now controlled the telephone industry in Canada, except for British Columbia. Bell Telephone therefore had about 150 employees and 2,165 telephones (see Table 2). The cities covered by Bell were not linked to each other; what is more, when two exchanges coexisted in one city they could not communicate with each other. In general, telegraph companies had treated the telephone as a purely local instrument of communication, while long-distance communications were considered the domain of telegraphy.

Charles Sise's intervention ended the first experiment with competition in the telephone industry. But how was he able to construct a monopoly without provoking the slightest reaction from the government? The answer, in a word, is speed. All of Sise's ingenuity was expressed in the few short months of 1880 when he stole a march on his adversaries, his own parent company, and the Canadian government, which had given him an unlimited charter. No one in business or political circles had seen or understood that the telephone

Table 2
Distribution of Telephones in Canada, 1882

City	Edison telephones	Bell telephones
Montreal	300	250
Toronto	50	200
Hamilton	50	300
Quebec City	40	75
Ottawa	50	50
Other	200	600

Source: Fetherstonhaugh, *Charles Fleetford Sise*, 119.[18]

would become a public utility and that Sise would lock up the market. By the time they caught on, Bell's monopoly was a *fait accompli*, executed in a lightning-quick campaign.[19]

The brief experiment with competition, which lasted from 1878 to 1880, had left a bad taste in the public's mouth because of the proliferation of wires in the streets and the incompatibility of the systems. Sise now turned his attention to merging the different companies he had acquired. When there were two exchanges in the same city, one had to be eliminated and its subscribers convinced not to abandon the telephone entirely. Free two- or three-month subscriptions were handed out, and no dissatisfaction was registered. The merger was a definite success.

Ownership of a few isolated exchanges, however, did not an industry make. Sise had created Bell Telephone from whole cloth in his own image – moralistic and penny-pinching to the point of miserliness. Control of operating costs would become a leitmotiv. Sise knew how scarce money was, and he played the card of self-financing to the hilt; it gained him the necessary independence from Boston, and it was the best way to keep rates down and answer the criticism of subscribers who were always ready to rise up against a company in a monopoly position. As soon as he noticed an anomaly in the expenditure curves – a disparity between two cities or between two consecutive years – he would write personally to whomever was responsible to demand an explanation. He thus managed to provide telephone service comparable to that in the United States but at a cost that was 23 percent lower.[20]

Sise, like Vail, had a "network" vision of the telephone. In the spring of 1881, he ordered construction of the first long-distance line in Canada, between Toronto and Hamilton. The cost almost bankrupted the company, and the experiment was not repeated; Sise had to advance money from his own pocket to defray part of the cost.

An underwater cable was laid between Windsor, Ontario, and Detroit, Michigan, in June, 1881. The rivers were separated only by a river, but it was still the first international link in Canadian telephony. Courtesy Bell Canada.

At the time, long distance was not profitable. Weak voice transmission and interference made all conversation chancy, and long-distance calls were routed from exchange to exchange, which caused delays in establishing communications. The concept was ahead of the technology and administrative organization of telephone companies, and the high rates for long-distance calls reflected the exceptional aspect of this service. What's more, were users scarce. People had to go to the exchange to place calls, while at the other end of the line a messenger had to be sent to inform the person receiving the call. In short, long-distance telephone calling followed the model of the telegraph, with which it found itself in unequal competition.

In June 1881 an international line was constructed between Windsor and Detroit, but it could hardly be called a long-distance line since the two cities were separated by a river. Nor was it a confirmation of the north-south trend, since the link did not connect a Canadian network to a dominant American network, as was the case with telegraphy. There was no telephone network either in Canada or in the United States. The Windsor-Detroit line was simply the connection point for two neighbouring exchanges.[21]

A LACK OF SOCIAL VISION

Sise had the gift of attracting the fidelity of a small group (all his life, he cultivated a taste for secrets) and excelled in the art of settling quarrels. When Vail and Forbes were at loggerheads in 1887, Sise managed not to take sides and remained on good terms with both men. He didn't seem to have any personal friends, however, and none of his letters betray any personal feelings – except, on rare occasions, anger, and then the consequences were disastrous. Also unlike Vail, he had no social vision. He had an eye for detail and advanced his career through hard work, but he was limited by his inability to see the big picture and by his lack of generosity.

Today, telephone companies pride themselves on always having favoured a policy of low rates with the goal of propagating service. To this end, Sise's speech to businessmen in Montreal in September of 1880 is often cited; he said that Bell Telephone "is at pains to give the public the best service possible, at the lowest rate compatible with shareholders' interest." However, everything was relative, and "low" rates didn't mean "affordable" to all. Between 1880 and 1887, the average annual cost of a subscription to telephone service was the equivalent of one-tenth of the average annual salary of a Canadian worker. So it was hardly surprising that the telephone was limited to companies, governments, and those in the upper echelons of society.

In any case, Bell Telephone's constant need for fresh capital forestalled any social policy, although Sise never applied the short-term policy of maximum profits that reigned in the United States after Forbes took over from Vail. For the period ending in 1900, the price of basic service in Montreal and Toronto varied between thirty and thirty-five dollars a year in the residential market and between fifty and fifty-five dollars in the business market, while in New York at the same time, the average rate was $240.

The reason for this was that when, in 1885, the Canadian government had cancelled Bell's patents, the resulting competition in the telephone industry had exerted some downward pressure on prices. It must be emphasized that although Sise always refused to lower rates, he froze them and let inflation do its work. In the United States, the patents remained in effect until their natural extinction in 1893–94. This explains why, at the very beginnings of the telephone industry, the Canadian rates were lower than those in the United States. Sise was forced to maintain a balance between the need to pay out scarce capital and the desire to hold onto subscribers who were always threatening to set up parallel telephone exchanges.[22]

Given the size of the territory to be covered, Sise favoured the business market; his motto was "Business first, homes after." (This approach has characterized Canadian telecommunications up to the present: the business market is called upon to amortize the initial costs of new techniques.) The number of business phones was constantly higher than the number of residential phones until the great waves of installation of rural phones, which took place between 1906 and 1920.[23]

In spite of careful, even conservative, planning, the lack of capital continued to be a heavy drag on attempts to expand the company. The first public stock issue, in December 1880, was very poorly received by Canadian investors. "The shares are selling slowly to serious men," Sise wrote to top management at American Bell in Boston. This was putting an optimistic face on it. That winter, Sise had to dip into his own pocket to pay Bell Telephone's operating expenses; this may seem unusual, but Sise considered the company to be his personal property. He also diversified Bell into alarm systems in order to find new revenue.

The lack of telephone hardware was a persistent problem. The death of James Cowherd at the beginning of 1881 had disrupted market supply. His replacement, another man from Williams, did not measure up to the task. No serious attempt was made to have Canadian third parties manufacture the apparatuses. As we saw in the Duquet affair, the patent law required Bell Telephone to make telephone equipment in Canada. However, Sise waited fifteen long months after Cowherd's death to make the decision that should have been made right away: to create a manufacturing division. It seems that he had not really grasped the nature of Canadian nationalism. Since 1879, however, Prime Minister Macdonald had been very busy putting his National Policy in place, and no one could have the slightest doubt about the Canadian government's intention to make firms respect the policy of manufacturing in Canada.

Finally, in July 1882, a telephone-manufacturing plant was set up on Craig Street in Montreal. Again, it was a Williams employee who came from Boston to set up the operation. But, in contrast to the unhappy attempt in Brantford, the employer was the Canadian company (Bell), not the American company (Williams). And it wasn't simply an assembly line but a true manufacturing plant, the ancestor of what would become Northern Telecom. Nevertheless, it was a modest workshop that had to appeal to Boston more often than was seemly for spare parts.

STILL A RUDIMENTARY INSTRUMENT

In the 1880s, using the telephone was not simply a matter of picking up a receiver. First, one had to check that the carbon microphone contacts had recently been sanded and adjusted. Then, one had carefully to attach the ammonia battery. Here too it was necessary to check whether the battery had to be renewed, which was a delicate operation since one had to take care that no liquid spilled. Besides, it smelled terrible and could burn holes in the carpet. Then one had to crank the magneto crank to warn the operators at the exchange that one was on the line. The weather was also important, since both

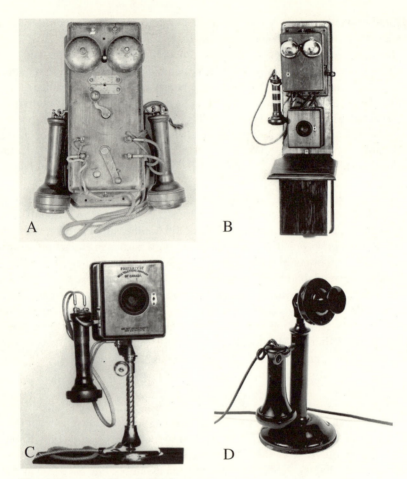

The telephone was still a primitive instrument.
a) Wall set in 1879. The crank was used to call the operator.
b) Blake telephone (1878–1900), wall set. The battery was inside the apparatus.
c) Blake telephone (1878–1900), desk set.
d) Desk set, 1900. The battery was located in the exchange, so the
    apparatus was smaller.

the single wire that routed the communication and the insulators had to be
dry. If there was an active telegraph or electrical wire nearby, all telephone
conversations became inaudible. The receiver was the only relatively reliable
element of the system. In short, the telephone required constant maintenance, which helped to limit its use to offices, where employees could take
care of it, and kept it out of the homes of all but a few innovative – and
wealthy – handymen.[24]

When telephone poles appeared along streets, municipalities protested strongly. Shown, linemen use muscle power to erect a pole. Courtesy Bell Canada.

In spite of all this, on the whole the public was receptive to the telephone, as the increase in subscriptions during those years attests. Ontario had a higher rate of telephone penetration than Quebec did, a discrepancy that lasted until the Quiet Revolution in the 1960s, when the Quebec market in its turn reached the saturation point. Sise noted in his log in 1880, "The French are not adopting the telephone, and will do so only in very low numbers unless they are forced by the necessities of their work." The true reason was that there were very few French-speaking members of the urban middle class.

Pressure groups did emerge to oppose specific aspects of the new technology. In 1881, for example, the Toronto Sacerdotal Association, strongly supported by the *Toronto Globe,* demanded loudly that telephone exchanges be closed on Sundays. More serious was the issue of telephone poles. Throughout the start-up phase of the telephone industry, some municipalities tried to stop Bell's progress by regulating the erection of poles.

The end of competition in the telephone market had momentarily reduced the number of poles, but it had not got rid of them. In 1881 the Quebec City *Daily Telegraph* launched a campaign against putting up telephone poles in the streets; the editor of the newspaper went so far as to chop down the pole

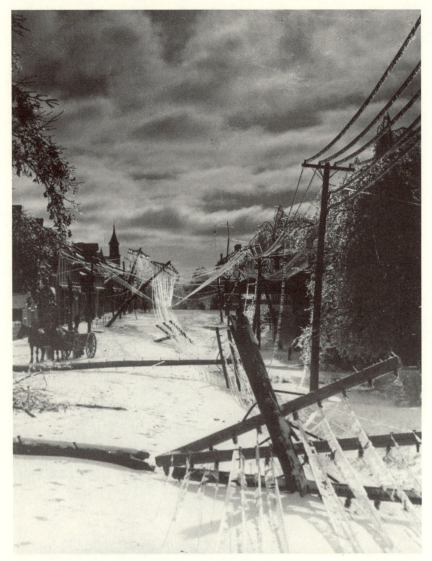

Telephone wires after an ice storm in 1893. In many cases, municipalities opposed Bell and tried to regulate the placement of poles. Courtesy Bell Canada.

in front of his office with an axe. The Halifax *Chronicle* led the way in demanding that the wires be buried. For its part, the Montreal city council required elimination of telephone poles altogether.

Today, it is difficult to understand this hostility to wards the poles: we are so used to them that we don't even see them. Sometimes, we even view their disappearance with nostalgia, as they are the last evidence of an era when wood and electricity worked together. But at the end of the nineteenth century, poles represented an intrusion of the public community into the private sphere.

Municipalities began to use their power to authorize the erection of poles as a means of regulating all of the telephone companies' activities, starting with rate setting. Bell could not tolerate such a constraint; if extended to all municipalities, it would compromise the very notion of the network. The Quebec City affair therefore went to the courts, which ruled that Bell Telephone's federal charter was invalid, since the company fell under provincial jurisdiction. Indeed, there were no interprovincial telephone lines at the time. Sise quickly pushed the legislative assemblies of Quebec (in 1881) and Ontario (in 1882) to vote in laws of application designed to reinforce the federal charter. This did not stop the opposition, and there was much more debate before 1904 when the Privy Council in London ruled, once and for all, that the federal charter had always been applicable.

Two defences were better than one, and Sise went back to Parliament in 1882 to broaden the powers of Bell Telephone's charter. The key phrase in the new document defined the company as "a work for the general advantage of Canada." Some senators grumbled about "purely local undertakings" that had the impudence to claim to be in the public interest, but the law passed with no great difficulty.[25] This innocuous-looking amendment would put Bell beyond municipal harassment and threats of nationalization by the provinces, thus proving to be an important asset in the trials to come.

BRITISH COLUMBIA: RAPID DEVELOPMENT

Bell's domain did not extend throughout Canada: development of telephone services in British Columbia was never part of its mandate, and Newfoundland, isolated from Canada by its status as a British colony, was a special case.

It was thanks to the initiative of an adventurer named Robert Burns McMicking that telephone service arrived in Victoria in March 1878. The thirty-five-year-old McMicking had been a gold prospector during the 1860 rush and had then worked on construction of the Overland telegraph project. After the project was abandoned, he became general manager of the British Columbian government's telegraph company, which came under federal control when the colony entered Confederation in 1871.

In 1878 McMicking, now general manager of the Dominion Government Telegraph Company, wrote to offer his services to Melville Bell and Thomas Henderson, who responded by naming him Bell Telephone's representative and sending him a pair of telephones. He promptly connected his office to that of the *Colonist* newspaper and invited Victoria notables to use the equipment. An article published in the *Colonist* on 26 March 1878 showed that the demonstration had been a success, as it had been everywhere else: people sang and whistled into the instrument and were stunned to recognize the voices of their friends issuing from it.

Encouraged by this enthusiastic reception, McMicking began to advertise an offer to rent telephones by the pair. No one took the bait, however, since potential customers went to San Francisco, where telephones were sold more cheaply than McMicking could rent them – when he had them. Indeed, instruments from Montreal had to detour through the United States on their journey to Vancouver. Henderson convinced McMicking that the solution was to set up a telephone exchange.[26]

McMicking, however, was still general manager of Dominion Government Telegraph, which had no plans to diversify into the telephone business, so he was working under the table. According to the official historiography, he resigned from his government position to launch a telephone company. In fact, he seems to have purchased the first telephones with Dominion Government Telegraph money when he was still in its employ; when its accounts were audited, there appeared to be some irregularities and he was sacked.[27]

McMicking did not let such details get in his way. He immediately ordered equipment from Bell Telephone in Montreal. A series of quid pro quos, errors, and bits of bad luck ensued, much of it caused by Melville Bell and Thomas Henderson's profound incompetence. Telephones were lost, some were missing parts, others failed to clear Customs; in some cases the instructions were delivered without the phones. To top it all off, the transfer of power at Bell from Melville Bell to Charles Sise was effected without the necessary instructions, and Sise was quick to question McMicking's claim to be Bell's representative in British Columbia.

In spite of these difficulties, the first telephone company in British Columbia, McMicking's Victoria and Esquimalt Telephone, received its charter from the provincial legislative assembly on 8 May 1880, nine days after Bell Telephone received its charter. How had McMicking convinced the provincial legislature to grant the charter? He was no longer working alone – in fact, he may no longer have been in charge, since he was receiving assistance from Edgar Crow Baker. Baker was an ex-marine officer turned financier who was influential both in Victoria's small-business community and in political circles (he later became a senator). A hard-bitten, business-driven man, he took the position of secretary-treasurer of Victoria and Esquimalt and re-established relations with Sise: perhaps the two old sailors spoke the same language. In July 1880, the equipment finally arrived and Victoria had a telephone exchange, one of the first cities in Canada to do so.

In the same year, the first phones on the mainland of British Columbia were installed by an Anglican missionary in a native fishing village called Metlakatla. The missionary laid a line between his store and the sawmill and connected a few cabins to his embryonic network. Today, the village has become Prince Rupert and it is the only city in British Columbia with a municipal telephone company.

Although Bell Telephone never did business directly in British Columbia, it had some influence on the beginnings of the telephone industry through Victoria and Esquimalt Telephone, which acted as its agent. This tenuous connection was broken in 1889, when the Vancouver Island company purchased rights to the telephone from Bell. Nevertheless, personal relations between Charles Sise and Edgar Baker were such that informal links between the two companies were maintained as long as the latter remained in his position in Victoria.[28]

But things were changing on the mainland with the planned construction of the Canadian Pacific railroad terminus at Port Moody, at the bottom of a bay called Burrard Inlet. A line was built between the town of New Westminster and Port Moody in 1883, a telephone exchange was installed at New Westminster, and a company was incorporated as New Westminster and Port Moody Telephone. The new company almost went bankrupt, however, when CP changed its route to end some twenty kilometres away, at Granville, near the mouth of Burrard Inlet. In 1885 the telephone line was hastily extended to serve the new terminus. An exchange was built there, and on 6 April 1886 the company changed its name to New Westminster and Burrard Inlet Telephone. On the same day, Granville received its municipal charter and changed its name to Vancouver. From then on, the development of Vancouver, and of British Columbia, was intertwined with that of CP – and of the telephone.

A few weeks later, a fire destroyed most of the town, including the bookshop where New Westminster and Burrard Inlet Telephone's exchange was located. Saved from the flames at the last moment, the historic exchange is now on display at the head office of British Columbia Telephone. Legend has it that one of the men who helped to save the exchange was a Canadian Pacific surgeon named James Matthew Lefevre, who subsequently invested in the cash-poor company. In his wake came CP managers, and CP was soon represented on the telephone company's board of directors, while Lefevre became manager of the Vancouver office. This alliance no doubt facilitated the telephone company's acquisition of CP's telegraph line from New Westminster to Snohomish, Washington. The line was converted to telephone use, and in December 1894 Vancouver was linked to Seattle.[29]

In April 1891 New Westminster and Burrard Inlet Telephone created Vernon and Nelson Telephone to serve the interior of the province. Independent companies were springing up all over, and Lefevre wanted to keep them in check with the support of local businessmen. After a series of financial reversals, the independents sold out to New Westminster and Burrard Inlet Telephone and to Vernon and Nelson. In June 1898, involved in a struggle to the finish with their competitors in the interior and short of money, the owners of New Westminster and Burrard Inlet Telephone sold the company to British interests (against the advice, it seems, of Lefevre, a principal but minority shareholder). The man who completed the transaction on behalf of the British consortium was an English immigrant named William Farrell.[30]

Meanwhile, Victoria and Esquimalt Telephone was languishing in Victoria. Since CP had arrived in Vancouver, the economic axis of British Columbia had shifted to the mainland. By 1899, Vancouver had outgrown Victoria. In August of that year, New Westminster and Burrard Inlet Telephone purchased Victoria and Esquimalt Telephone. Robert McMicking stayed on in Victoria as general manager, but it seems that this position was mostly honorary; the power had passed to the men in Vancouver.[31]

At this point, Lefevre, who had remained vice-president of New Westminster and Burrard Inlet Telephone and had never accepted the sale to the British interests or the loss of his own power, staged a coup. In 1902 he went to Great Britain and managed to divide and neutralize the owners of the company. He then made a completely hostile offer to purchase, acquiring a majority of shares. When he returned to Vancouver, at the beginning of 1903, he amalgamated all the telephone companies that he owned under Vernon and Nelson Telephone. Why the subsidiary and not the parent company, New Westminster and Burrard Inlet Telephone? Simply because the charter of Vernon and Nelson Telephone was broader and gave him a freer hand. In any case, he did not keep this name long; in July 1904 he renamed it BC Telephone (or BC Tel).[32]

All, or almost all, of the telephones in British Columbia were now under one umbrella. Curiously, the man chosen to head the new company was William Farrell. The BC Tel historian hypothesized that the sale to British interests had very likely been planned, probably by Lefevre himself. It was one way to raise fresh capital to continue the fight against the independents, which had powerful support from the United States. Once the threat dissipated, the crafty Lefevre regained control of the company, which had meanwhile doubled in size.[33]

In December 1904 a cable was laid between Victoria and the American town of Bellingham on the mainland. Because the law in Washington State prevented foreign interests from owning a public utility, Lefevre set up a firm, called International Telephone Company, to be owned by an American friend of Farrell's. In a few years, with no capital but with boundless ingenuity, Lefevre had created a true telephone network in British Columbia.[34]

NEWFOUNDLAND: TRIUMPH OVER ISOLATION

In Newfoundland, the situation was completely different. First, Melville Bell had failed to file a patent application in the British colony. It is said that the first telephones were installed in St John's between the residences of the postmaster and the meteorologist in March 1878, but it seems that it was a private link with no commercial application.[35]

When Bell Telephone tried to obtain an exclusive operating permit for telephones on the island, it was blocked by Anglo-American Telegraph. This company, founded by Frederick Gisborne and then renamed by Cyrus Field when the transatlantic cable was laid in the 1850s, had received from the government an exclusive right to operate a telegraph service for fifty years. Alexander M. MacKay, managing director of Anglo-American since the pioneering era, believed that the telephone would compete with the telegraph in transatlantic communications. He proposed to the government of Newfoundland that Anglo-American's monopoly cover all electrical communications, telephone as well as telegraph, and his argument was convincing.

Bell thus had to deal with Anglo-American Telegraph. Sise convinced MacKay that the telephone did not threaten telegraph service in the northern Atlantic, and he ceded his patents against payment of royalties. The first telephone exchange in Newfoundland was opened in 1885 in St John's by Anglo-American Telegraph.

5 Bell Comes Out Fighting

In Bell Telephone's own territory, crisis was looming. With its chronic lack of capital and selective penetration of markets, it had been neglecting the rural regions. There was growing impatience, which manifested itself in challenges to Bell Telephone's monopoly. Left waiting for telephone service, many villages took matters into their own hands.

Country doctors often played a key role in this. Usually the only prominent citizens in their community to have benefited from scientific training, they would have lines installed between their office, the pharmacy, and their patients' homes and sometimes ended up creating companies. Elsewhere, the municipality itself set up a public telephone utility, similar to those for natural gas and running water.

CANADA CANCELS BELL'S PATENTS

The Toronto Telephone Manufacturing Company opened up to handle the backlog that had built up in the telephone industry. Not content with making telephones without a patent, this small company attacked Bell Telephone before the federal minister of Agriculture on two points. The first had to do with manufacturing the apparatuses. The Patent Act of 1872 required that all patented products be manufactured in Canada no more than two years after the patent was registered. As we have seen, Bell had continued to import parts from the United States until its manufacturing division opened in 1882, well after the grace period expired. The second point concerned free access to products covered by the Patent Act. According to Toronto Telephone Manufacturing, Bell Telephone should have agreed to sell its telephones so

that the public could have full access to them, as the law stipulated. Bell refused to sell the instruments, agreeing only to rent them.[1]

This challenge was more successful than Cyrille Duquet's had been five
years earlier. The Department of Agriculture ruled in favour of Toronto Telephone Manufacturing, and Bell's patents were cancelled in January 1885. The
decision was beyond appeal. Sise's reaction was surprisingly weak: "And I
suppose that I will come in for all the blame, although – as you know – I was
much the mouthpiece of the Am[erican] Bell in the matter, and acted throughout under advice of our Counsel."[2]

One of the first consequences of this decision was Bell Telephone's withdrawal to the most populous sectors of its territory, Quebec and Ontario, and
its westward expansion to Manitoba and, starting in 1883, the districts of
Alberta and Saskatchewan, which were still called the Northwest Territories
at the time.

The West was the key to the future. Meanwhile, over a three-year period,
Prince Edward Island, New Brunswick, and Nova Scotia asked to leave
Bell's fold.

THE MARITIMES GAIN FREEDOM FROM BELL

The disengagement of the Maritimes started in Prince Edward Island immediately after the Department of Agriculture's ruling. Bell had been a presence on the island for barely one year, in the person of an agent named
Robert Angus. This eminently practical Scotsman immediately set out in
search of the twenty-five or thirty subscribers needed to justify the opening
of a telephone exchange, a goal he accomplished in December 1884. Prince
Edward Island businessmen were so interested in the telephone that they decided to build a network covering the entire island themselves. Angus lent an
accommodating ear to this plan and a purchase offer was made to Bell Telephone, which agreed to sell its on-site equipment.

The deal was concluded in July 1885, with Bell getting $1,500 and forty
shares in the new Telephone Company of Prince Edward Island, as well as
the contract for supplying telephone equipment. Angus was appointed general manager of a company that consisted of eleven telephones connected to
the Charlottetown exchange. In a few months, the island's main towns were
linked in a rudimentary fashion: single naked wires were strung from house
to house, without benefit of poles. Needless to say, a few drops of freezing
rain were enough to cause a complete breakdown in the system, and funding
problems intensified.

Effective pressure for assistance was exerted on the provincial government: country doctors encouraged their patients to write to their representatives in the provincial legislature. Finally, in July 1895, the Prince
Edward Island legislative assembly voted a credit of $250 per year for

fifteen years for the Telephone Company of PEI. It was a token sum, but it meant official recognition.[3]

The first pair of telephones were introduced into Nova Scotia in the summer of 1877 by Gardiner Hubbard, Graham Bell's associate and future father-in-law. Hubbard had been a member of the board of directors of Caledonia, one of the many Cape Breton mining companies in Glace Bay. On one of his visits to the company's offices, he took telephones, which were set up to connect the bottom of the shaft with the surface. It was, no doubt, the first commercial application of the telephone in Canada.

In Nova Scotia, as elsewhere, systematic introduction of the telephone really began with competition between Dominion Telegraph and Western Union. Two exchanges were installed in Halifax in 1879: Western Union took the first shot in November with a system designed for Edison instruments, for which it held the patents, and Dominion Telegraph returned fire one month later with an exchange for Bell instruments. It seems that Western Union also constructed a private network in Yarmouth (the subscribers were connected to a central line and everyone could talk together, conference-call style).

Bell Telephone's Nova Scotia "blitz" was unleashed in July 1880 with the purchase of Dominion Telegraph's installations and was extended the following year with the purchase of Western Union's equipment. In 1885 Bell built a sixty-kilometre long-distance line between Halifax and Windsor, a small town on the Bay of Fundy. Most of Bell's efforts, centred on Halifax, however, while the rest of the province was left to local initiatives.

In May 1887 a group of Halifax businessmen created the Nova Scotia Telephone Company, with an authorized capital of $50,000, to link Halifax, Truro, New Glasgow, Pictou, and Amherst, thus serving the central part of the province. Construction of the Halifax-Truro line quickly got under way and was completed in July, for Bell absolutely could not be given time to react. The new entrepreneurs then mailed a circular offering their service to the three hundred or so Bell subscribers. Two small local companies were purchased in rapid succession, Hants and Halifax Telephone and Parrsboro Telephone, but the purchases seemed to be symbolic gestures since neither company had been very active. The authorized capital was raised to $100,000 and the stage was set for installing exchanges in Nova Scotia's main cities – and for a major confrontation with Bell.

A letter out of the blue from Sise that November changed the situation. Sise was offering to sell Bell's installations in Nova Scotia and, curiously, New Brunswick for $50,000 in cash, $65,000 in shares, and the commitment to purchase, at equal price, Bell's equipment rather than its competitors'. Moreover, Bell would have the right to two representatives on the board of directors of the new company. The proposal was quickly accepted and the telephone war never took place. In February 1888 Nova Scotia Telephone took possession of Bell's installations in both provinces, representing 539 subscribers served by four exchanges.[4]

Over the years, Nova Scotia Telephone increased its capital, which reduced Bell's capital share; by 1905, Bell's share in NS Telephone had slid to 14 percent. Sise deserves credit for having effected a gentle "decolonization": the goodwill between the parties was such that Bell's auditor went to Halifax to help the new company set up accounting and payroll procedures similar to those in Montreal. The relationship was summed up by a NS Telephone manager: "We never claimed to manage the business better than the Bell Company did; but we did claim that our local position enabled us to increase the business in this Province, and I think that our late statement fairly bears out that expectation."[5] Bell Telephone and NS Telephone thus began a cooperative arrangement that, despite a few hitches, has lasted up to the present.

On the other hand, business circles in New Brunswick reacted badly to the announcement that their installations would be taken over by NS Telephone. They did not want to be towed along in Halifax's wake anymore than in Montreal's.

In New Brunswick, the first telephone exchange, a system using Edison instruments, had been installed in Saint John in December 1879 by Western Union. A few days later, Dominion Telegraph installed a Bell system so that, just like Halifax, Saint John had two telephone exchanges with incompatible standards. In 1881 Bell Telephone inherited both companies and dispatched Lewis McFarlane to Saint John to handle the merger. (This was the first Bell Telephone assignment for the man who had convinced Dominion Telegraph to let its lines be used for the first long-distance telephone call, in 1876.) McFarlane was immediately arrested for conducting commercial activities without a permit. The future president of Bell Telephone spent one night in jail and then hastened back to Montreal.[6]

At the beginning of 1888, a group of businessmen founded the New Brunswick Telephone Company; in March, the New Brunswick legislature passed a law giving NB Telephone exclusive operating rights in the Fredericton-Saint John-Moncton corridor. Nova Scotia Telephone reacted quickly, applying for a federal charter, which was indispensable for operating future interprovincial lines, under the name Nova Scotia and New Brunswick Telephone. It seems that the federal authorities put things off, since the charter was never accorded in the end. During this time, NB Telephone opened exchanges in prime locations, started a price war, and hired away a number of NS Telephone's agents.

The situation quickly got out of hand and NS Telephone appealed to Sise, who went to New Brunswick in September. He explained Vail's theory of occupation of the territory and suggested construction of an Amherst–Moncton interprovincial line. Nothing worked. The directors of NB Telephone spread

Table 3
Number of Bell Subscribers and Exchanges, 1885–1905

Year	Number of Bell subscribers	Number of Bell exchanges
1885	10,200	126
1890	20,437	212
1895	30,908	345
1900	40,094	343
1905	82,351	526

Source: Fetherstonhaugh, *Charles Fleetford Sise*, 223.

a rumour that they had CP's financial support through Federal Telephone.[7] At the time, CP was the giant of British and Canadian capitalism and represented a sacred union between capital and political power. NS Telephone had no choice but to sell.

The sale took place in two phases, since antagonism was running so high that NB Telephone's directors refused to negotiate directly with NS Telephone's. Bell bought up NS Telephone's installations in New Brunswick and resold them to NB Telephone for $50,000, half in cash and half in stock, for a holding of 31 percent. As NS Telephone had, NB Telephone named two Bell representatives to its board of directors. The Bell–NS Telephone network in New Brunswick had consisted of four exchanges serving 520 subscribers. As with NS Telephone, Bell's participation in NB Telephone diminished over time, but Bell continued to be represented on the boards of directors of both companies.[8]

In total, the loss of Nova Scotia and New Brunswick cost Bell Telephone 1,200 subscribers. Throughout this period of technological adolescence, however, Bell expanded rapidly, and the loss of the Maritimes market had no effect on its results, as Table 3 shows.

Bell Telephone's withdrawal to Quebec, Ontario, and the West was thus relative, since most of the Canadian population was concentrated in the two provinces of central Canada.[9] Moreover, Bell continued to supply equipment to the new companies in Prince Edward Island, Nova Scotia, and New Brunswick. But the causes for the withdrawal persisted. Lack of capital continued to be a discriminatory drag on telephone penetration. Bell Telephone concentrated its activities in large cities, where the return was better for investors, and neglected the countryside, where it was less so. The consequence was fury in much of Canada. Furthermore, the loss of telephone patents threatened Bell's monopoly even as its own territory was melting away.

BELL CONFRONTS THE COMPETITION

Bell Telephone did not simply withdraw in reaction to the 1885 decision to cancel its Canadian patents. The company intended to preserve its monopoly in central Canada and the Prairies, although this monopoly had no legal basis – anyone could make telephones and put together systems. Already, competition was rearing its head. The response was planned at the highest level in Montreal and Boston.

For Theodore Vail, the Canadian situation was a dress rehearsal for what would happen in the United States when the American patents expired in 1893–94. He went to Montreal in February 1885 to lay out a battle plan based on construction of long-distance lines. The scene has become part of Bell's legend. Vail was leaning over a map of Quebec and Ontario and drawing blue lines between Montreal and various towns within a five-hundred-kilometre radius:

"Build long-distance lines at once to connect all the exchanges within this territory right away," he said.

"But they won't pay," protested the general manager, whose longest line was just 20 miles.

"I didn't say they would," Mr Vail observed, "but they will unify and save your business."[10]

He then generalized this plan into one of his patented pithy expressions: "In this business, the possession of the field is of more value than a patent."

Vail and Sise's plan would involve agreements with the main railway companies, which owned the land and installations on which Bell Telephone could run its long-distance lines. Sise offered the railway companies free telephone service in return for the right to hang Bell's wires on their telegraph poles. One by one, the railway companies signed on the dotted line, providing Bell with a guaranteed right of passage on all future long-distance lines. As a corollary, Sise asked for and obtained exclusive telephone service in train stations. This seemingly innocuous clause would later compromise Bell's future, since it provoked great animosity both in rural regions and in the West.

Sise then turned to the municipalities, which conducted an incessant guerrilla war against the proliferation of telephone poles and wires. Up to then, Bell had refused to bury its wires, arguing that the cost was too high. However, the appearance of the first electric streetcars in the 1880s made aboveground wires less attractive because of the interference with transmission. Sise decided to transform an inconvenience into an asset. Using Toronto as a testing ground, because of the virulence of the opposition there, he negotiated with the city a five-year contract giving Bell a monopoly on telephone service in exchange for running its downtown wires underground, fixing an annual ceiling for residential and business rates (twenty-five and fifty dollars, respectively), and paying a 5 percent royalty on revenues. Similar agreements would be signed with thirty-six other municipalities in subsequent years; in this way, Sise provided an empirical form of regulation to municipal authorities.

Also in 1885, Bell Telephone froze its rates to nip competition in the bud. In isolated cases (Dundas, Peterborough, and Port Arthur), Bell offered free telephone service for as long as it took to eliminate a rival, then returned to previous rates. This predatory policy was crushingly effective, and all commentators rightly decried it. They neglected to mention, however, that in every case documented during the inquiry into the telephone industry organized in 1905 by the Select Committee of the House of Commons, it was Bell's reaction to opposition by the municipal companies, which were using taxpayers' money to discount their service. Overall, however, the practice was rare.[11]

Sise boasted that his rates policy was paired with a quality policy. Like Vail in the United States, he was convinced that the quality of service, even more than low prices, would ensure customer loyalty. The expanding use of electricity, however was causing more and more interference with telephone transmissions. A great deal of money was devoted to the addition of a second wire to subscriber circuits and the elimination of the earth returns that had been used up until then, a response that proved to be a formidable engine of war against the competition.

The companies that sprang up when Bell Telephone's patents were cancelled could be put into one of two categories: rural companies that were taking up the slack in Bell's absence, and urban companies that wanted to take over parts of markets that were already being served. An example of the first type of competition was the Compagnie de téléphone de Métis, founded by J. Ferdinand Demers, a physician in Saint-Octave-de-Métis, a small village in the Lower St Lawrence region near Rimouski. In 1897 the young doctor, freshly graduated from Laval University, laid a line between his house and the railway station. He then extended the line to the neighbouring village of Sainte-Flavie, where he faced competition from a colleague, Dr François-Xavier Bossé, who had opened the first exchange in the region the year before. The two adversaries went before the municipal council to find out which would receive the exemption from municipal taxes.

Demers won the fiscal battle and set up the Compagnie de téléphone de Métis. His powers of persuasion must have been great, for Bossé, far from holding a grudge, became a stockholder. It was a good investment, for Demers soon extended his lines east to Matane and west to Rimouski.

Bell Telephone had been established in Rimouski since 1890. Would it be able to stop Demers's juggernaut? Demers told the story years later, during hearings of the 1905 Select Committee on the Telephone: "Their system covered only the town [of Rimouski], and we spread the rumour that if they didn't agree to sell it to us, we would build a parallel line that would take away all of their subscribers. I don't know what they thought, but they decided to sell for two thousand dollars, and we paid them."

The chairman of the special commission could not refrain from irony:

"I suppose that Bell Telephone Company had to lay down arms before its powerful competitor."

"So we told them that they had to sell their installation or we would ruin their company," Dr Demers repeated, and one could see his pleasure with the effect produced.[12]

The battle of Rimouski was a total victory for Demers – although one that had to be put into perspective: Rimouski, in 1899, had thirty-three subscribers. Yet David had conquered Goliath. How? Of course, Métis Téléphone's rates were lower than Bell's. But the deeper reason was the company's roots in the community, with which it had a symbiotic relationship. Telephone operators were paid on commission (ten percent of subscriptions, twenty percent of long-distance calls).

Demers's unstoppable juggernaut rolled on. In 1900 he bought Compagnie de Bellechasse, which served the rest of the Lower St Lawrence region, from Lévis to Matane, and dropped the name Métis Téléphone in favour of Bellechasse. With these moves, Demers's company joined the ranks of full-sized telephone companies, with a long-distance network that stretched from Lévis, at the gates to Quebec City, to New Brunswick, where it interconnected with Central Telephone. In 1904 Demers gave up medicine entirely to devote himself to his telephone business. It was the right career move; his company still exists today under the name Quebec Tel.[13]

Rural telephone service provided by independents would never have made such great strides without the phenomenon of the party line. When Demers lowered his rates below Bell's and remained profitable, it was because one party line could serve up to thirty subscribers. It was impossible for two subscribers to make a call at the same time, and privacy was out of the question, but this choice on the part of the independents was in the purest North American technological tradition: quality was traded for accelerated propagation of the telephone. At first, Bell deplored the low quality of service offered by the independents, but later it also offered party lines, at a lower price, shared among an average of twelve subscribers.

Another element common to many independent companies was their link with municipal authorities; the role of the town of Sainte-Flavie in the establishment of Métis Téléphone is an example. In some cases, the municipality itself created a phone company. Things generally happened as follows: a group of farmers not served by Bell sent a petition to City Hall, which then issued bonds guaranteed by the farmers' land to cover the cost of constructing the network. Then a special levy was simply added to the municipal tax account to reimburse the cost of the bonds and pay for network maintenance.

This system meant that poor small-holding farmers could have access to telephone service, while non-landowners could have a telephone by paying local rental fees. The non-owners were called "renters"; the term "subscribers" was reserved for owners. Thus, thanks to independent companies, telephone service in Canada, as in the United States, very quickly stopped being the preserve an urban elite.

COMPETITION OR OLIGOPOLY?

Many private and municipal phone companies came into being more or less haphazardly. At first their status was uncertain: should they be considered competition for Bell Telephone or independent companies exercising a monopoly on their territory, as Bell did on its own? Until the great public debate of 1905 on the telephone industry in Canada, Bell Telephone saw independent companies as competition, refused to interconnect their local systems to its long-distance network, and attempted to buy them out the moment they ran into financial difficulties. Two official reasons were given for the refusal to interconnect: the inability of the independents to collect the fees for long-distance calls, and the technical incompatibility of the exchanges.

Nevertheless, at the beginning of the century there were some twelve hundred independent phone companies in Canada.[14] In 1905 a British-born engineer called Francis Dagger brought many of them together in the Canadian Independent Telephone Association, which led the battle against Bell. At first, the group comprised mostly Ontario members, but Demers's Bellechasse also joined. CITA still exists today, but it has abandoned its militant origins and collaborates with Bell rather than confronting it.

The fate of companies that attacked Bell Telephone on the urban front was altogether different. The most determined onslaught was Canadian Pacific's offensive in Montreal, at the very heart of Bell's territory.

Since its incorporation in 1881, the powerful railway company had been eyeing the telephone market. Its charter allowed it to offer telephone service as well as telegraph service.[15] However, it conducted an indirect offensive. A company called Federal Telephone was incorporated. It had no corporate link with CP, but its board of directors, composed of Sir Donald Smith, William C. Van Horne, R.B. Angus, and C.R. Hosmer, was identical to CP's. In April 1888 the city of Montreal gave the newcomer permission to lay lines in the streets. Like all independents, Federal Telephone lowered prices, but unlike most, it had the means to persist with this policy. Bell had to lower its rates by 30 percent and supply telephones at a loss to stay in the race.

To understand what was happening in this price war, one must note that at the end of the nineteenth century, the notion of depreciation was utterly unknown: all independents committed the error of not planning for a fund to replace used equipment. Thus, they offered telephone service at a lower price than Bell's, but after a few years, when their equipment began to age, they were obliged to raise rates again, with catastrophic results. It was generally too late by then; subscribers didn't understand why they had to pay rates as high as Bell's for inferior service, and they refused to pay. The independents could do nothing but fold their tents ... or let themselves be bought out by Bell. Charles Sise summarized this strategy in a letter to one of his managers in the West: "The Federal Company, starting with an entirely new equipment, will for two years have little or no maintenance and no depreciation; they will therefore be able to work for those two years, as you are well aware, at a much lower price (than what I have quoted), but when the changes which are inevitable begin to come, they will increase, as we did ... Before they have obtained that experience however they will think there is more money in the business than there really is and will be inclined to go into other cities."[16]

After a year and a half of heated competition, Federal Telephone had between 1,250 and 1,300 subscribers in Montreal, while Bell had 4,500. Both companies were operating at a loss, but neither had the means to eliminate the other. And neither gave up, because, beyond the companies themselves, their battle represented a confrontation between two distinct brands of capitalism. Federal Telephone and its backer, CP, represented the British capitalism that had founded the empire; Bell Canada symbolized North American capitalism, still awkward but growing fast.

Sise seems to have been aware of the national and international dimensions of the stakes; in a letter to the president of AT&T, he defines the leaders of the enemy group as "these people, who *to all lookers-on appear to own Canada*."[17] Sise wasn't the kind of man to let himself be intimidated, but he understood the nature of the adversary; it was the fear of seeing Federal launch raids on other cities that prompted him to sign a first municipal contract with Toronto. As predicted, Federal Telephone lost money. The affair ended in June 1891 with Bell acquiring the majority of Federal's shares, under conditions set by Federal. Bell had bought peace at a high price, but it gained ultramodern installations. What's more, it finally concluded an agreement with CP that was analogous to those concluded with the other railway companies.

British capitalism at the pinnacle of its glory had signed a peace with young North American capitalism. The battle for Montreal was only one skirmish in an ongoing war, but for Sise it sounded an alarm. "In 1891, we got no advantage – or very little – from the collapse of Federal [for] the battle was of the worst sort and very expensive."[18]

Still in the urban arena, another type of competition sprang up, based on linguistic criteria. In September 1892 a group of French Canadian business-men in Montreal decided to launch their own company, Merchants Tele-phone Company. Each subscriber would also be a stockholder. The newcomers began to erect their poles the following September with no mu-nicipal permit, which caused them some problems at City Hall. Further, the timing of this enterprise was unfortunate: it was the very moment when Bell was, at great cost, laying its underground cables. The City did grant a permit to Merchants in November 1893, but it was conditional on case-by-case ap-proval by the municipal inspector.

Having learned nothing from the recent failure of Federal, Merchants sys-tematically reduced its rates. At first the company was so successful that its directors considered opening up shop in Toronto.[19] The telephone giant re-plied with an advertising campaign using the slogan "Two Bells Means Two Bills." Indeed, users of Merchants' services had to continue to subscribe to Bell to communicate with everyone else and thus paid two bills for the same service as before. Merchants had 1,546 subscriber-stockholders in 1905, most of them owners of small businesses, but it never managed to recruit outside of this initial circle. Bell eventually purchased the company in April 1913.[20]

All independent companies repeated the losing scheme, which led from the illusion of a promising start to progressive financial strangulation. Aside from ignorance of the basic tenets of accounting, they all had something in common: an eye for the short term. None seemed to have foreseen the scope of capitalization needed to build a telephone network. And Canadian banks continued to give the telephone industry the cold shoulder, for its return was deemed insufficient.

Bell Telephone had to confront the same problem at the same time. But it had a strategic asset: its technical and financial links with the United States. Each move it made was part of a carefully weighed strategy aimed at protect-ing the monopoly. In total, in spite of some unhappy exceptions that would be driven home hard in the following decade, Bell proved to be relatively accom-modating to the rural companies, a number of which have survived to this day. On the other hand, the company was vicious when it came to eliminating competition in the cities and managed to preserve its monopoly through a ju-dicious combination of brute force and negotiation. It had to yield on the pe-riphery to keep the heart of its territory – cities in Quebec and Ontario. Bell's monopoly was acquired by audacious surprise attacks and maintained with difficulty, given the indifference of financial circles and the government.

The main opposition came from the municipalities, which were not at all happy with Bell's fluctuating rates, its slowness and occasional refusal to in-stall telephones, or, of course, its poles. The increasing pressure on Bell was ultimately played out on the federal political stage, the only forum in which

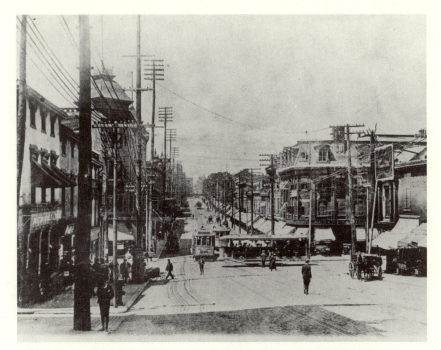

Montreal in 1904 (St. Lawrence Boulevard seen from St. James Street). The Merchants Telephone Company's telephone wires were competing with Bell's. In addition, there were electric and tramway wires. It is not surprising that the public was unhappy with such an unsightly tangle.

to confront Bell, whose protective charter declared it to be "a work for the general advantage of Canada."

In 1892 Bell Telephone asked the federal government for permission to increase its capital. Rapid development of streetcar systems was forcing it to switch to metallic circuits, and it therefore had to make considerable investments over a very short period. The legislation quickly passed through the House of Commons but caused a storm in the Senate, where Bell was criticized very severely for its attitude towards municipalities. The debates over rights of way and rates were in full swing in Toronto, and municipalities were manoeuvring so well that it was urgent for substantial modifications to be made to the law regulating Bell.

Finally, on 9 July 1892, a special law was adopted stipulating that all increases in telephone rates had to be approved by Cabinet. In exchange, the increase in capital was granted and Bell sought the sums it needed on the market. No one seems to have realized the importance of the decision at the time, but Toronto's vehement pressure and the Senate's hesitation had started the era of telephone regulation.[21]

POPULISM HITS THE TELEPHONE INDUSTRY

The affair couldn't end there. As in the United States, a wave of populism was sweeping through Canada. This time, it was the very concept of a private company providing telephone service that was at issue. There was talk of municipalization and nationalization, the British model was evoked – in short, Canada had telephone sickness, and Bell was accused of being the cause.

The movement for nationalization of telephone service in Canada did not originate with socialist ideology but with a typically North American phenomenon: populism. The urban branch of the movement was led by historian and poet William Dough Lighthall, an ivory-tower intellectual. Lighthall was mayor of the Montreal suburb of Westmount, and in August 1901 he founded the Union of Canadian Municipalities, which was responsible, as we shall see, for the telephone crises in the Prairies.

The biggest boosters of civic populism were owners of small businesses who were frightened by the rise of large capitalist consortiums, symbolized in their eyes by private companies providing public services. Their solution was simple: municipal freedom and government control. This minimal program attracted people from all walks of life, especially progressive and union circles. Christopher Armstrong and H.V. Nelles, in their excellent book *Monopoly's Moment* (1986), explained the success of this trend by its "sacred union" against big capitalism. Civic populism joined forces with rural populism, which wanted public services to be subjugated to the interests of farmers.[22]

Both civic and rural populism were steeped in a moralistic, almost religious mentality that was fed by the periodic announcement of financial scandals linked to corruption and acts of collusion between public utilities and the government. In Ontario, the populists ended up demanding nationalization of the electrical utility; in the West, CP's monopoly was at the centre of the storm.

It was in Toronto that "civic populism," an ideological quasi-religious movement whose battlehorse was municipalization of public utilities, broke out most virulently. The pretext for the crisis was supplied by the failure to renew Bell's exclusive contract in Toronto in 1896. The City wanted a reduction in rates and a rise in the royalties Bell paid it in exchange for maintaining its franchise. Bell refused outright. For Toronto, it was a question of principle: who fixed the rates, Bell or the city?

Under the special law of 1892, Bell's rates were subject to approval by the federal government. In January 1897 the company applied to the government to raise its rates. Hearings were organized by the Privy Council's Railway Committee. In Toronto's systematic guerrilla war against Bell, it found support even within the federal bureaucracy. Sir William Mulock, postmaster general and senior minister for Ontario, a respected politician known for his progressive ideas, declared in the House of Commons, "I cannot see why it is not as much the duty of the state to take charge of the telephone as it is to conduct the postal service."[23] After a year and a half of hearings, debates, and negotiations, Bell's request was turned down. The divided Cabinet was incapable of making a decision on the highly technical and politically explosive issue of rates. Government regulation was proving to be a complete fiasco.

Bell made no further requests for rate increases; to no one's surprise, Charles Sise decided instead to find a way around the 1892 law. He used a loophole in the legislation to raise rates for new subscribers, arguing that the government regulation concerned only subscribers from before 1892. This interpretation, which was manifestly contrary to the spirit of the law, still received approval from the minister of Justice in 1901, which did nothing to quiet the controversy.

Strangely, it was the unelected Senate that proved to be the most attentive to the interests of the public and the municipal opposition. In 1902 Bell asked again to increase its capital, and the Senate took advantage of this opportunity to take regulatory matters away from Cabinet and entrust them to a Supreme Court judge or to the old Court of the Exchequer. Needless to say, this modification remained a dead letter, since Bell never again asked to increase its rates but simply did so on its own.

More significantly, the 1902 amendment required Bell to provide service to anyone whose residence was located less than two hundred feet from a public right of way where a telephone line was located. The goals of this clause, which is still in force, were to force Bell to provide service outside of the densely populated zones it had favoured up to then and to prohibit discriminatory practices. Although the amendment had some effect, it fell far short of the demands of civic populists.[24]

What Toronto wanted, in fact, was the power to regulate Bell itself. In 1900, after it lost the rates battle, the City moved the point of contention to the issue of poles, forbidding Bell to erect them without its permission. The City's basic argument was that local service fell within provincial jurisdiction. The company replied that its federal charter took precedence over municipal and provincial jurisdiction. In 1904 London decided in favour of federal competence: the Privy Council refused to distinguish between local service and long-distance service, taking the position that the same installations were used for both types of service. The decision ended the multiple quarrels that had always impeded development of telephone service in the cities, not only in Toronto but across Canada.

But this didn't keep Bell from attracting powerful enemies in Toronto. In January 1900 Mayor E.A. Macdonald asked Parliament to nationalize long-distance telephone and telegraph lines. For his part, Conservative MP

William Findlay "Billy" Maclean, editor of the tabloid newspaper *Toronto World*, used his position to fight Bell and promote municipal ownership of telephone service.

It was in fact an initiative by Maclean that took Bell almost all the way to nationalization. When Wilfrid Laurier's Liberal government created the Railways Commission in 1903, Maclean led a bipartite coalition to broaden the telephone law. If the railways were regulated, he argued, why not the telephone companies, and especially Bell?

The timing was propitious. North of the Great Lakes, the towns of Port Arthur and Fort William (which later merged to become Thunder Bay), the site in the 1880s of one of the most hotly contested episodes in Bell Telephone's struggle against the competition, had created a municipal telephone network. The towns approached the Railways Commission to ask permission to serve the train stations, in spite of past exclusive contracts between the railway companies and Bell, but the commission turned them down in March 1904.

The affair didn't stop there; a veritable civic rage built up against Bell. The town councillors began a campaign asking people to cancel their subscriptions to Bell's service. Forms to break the subscription contract were printed and distributed. To convince people to subscribe to the municipal system, the councillors promised free telephones. Public utilities employees in both towns tangled and even cut Bell's lines. Employers using Bell's service were boycotted. All of this was completely illegal.

Charles Sise responded by sending William C. Scott to the site. Since 1895 Scott had been head of the Special Agents Department, which was responsible for a wide range of activities, from advertising to producing phone books to "special missions" (spying). When Scott tried to turn free service, the weapon of the disloyal municipalities, against them, it was tantamount to a declaration of war.

Whipped into a populist frenzy, the public subscribed *en masse* to the municipal telephone service: of the one thousand households in the two towns, 763 had a telephone. This was an obvious sign that the telephone service was no longer reserved for economic élites; from now on, through competition and political struggle, it would spread through all of society.

PRAIRIE FIRES

News that the Railway Commission had rejected the request of Port Arthur and Fort William eventually arrived in Edmonton, where it found fertile ground. Edmonton didn't even have the status of a municipality, but it had had a telephone company since July 1893, so great was the need for communication among people in the West. Public opinion saw in the decision by the Railways Commission approval of "the consortium of the tyranny of large companies." A powerful movement in favour of a farmers' phone service quickly developed in Alberta, with Charles Walter "Pete" Peterson as its main proponent. This Danish-born man had worked in the territorial government, becoming the apostle of the farmers' movement. In 1904 he wrote in the *Toronto World*: "I have long since maintained that the evolution of the ranch and farm requires the generous use of all modern inventions. The telephone, I believe is the most useful of the many modern devices for rendering farming easy and profitable. In the past few years, I have called many meetings of farmers to discuss the subject of cooperative work in this direction but in every instance we have been blocked by the Bell people. That we are being damaged, directly and seriously, by the Bell people, I am certain."[25]

Paradoxically, opposition by farmers to Bell was even stronger because farmers liked the telephone. When Edmonton became a municipality in December 1904, its first move was to organize a referendum on municipalization of the telephone. More than ninety-one percent of the population voted in favour of municipalization, which came into effect on 1 January 1905. The four hundred phone subscribers in Edmonton did not carry much weight compared to Bell's seventy thousand, but their actions were symptomatic of the overriding sentiment in the Prairies.

The fundamentally anti-Bell attitude had a theoretician in the person of Francis Dagger, the British-born engineer who had founded the CITA. Dagger tirelessly travelled across Canada giving speeches and writing articles favouring the independents and railing against Bell's poor service, high rates, and obsolete equipment. He had worked in a number of telephone companies in Great Britain, where he had already made his pro-municipalization feelings known. Since he couldn't put his ideas into practice in England, he came to Canada in 1899 and began by working for Bell in order to familiarize himself with the Canadian telephone industry. He quit his job eleven months later, his opposition to gigantism as firm as ever, and began a career as a telephone consultant, first with the federal government, then on the Prairies.

Dagger divided the telephone industry into three parts: long-distance lines, local service, and rural networks. He saw right through Vail and Sise's strategy of developing long distance to the maximum to control the industry as a whole and recommended nationalization of the long-distance network by the federal government or the provinces. Under no circumstances, he felt, should this strategic tool be left in the hands of a private company that also had an interest in local telephone service: "From a public standpoint it is most undesirable that the same private interests which own and control the long distance lines should also own and control the local exchange systems ... In other words, a company operating both long distance lines and local exchanges in any territory has by reason of its control of the long distance service a virtual monopoly of the business to the disadvantage of the general public."[26]

On the other hand, Dagger opposed nationalization of local service. He felt that any centralized telephone system was by nature too rigid to satisfy the needs of subscribers. Local service should fall to the municipalities, "like water, natural gas, public lighting, electricity, and streetcars." Nevertheless, as the example of the existing independent companies proved, the efficiency of municipal service often left something to be desired. The government therefore had to set standards and offer its expertise at cost price in order to guarantee quality.[27]

Finally, the rural network was looked at most carefully, since it was there that the problem was most acute: "This branch of telephony has in the past been absolutely neglected and discouraged in Canada by the existing companies, for the reason that it does not prove such a lucrative business as the exchanges in the towns and cities."[28] Dagger hoped that rural telephone service would be provided either by agricultural cooperatives or by rural municipalities with government support, as circumstances dictated. All the same, he sometimes descended to the most simplistic demagogy: when he appeared before the 1905 Select Committee, he demanded that farmers be permitted to use telephones for free in town.

Dagger failed to achieve his lifetime goal – nationalization of Bell's long-distance network – and his civic and cooperative dream was only partly realized. However, he managed to drag Bell into a parliamentary commission where a number of abuses were exposed in a public forum, and his ideas triumphed in the Prairies, where three provinces nationalized Bell.

Today lost in obscurity, Dagger was the most important figure, apart from Sise, in the Canadian telephone industry in the late nineteenth century. He was no saint, however: in his eagerness to "sell" his ideas to the governments he was advising, Dagger completely failed to take account of the depreciation of telephone equipment or to plan for funds to be laid aside for amortization. The lower telephone rates that he held up like a holy grail for his clients thus arose from accounting sleight of hand. It seems that Dagger was not cut out to be an empire builder.

On balance, through his role as critic, Dagger exerted a positive influence on the Canadian telephone industry, forcing Bell to end its most contemptible predatory practices. Ahead of everyone else in the industry, he promoted automatic telephones at a time when Bell was pitting all the weight of its monopoly against this new technology.[29] On the other hand, his role as advisor was spotted with manifest errors and omissions that cannot have been innocent. He was swept along by the great movement in favour of civic populism, but he could not manage to rise above the passing fad of marrying the public's short-term interest to long-term realities. Nevertheless, Dagger was one of the few Canadians to have placed the issue of the telephone industry in a political context.

Dagger took Bell to the brink of nationalization. At the beginning of 1905, he found an attentive ear, as we have seen, in William Mulock. Mulock convinced Prime Minister Wilfred Laurier to appoint a Select Committee of the House of Commons with a mandate to inquire into the telephone industry in Canada. It was a deliberately vague mandate that made no mention at all of nationalization.

BELL STANDS ACCUSED

Because of his personal relationship with Sise, Laurier was opposed to the expropriation of Bell. In general, Laurier's connections to the business world were as incestuous as Macdonald's had been, and Mulock was isolated within the Cabinet; nevertheless, municipal pressures were such that the government had to make a decision with regard to Bell. The Conservative opposition had made nationalization of public utilities a part of its platform. In this it was taking its political cue from Great Britain, which had nationalized its telephone system in 1901.

By 1905 the telephone could no longer be considered a technological curiosity or a luxury item; it had become a public utility. As service became widely available, Bell's monopoly became intolerable in the eyes of a growing portion of the population. The right of ownership was running up against a new notion, the public interest.[30]

The Select Committee began its work on 20 March 1905. After one week, it seemed apparent that Bell Telephone's monopoly was well and truly at the heart of the debate. The conjunction of civic populism and British imperial ideology, as the Conservative party interpreted it, was at its height; it was to lead to nationalization of electricity in Ontario, but would it triumph over the telephone at the federal level? Many people thought that it would. Charles Sise, on vacation in Italy, was called back to respond to the emergency, and the president of AT&T, Frederick P. Fish, came running from New York to study the situation.

Sise was Bell Telephone's main witness. His position was clear from the start: any state intervention, whether nationalization or regulation, must be opposed. He hired a Toronto Liberal lawyer, Allen B. Aylesworth, to assist him in front of the Select Committee, and a British telephone expert, who painted the public telephone utility in Great Britain in rather unattractive colours. The British example had often been cited by proponents of nationalization, but Sise himself vigorously led the counteroffensive. At the time, he was seventy-one years old, and no one in Canada outside of business circles knew who he was, so great was his taste for secrecy. The committee was surprised to have before it a witness who, while endowed with an uncommon memory for facts, was not loquacious, often responding with a simple yes or no.

Once Bell had been presented in a most sombre light, the Select Committee, with no specialized personnel, quickly found itself in an inferior position before the old man, who was concealing his growing deafness behind a double helping of arrogance. Of course, Sise had to face hostility from some committee members, most of them from Ontario and the Prairies; on the other hand, he could count on the support of members from Quebec. Although Mulock, the Select Committee chairman, favoured nationalization, he surprised everyone by remaining neutral.

Sise's main asset was his knowledge of the dossiers, which enabled him to refute most of the accusations of inadequate service in rural areas, too-high rates, the Bell/AT&T affiliation and the flight of capital to the United States, and the crushing of competition and creation of a monopoly.

Sise spoke without notes. He pointed out the paradox of asking at one and the same time for rates to go down and for service in the countryside to be improved. He revealed to the Select Committee that the Americans had always invested at a loss in Canada and only in response to Bell's requests.[31] Sise presented himself as the spokesperson for pure economic rationality over social, cultural, or human considerations. More debatable was his argument that Bell was not a monopoly because of the existence of a hundred independent companies, many of which had in fact come to speak against Bell at the hearings.

When the House of Commons went into recess in July 1905, the Select Committee ended its work without reaching a conclusion; the federal government was content to order publication of the transcript of the hearings. This exercise was not a simple admission of impotence. One of the federal government's main handicaps in its confrontation with Bell since the beginning of regulation in 1892 had been its total lack of information on the telephone industry; it had depended entirely on what Bell chose to reveal. The committee's work constituted Canada's first attempt to obtain complete and objective information on an industry that had until then developed behind a shroud of mystery.

The dissolution of the Select Committee did not, however, bring serenity back to the Bell camp. Nationalization had been avoided, but the issue of the telephone industry would return to the House of Commons during the following session. The main point of contention was still the question of independent companies, which Bell refused to interconnect to its network, although the spread of the central battery switchboard, starting in 1900, had removed the main technical obstacle to interconnection by standardizing the electrical currents employed.

It was during the summer of 1905 that Sise began to accept the idea of regulation in exchange for the many advantages of the monopoly that Bell Telephone would acquire *de facto* in the most populous regions of Canada. Again, this regulation had to be as "harmless as possible" – that is, limited to the field of interconnection – as his letters to the president of AT&T show: "The difficulty which I experience personally as a witness [before the Select Committee] is in explaining why – except that we propose to maintain our monopoly – we grant physical connection under certain restrictions to certain independent but non-competing Companies but refuse it to competitors."[32] In spite of this restriction, the idea that germinated in 1905 among upper management at Bell Telephone was that of a regulated monopoly.

In October 1905 Mulock resigned from the Post Office to become chief judge of the Ontario Court of the Exchequer. Laurier appointed Aylesworth, Bell's legal advisor, to replace him. Among opponents to Bell's monopoly, the disappointment that had greeted the fiasco of the committee turned to rage: the collusion between Bell and the federal government was now out in the open. The historian Donald Grant Creighton termed this period the "Laurier plutocracy."

FEDERAL REGULATION TRIUMPHS

When the House of Commons convened in February 1906, it could do nothing but legislate. An amendment to the law that created Bell placed the company under the control of the Railways Commission. The main provisions of the amendment stipulated that all rates had to be approved by the commission; that the commission had the power to authorize independent telephone companies to serve railway companies without having to pay compensation to Bell; that Bell had to submit to municipal regulation, though the municipalities could not unduly delay the laying of new lines; and that any request by an independent company for interconnection to the Bell network had to be approved by the commission, which had the power to decide on the amount of compensation Bell was to be paid as well as the technical standards to be imposed on the independent company.

It was a compromise solution. The proponents of nationalization were disappointed, but the independent companies obtained satisfaction on their main points. Bell saw regulation imposed by a neutral and permanent agency, a guarantee of stability with regard to parliamentary regulation, and a tacit exchange of monopoly for regulation – even a legitimation of the monopoly. Indeed, in the government's view, independent companies were themselves monopolies within the territories they served. This was a clear and clean response to Sise's ruminations on the difficulty of deciding between the independents and the competitors.

Bell, which had rallied to regulation as the lesser of two evils, was soon unreservedly on board, with the enthusiastic support of top management at AT&T, especially once Theodore Vail returned to the presidency of the company the following year. In July 1906 Charles Sise was able to leave for Europe and continue the vacation that had been interrupted the year before.

He had managed to avoided nationalization, but he lacked an understanding of the new complexity of a world that was beginning to set up counter-weights to unbridled capitalism. He had reacted to events rather than create them. The Sise era was coming to an end.

The question how the commission would regulate the company came to a head when it was time to set telephone rates. A first series of hearings on rates took place in 1907; in effect, the 1906 amendment had required Bell to submit its entire rate grid to the Railways Commission.

There was something surreal about the hearings. At the outset the general manager of Bell's long-distance lines – none other than Charles Sise, Jr, the son of Bell's founder in Canada – had stated in a challenging tone, "Many of our rates are not reasonable." Questioned closely by the commissioners, Sise, Jr, explained things this way: "Really, [the rates] have not been de-vised, but have just grown ... The original business started with private lines and no exchanges. But the rates were fixed, and as the business increased to its modern proportions, rates have never been changed, so far as I know. Then we got to the point where we should have to put in metallic circuits, with two lines instead of one to each subscriber, but we could not change the rates as we wished, because of the amendment to our charter. As a result we have regarded our rates as a fixture, and have paid little attention to them. But we have equalized the rates at various places, some up and some down."[33] Sise, Jr, then obstinately refused to separate long-distance-network from local-network expenses. Not only had the calculation never been made but it would be impossible, he argued, because the various elements of the network were so intertwined. When the commissioners, manifestly frus-trated by his evasions, pressed further, Sise, Jr, finally conceded that "the long distance service was profitable, because every call was paid for, and was, at any rate, more profitable than the local business in many places."[34]

Try as he might to gloss it over, Size, Jr, had made a declaration of the ut-most importance: it was the first time anyone from Bell had admitted that the company's long-distance service was profitable. Up to then, the telephone industry had claimed that long distance was a bottomless pit whose revenues were purely hypothetical, since it was really the domain of telegraphy. By May 1907, thanks to technological progress, things were starting to change, but the company was still very reluctant to raise rates because of elasticity of demand – if the rates went too high, the number of users would diminish. Basic service was much more stable; rates in this sector could thus be raised by a larger amount than could long-distance rates without provoking people to cancel their subscriptions.

The commission's examination of Bell's rates revealed both the arrogance of top management towards the public authorities and the company's cre-ative accounting practices. Bell required that its rates be approved in their entirety so that it could achieve an acceptable rate of yield. But what was acceptable? Sise refused to give an evaluation of the company's assets and insisted that the rate of yield be calculated taking into account the rise in the cost of living. Things were at an impasse, when the chief commissioner died. The work of the Railways Commission was halted and Bell's rates were accepted as a unit.[35]

This anticlimactic conclusion has often been cited as damning proof of collusion between the Railways Commission and Bell, or, at the very least, of the commission's ineffectiveness. It is more likely, however that the difficulties with the new regulations were due to a search for tools and a tradition. With the exception of those of the independent companies, Bell's rates were no doubt the lowest in the world – even lower than AT&T's – so there were no fires to put out. In any case, the application had a positive effect in that it obliged Bell to establish global rates: long-distance calls would be calculated in a uniform way throughout the company's territory and local rates would be identical for population centres of similar sizes. This was an appreciable simplification.

Finally, the 1907 hearings were historic because they raised the issue of Bell's non-telephone revenues. The company's lawyer, Aimé Geoffrion, objected to any examination of revenues resulting from real estate and financial investments: "If the company had made wise real estate investments that should have no bearing on the telephone tolls. The company was entitled to its full legitimate profits on its real estate and other holdings, quite irrespective of the telephone's per se, otherwise the company could reimburse itself for possible losses on real estate by increasing telephone rentals."[36]

Geoffrion's argument eerily presaged that of Bell CEO and president Jean de Grandpré's in 1983, when Bell Canada Enterprises was created. The chief commissioner's reply was just as trenchant: he retorted that if Bell's charter did not limit the company's activities to providing telephone service, then all of the company's activities were open to examination. De Grandpré cut the Gordian knot by separating Bell's regulated telephone activities from its other operations. It was in fact the old Railways Commission, decried for its laxness, that opened hostilities in 1907 by examining all of Bell's investments, including the relationship between Bell and Northern. The commission specified that it did not intend to regulate Northern but only the Bell–Northern relationship – a significant nuance.

In 1911, 1912, and 1915 the cities of Toronto and Montreal in turn contested Bell's rates before the Railways Commission. Through press releases, subscribers who felt they were paying too much for their telephones were asked to come forward; then, backed by this popular protest, the municipal governments built their case. Each time, the commission refused the City's demand for lower rates. In 1912, however, Bell seized the opportunity on the fly and presented a counterproposal for a rate hike in Montreal. The commission refused in the same breath to raise Bell's rates, citing the company's good financial health.[37]

INTERCONNECTING THE INDEPENDENTS

The other major area to fall under regulation because of the 1906 amendment concerned the interconnection of independent companies with Bell's long-distance network. As we have seen, it was Bell's refusal to hook up the independents that had sparked the crisis. Once the amendment to its charter was adopted, Bell negotiated 378 connection contracts and refused eleven. The latter involved disputed territories between Bell and the independents, where there was competition for local service and access to long distance was a coveted bonus.

Ingersoll Telephone, an independent located in a town of the same name near London, Ontario, led the battle for interconnection. In May 1911 it applied to the Railways Commission in May, 1911, which made a historic decision that aimed to suppress competition. The head commissioner, J.P. Mabee, clearly laid out the regulating body's position: "Many people think competition is desirable. It is in most things; but competition in connection with telephones never appealed to me."[38]

The Bell network was obliged to interconnect with those of the eleven independents but was given the right to charge higher-than-usual rates, fifteen cents per long-distance call, from these small companies. Two years later, the commission approved a standard agreement between Bell and non-competitive independents. Once it had this victory, Bell could afford to be gracious. After a series of negotiations with the independents, it stopped surcharging on long-distance calls, for competition gave way to a long period of cooperation, which endures to this day.[39]

THE CANADIAN MODEL

The 1906 amendment to Bell's charter, which was the basis for Canadian regulation of the telephone industry until the 1992 decision to allow long-distance competition, inspired the 1908 amendment to the Railway Act extending the regulatory agency's powers to all telegraph and telephone companies under federal jurisdiction. It placed a private industry under the authority of a neutral regulatory agency, thus limiting the abuses of unbridled capitalism while avoiding political interference. Like any compromise solution, the 1906 amendment didn't make everyone happy. Some thought that regulation served only the interests of the phone companies, others, that it served only the public interest.

In Part Two of this book, we will see how Canadian regulation addressed both sets of interests at once. It allowed efficient and socially responsible companies to flourish and served the public interest better than any other economic system in the world. Indeed, by submitting a private enterprise to the control of an independent public agency, regulation imposed an early form of *glasnost* on Bell. To my knowledge, there is no other economic system that forces a company to lay open its books, management practices, and, to a certain extent, internal policies – and much more effectively than nationalization could. Beyond the content of the rulings made over the years by the Railways Commission, the Canadian Transport Commission, Transport Canada, and the Canadian Radio-Television and Telecommunications Commission, and beyond the nature of the controls imposed on Bell, the simple fact of getting information out of the closed rooms of the board of directors represented an enormous step in the direction of economic democracy. It worked marvellously for seventy years.

In the 1970s, once the goal of universal service had been reached, the regulatory process lost its *raison d'être*. With the rise of competition and the multiplication of pressure groups, it got weighed down and legal procedure took precedence over fundamental issues. Reality sometimes rode roughshod over the CRTC and it erred in some decisions, going so far as to compromise the industrial development of telecommunications in the name of a now-irrelevant social principle. Overall, the Canadian model has been an economic and political success: universal service.

This model was picked up by the United States when it decided to regulate AT&T in 1910. The far-sighted president of ATT, Theodore Vail quickly favoured regulation of the telephone monopoly. For the industry and the American legislators, the Canadian precedent served as a paradigm. It was one of the few times in the history of the two countries when the mouse influenced the elephant.

6 Balkanization
of the Telephone Industry

While the federal government came up with the solution of government regulation of a private monopoly, the Prairies nationalized the telephone industry. Bell Telephone's divorce from the West was fraught with *sturm und drang*.

In the 1880s and 1890s, Bell had satisfied the demand for telephones in the Prairies without too much difficulty. It had opened an office in Winnipeg in 1881 and was serving a territory stretching from Manitoba to the Northwest Territories to the Rockies. Bell's *modus operandi* was to give priority to cities while letting rural markets languish. On the other hand, long-distance lines were still rare; at the beginning of the twentieth century, Manitoba still did not have a telephone link to the rest of Canada (though it had a line to the United States). The reason for this was economic: capital invested in eastern Canada yielded a better return than investment in the West. Respect for this economic orthodoxy was to cost Bell dearly later on.

In the autumn of 1885, a group of Winnipeg businessmen launched the Wallace telephone cooperative. A machiavellian Charles Sise replied by creating, under a heavy cloak of secrecy, the paradoxically named People's Telephone, which was to compete not only with Wallace but with the local Bell agent, who wasn't in on the secret. Wallace sold his assets to People's in April 1886 and order was reestablished.[1]

Table 4
Number of Telephones in Manitoba and the Northwest Territories, 1896–1907

Year	Number of telephones
1896	2,000
1903	4,000
1905	8,000
1907	19,000

Source: Armstrong and Nelles, Monopoly's Moment, 175.

BELL LOSES THE BATTLE OF THE PRAIRIES: 1885–1908

The Beginnings of Discontent

What upset the apple cart was the massive influx of immigrants in the early 1900s. Populations in the Prairie provinces skyrocketed: between 1881 and 1911, Manitoba's went from 62,000 to 461,000; Saskatchewan's, from less than 30,000 to 492,000; Alberta's, from 1,000 to 374,000. There would have to have been a massive infusion of capital and energy into the telephone infrastructure, and the return was far from certain. Bell made a substantial effort at the beginning of the century, but it was too late to check the anti-monopoly wave that was sweeping the Prairies. And the potential was enormous, as Table 4 shows.

This spectacular growth was even more remarkable because the telephone was the first industrial technology to make inroads in the West without government assistance, whereas the first telegraph line to serve the Prairies, opened in 1876, had been built by the government. The first transcontinental telegraph line, opened in 1886, had been CP's and it had received indirect government assistance. Compared to the telegraph, the telephone was a poor relative. Bell had to raise the funds it needed on the stock market, but no one wanted to invest in the Far West, which was the stuff not yet of legends but only of strange rumours.

On top of these economic and demographic facts, there was a cultural animosity towards anything that came from the East. Villages with telephone service complained that the rates were too high, and those with little or no service complained about being neglected. The telephone historian Tony Cashman, in his retrospective of the telephone industry in Alberta, summarizes the situation on the Prairies very well: "This unappreciated company was able to achieve unpopularity in two ways: a) by coming into a town; b) by staying out."[2]

As Manitoba threatened nationalization, Bell increased its system-building efforts. Here, a team of technicians buries a cable on a Winnipeg street in 1903. Courtesy Bell Canada.

For Bell, the Prairies were a bottomless pit for investments, with a singularly low rate of return. Subscribers in the West tended not to pay their telephone bills and Bell employees did not display an excessive zeal for pursuing its accounts receivable. Bell's correspondence with its local managers reveals Montreal's incomprehension regarding the baffling "Far West." The company's treasurer wrote to the manager in Calgary in 1889, "I do not think our auditor will accept your explanations of why outstandings are unpaid. Out of 30, the first on the list has 'skipped out' and the other 29 are 'away from home.' You do not appear to understand our system."[3]

On the other hand, Prairie subscribers felt completely justified in not paying for a service that was out of order more often than it worked. What's more, the minimum of twenty to twenty-five subscribers that Bell required in one location before it would install an exchange irritated people, who considered the telephone to be a key part of rural development. Yet populations moved around so much in the West that Bell had already had to close exchanges that it had just opened, and it wanted to be more prudent.

Unlike in Ontario and Quebec, few independent telephone companies set up shop in the Prairies before the industry was nationalized, so the safety valve of a dynamic independent telephone industry was missing. With the arrival and geographical stabilization of a large population, this dearth might have been corrected. At the beginning of 1905, two telephone companies filed with the Manitoba legislature charters specifying that their common goal was to stem the abuses resulting from monopoly. But the Private Bills Committee rejected the applications, arguing that the government had to intervene, either by purchasing Bell's installations or by creating a parallel network.

In July 1905 the Union of Municipalities, spearhead of Canadian populism, organized its annual general meeting in Winnipeg. Francis Dagger, the leader of the anti-Bell movement, was shifting his focus to the more fertile battlefield of the Prairies.[4] The Conservative premier of Manitoba, Rodmond Palen Roblin, used this movement to launch a campaign promoting a public telephone network. Roblin's early political career had been built on opposition to CP's monopoly. When Bell signed an exclusive contract with CP to provide telephone service in its train stations, the manoeuvre, seen as a smart move in the East, seemed in the West to be a provocation by the large monopolies. Bell's excessive centralization had cut top management off from the sociopolitical realities of the West, and CP's financial and political situation made it an impregnable fortress to the new Prairie provinces. Moreover, the telephone issue had just hit the headlines thanks to the Select Committee of the House of Commons, and the neighbouring province of Ontario was flirting with the idea of nationalizing the electricity industry (Ontario Hydro was created in 1906).

Charles Sise thus felt, with reason, that "the movement in Manitoba [was] a purely political matter."[5] For some time, he considered using the Maritimes model, creating a company with local capital and some degree of ownership by Bell Telephone. But at the turn of the century, the Prairies didn't have the capitalistic tradition of the Maritimes, and local businessmen saw real estate and commerce as vastly more interesting investments than the telephone, where the rate of return was around 7 percent year in, year out. In any case, the situation was far too sensitive politically for Bell to play a key role. The Prairies as a whole, not just Manitoba, were rising up against Bell. Sise even considered pulling off the People's Telephone scheme a second time – launching a subsidiary disguised as an American firm – but he dropped the idea when he found that it would involve paying bribes to Manitoba politicians. He then committed a tactical error: he sent out his Chief of Special Agents, W.C. Scott, to organize a counteroffensive. Scott launched press and lobbying campaigns based on economic rationalization. But public opinion was not as easily swayed as a parliamentary committee; political passions could not be fought with figures.[6]

In November 1905 Premier Roblin announced his intention to go ahead with the creation of a jointly owned municipal-provincial telephone network, and a special committee was formed in January 1906 to study how to make this a reality. Sise then presented Bell's legal argument: a province could not expropriate a company that had been declared "for the general advantage of Canada." This was another error. In Canada, a company that was also a public utility could not allow itself the luxury of confronting a provincial government. Ever.

The Manitoba government tabled two bills, the first to tax private telephones and the second authorizing the establishment of the jointly owned company, on condition of taxpayer approval. During this time, the Bell teams redoubled their efforts to improve service: a long-distance line from Winnipeg to Regina was completed in 1906. But it made no difference.

Town by town, a series of referendums was organized. The Manitoba government hired Francis Dagger as a consultant to inform the electorate on the telephone issue in what was supposed to be a non-political manner; Dagger invited Dr Demers, the telephone pioneer of eastern Quebec, to address francophone Manitobans. Official propaganda promised to lower the basic rate by half and long-distance rates by two-thirds. For its part, Bell organized a countercampaign designed to show the real choice: either the rates would drop and the majority of taxpayers would have to finance the minority of users of a public telephone network, or the rates wouldn't drop, and then what would be the good of nationalizing?

The results of the government propaganda were mixed. A majority of municipalities, sixty-seven to fifty-five, rejected the government proposal. In the popular vote, however, a slight majority (13,688 to 11,567) was in favour of the proposal (60 percent of the vote was required for adoption in each municipality). Not at all discouraged, Roblin called a general election on the telephone issue and was returned to power in March 1907. He immediately wrote a letter to Sise stating his intention to purchase Bell's installations in Manitoba, but the answer was a flat no. Roblin then made plans to create a parallel network, as the law authorized him to do. Bell was unprepared to face this eventuality. In his daily log, Sise wrote laconically that the company had a choice: "to lose the territory with the money or without it."

Bell lost Manitoba but made money, lots of money. Sise was as talented in business as he was poor in politics. Without making even a cursory evaluation, he demanded $4 million and was only too happy to close the deal at $3.3 million plus $100,000 in construction materials in December 1907, despite Dagger's indignant clamouring. From Regina, where he was working for the Saskatchewan government, Dagger was still watching closely; it was "a present of one million dollars over and above the actual value of the plant,"[7] he shouted at the top of his lungs – and from a purely accounting perspective, this wasn't far from the truth.

Another factor that certainly influenced this outcome was a crisis at AT&T. John Pierpont Morgan had executed a financial *coup d'état* that ended with Theodore Vail's return to the business. As he was attempting to get AT&T back on its feet, Vail advised his old friend Sise to play for time and seek compromises. But once again, Bell Telephone had failed to find in Canada the funding it needed to develop. By depriving Bell of its usual recourse, events at AT&T forced it to step aside from the competition in Manitoba, but also enabled it to turn a negative situation into an extraordinary source of revenue.

On 15 January 1908 the provincial government officially took possession of a network with seven hundred employees serving fifteen thousand subscribers. One clause of the contract between Bell and Manitoba specified that Bell employees in the province were to be automatically hired by the new government administration.[8]

Alberta Slams the Door

With the upheaval in Manitoba, people in Alberta started to pay attention to the telephone. Since the beginning of the century, conflicts had been multiplying between Bell and Albertan businessmen who, unlike those in Manitoba, were launching phone companies in growing numbers and trying to dissuade the Montreal giant from opening exchanges in their localities. Everywhere, Bell invoked its federal charter to go ahead and install a system, nipping local initiatives in the bud. The core of the Bell strategy in Alberta was, as usual, to promote long-distance lines. It built the Edmonton–Calgary line in 1903.

In Red Deer, a small town halfway between Calgary and Edmonton, the power struggle resulted in the installation of two competing exchanges. When Alberta became a province in September 1905, both main political parties had on their platforms nationalization of public utilities in general.

Once elected, the Liberal government quickly tabled a bill that would allow municipalities to construct and operate local telephone networks; the provincial government planned to link these local networks with a long-distance one. During the first session of the first legislative assembly in March 1906, the Telephone Act was adopted, making Alberta the first province to create a public telephone network.

The first government long-distance line was built, with great difficulty, between Calgary and Banff in the winter of 1906–07. The new government company, Alberta Provincial Telephones, set its rate 30 percent lower than Bell's over an equivalent distance. Bell reacted by doubling the capacity of its Calgary–Edmonton line. Premier Graham Cameron Rutherford's Liberal government realized that the cornerstone of Bell's monopoly was long-distance service. In February 1907 it adopted a new telephone law that specified the modalities of executing the 1906 law. The debate in the assembly reached a level of verbal violence unequalled until then in the history of the Canadian telephone industry: Bell was called an "octopus" and Sise was held up to obloquy. "He may be a Napoleon," thundered one member of the assembly, "but fortunately Alberta has a Wellington, and today the Bell Telephone Company meets its Waterloo."9 The member in question was the owner of the Red Deer telephone company and he had personal reasons for being angry at Bell. Even so, the tone was set.

After this debate, the provincial government offered to purchase Bell's installations in Alberta. Charles Sise proposed the creation of a joint enterprise, which would have allowed the province to avoid paying to acquire the existing installations. The government refused the offer, although it was advantageous in all ways for the province – except for Albertan "patriotism." Sise then committed an error that was all the more unpardonable because it was so unlike that austere, cold man. He lost his temper in his response to Alberta's minister of Public Works: "In the twenty-seven years in which this Company's stock has been offered for sale I am not aware that one Citizen of Alberta has ever taken or offered to take one share of stock in the Company or in any way aided the Bell Telephone Company in prosecuting its business."10

Once again, the dialogue between Bell and the government authorities broke off. The year 1907 was marked by a race between APT and Bell. APT built or repaired fourteen exchanges, and long-distance lines shot out in all directions – everywhere Bell's weren't. Construction of the Edmonton-Lloydminster line gave rise to an incident that revealed the enthusiasm with which isolated towns greeted the arrival of the telephone. Lloydminster straddled the Alberta–Saskatchewan border. When the long-distance line reached the part situated in Alberta, residents of the part situated in Saskatchewan demanded to be hooked up. APT was hesitant; it did not want to create a conflict with its neighbouring province. Premier Thomas Walter Scott of Saskatchewan sent a letter giving APT permission to construct a line wherever in his province it wanted.

When telephone service took too long to reach an Albertan town, local initiatives took up the slack. Small companies sprang up all over; the telephones were purchased wherever they could be found and the lines were made with iron fence-wire.

In this context, the result of the war between Bell and APT was never in doubt. Bell played the card of economic rationality, while the Alberta government and APT played the cards of western pride and opposition to large companies and distant eastern bureaucracies. The battle was not an equal one. In April 1908 Bell agreed to sell its installations to the Alberta government for $675,000. Its network comprised 1,100 kilometres of long-distance line, 18 long-distance exchanges, 19 local exchanges, 2,270 subscribers, and 150 employees (only the top managers stayed with Bell and went back to Montreal). As had happened in Manitoba, telephone rates immediately dropped.11

Saskatchewan Finds a Way Out

Saskatchewan had no choice but to follow the example of its provincial neighbours. The story was a familiar one: the government hired Francis Dagger as a consultant in 1907, and he recommended nationalization of long-distance service and municipalization of local service. Special attention was paid to farmers, who were the majority of Saskatchewan residents: rural service was entrusted to telephone cooperatives. This was an original solution, adapted to a province that would otherwise have been unable to finance an essentially rural, and therefore very expensive, network.

Saskatchewan's battle with Bell was less intense then Alberta's or Manitoba's. Lewis McFarlane, Bell's negotiator at the time, put it rather colourfully: "Having cut our arms and legs [Manitoba and Alberta] off, there wasn't much use keeping the body!"[12]

In June 1908 most of Dagger's recommendations were implemented in three laws establishing a system of municipal or agricultural cooperatives for the local network and a provincial system for the long-distance network, all headed by the new Department of Railways, Telephones, and Telegraphs. The provincial government planned to construct a telephone system jointly with the municipalities and farmers, supported by local initiatives, and not simply replace Bell with a nationalized company. Nowhere was Dagger's cooperative theory brought to life with such fidelity to his original model as in Saskatchewan.

The province's lack of ready cash delayed the outcome a little. Saskatchewan, like Alberta, had become a province only in 1905, and acquisition of a telephone system was something of a baptism by fire with regard to economic intervention for a young province with few resources.

Elections in August 1908 returned Scott's Liberal government to power, and an agreement was concluded in April 1909. Bell sold its Saskatchewan installations for $368,920.[13] Its network at the time comprised 365 kilometres of very long distance line (mainly the Saskatchewan part of the Regina-Winnipeg line opened in 1906), 11 long-distance exchanges, 13 local exchanges, and 2,020 telephones. After this acquisition, the provincial government purchased three independent companies and built its first two exchanges; a year later, the Saskatchewan network comprised 792 kilometres of long-distance line and 5,710 subscribers.[14]

Between start-up on Prince Edward Island in 1885 and start-up in Saskatchewan in 1909, Bell Telephone had lost half of its territory, but the number of telephones it served had increased from 8,000 to almost 120,000. The geographical losses had been compensated for by exceptional penetration in the heart of Canada, especially in the cities.

From a purely economic point of view, the loss of the Prairies marked the limits of the North American brand of private enterprise, with its goal of short-term profitability. It would have taken a long-term vision, which Bell didn't have, to invest in the West at a loss, or even at a lower rate of return than in the East.

Confronted with the same problems, the telephone industry in the United States remained unified within AT&T. But Theodore Vail's management philosophy was not characteristic of unbridled North American capitalism, while Charles Sise's was. Under these circumstances, intervention by provincial governments was seen as the only way to make up for the lack of capital on the Prairies, as the historian and philosopher of communications Harold Innis suggested: "The advantages of government ownership were shown in the immediate possibility of commanding tremendous capital resources at a comparatively low rate of interest and placing at the command of the community in the shortest possible time the conveniences of modern civilization which involve heavy capital investments."[15] For Innis, nationalization was dictated above all by the conjunction of three factors: scarcity of capital, low population density, and rapid economic expansion. In this view, the ideological dimensions of the controversy were secondary; nationalization was an efficient way of raising cheap capital when all other means had failed.[16]

THE END OF SISE'S REIGN

After thirty-five years at the top, Charles Sise let go of Bell's reins in February 1915. He was over eighty. His entourage had been complaining for a few years about his growing deafness, but no one would have dared ask him to leave. Bell Telephone had become Sise's baby. Now, he named his faithful assistant, Lewis McFarlane, to the presidency, and kept the chair of the board of directors for himself. Sise had built Bell in his image – authoritarian, severe, and moralistic, insofar as the moralism coincided with economic rationality. Ultimately, however, Sise's stamp on the company involved a stubborn will to build, structure, and create.

Sise's greatest feat was without doubt Bell's founding "blitz" during the summer of 1880. In just a few months, he had made the monopoly a *fait accompli* by manoeuvring masterfully between the parent company, the telegraph companies, and the federal government. When he arrived in Montreal in March 1880, he was a representative of American Bell with an ill-defined role. By June he was officially the number-two man in the new company; in fact, he was already number one. In October of that year, by acquiring the telephone installations of the telegraph industry's "big two," he ensured Bell's monopoly in Canada.

No less striking was the virtuosity with which Sise cut the American parent company's fetters on the Canadian firm. In two years, American Bell went from majority to minority shareholder – after having provided one hundred percent of the investment. In the short term, this hardly changed the relationship between the two companies since the Americans still held the controlling minority share, but it was the first step in the Bell group's long march to independence.

The successful "decolonization" of the Maritimes represented Sise's greatest achievement. Since Bell was incapable of continuing to operate telephone service in these provinces, he negotiated a gentle disengagement from the new companies in Prince Edward Island, Nova Scotia, and New Brunswick and maintained the Bell group's presence in these provinces as an advisor and a supplier of telephone equipment.

Top among Sise's liabilities was his lack of an overall vision of the "telephone phenomenon." Unlike Vail in the United States, Sise had no long-term plans. Nothing in his log revealed a grand design. He managed the company as tightly as possible, with neither errors nor inspiration. Not one sentence in his immense correspondence merits being passed on for posterity. (It is true, however, that there is a tendency to be even more severe with Sise in comparison to Vail.)

An absence of vision, along with a total absence of political sense, was behind Sise's greatest mistakes. During the Civil War, for example, the New Hampshire-born, pure New England Yankee must have been singularly lacking in judgment to join the ranks of the South. At the head of Bell, Sise did not grasp the importance of Canadian protectionism, which was at the heart of Prime Minister John A. Macdonald's "National Policy." He continued to import telephone equipment from the United States, losing Bell's Canadian patents. Nevertheless, Sise quickly realized the extent of his blunder, and after this setback he took every care to build an independent manufacturing sector in Canada.

Up to the turn of the century, in spite of some undeniable limitations, Sise's management report card was positive on the whole. Even without the vision of a Vail, he was undeniably an empire builder. After 1900, however, he began to backslide. He was ageing and going deaf, and his conservatism won over his energy. He was completely puzzled by civic populism, and he misunderstood the power relationships in the Canadian West. This double error in judgment lost him, in turn, Manitoba, Alberta, and Saskatchewan.

The rise of unionism brought confirmation of the old man's inability to adapt to the twentieth century. The Toronto strike caught him unawares, frozen in an attitude of primitive capitalism. He did not see the need for employee/management relations, still less the need to maintain relations with the government, and he displayed a contempt that was costly for his company. He was the patron saint of divine right.

What is more, Sise had worked for thirty-five years in a French-language city – Montreal – without letting it affect his management style at all. He had designed Bell as an Anglo-Saxon company in a French environment, which for companies at the time was not unusual. On the other hand, for a public utility, such an attitude was suicidal. His error was all the more egregious given that he himself was fluently bilingual.

Very early, as we have seen, Sise entrusted Bell's relations with the outside world to a "Special Agents" department, whose director, William Scott, was sent on spying missions. This was far from the transparency favoured by Vail during the same period. While Vail invented public relations out of whole cloth, Sise carefully cultivated a cult of secrecy, both personal and corporate. His log abounds in codes and more or less obscure phrases that connote an unshakable penchant for the shadows and for backstage ploys.

Sise was unable to see that the telephone industry had changed fundamentally by 1900. It was now a public utility and thus had new responsibilities. Bell's official policy persisted in denying any corporate evolution, as this turn-of-the-century oath of loyalty shows: "Telephone service is not universal in its character as are the systerms of Waterworks, Gas, Electric Light, or even the Street Railway Service."[17]

It took the 1905 Select Committee and Theodore Vail's return to power at AT&T for Sise to acknowledge that things had changed. The president's message in Bell Telephone's annual report for 1912 attested to Sise's sudden turnaround: "The general feeling now seems to be that the telephone service to be perfect must be universal, intercommunicating, interdependent, under one control ... and that rates must be so adjusted as to make it possible for everyone to be connected, who will add to the value of the system to others."[18] The text might have been signed by Theodore Vail. This surprising reversal, however, shows the adaptability of this nineteenth-century man. Sise did not have the inspiration to innovate, but he had the instinct to survive.

Sise's reign would have been better had it been shorter. Thirty-five years at the head of the company was too long. But Sise did not consider his role at Bell to be just another job. He had invested in "his" company the sense of belonging that one would normally have for one's country. Bell was Sise's homeland. The historian Graham D. Taylor said, more crudely, that Sise "colonized" Bell Telephone.[19] Indeed, in 1913 he sat his son, Charles Sise, Jr, on the board of directors, and Sise, Jr, was to succeed McFarlane as president in 1925. Sise's other two sons, Edward and Paul, were appointed to the board of directors of Northern Electric in 1911. Edward became the president of that company in 1914, and Paul succeeded him in 1924. Although they held only a tiny fraction of the company's stock, the Sise dynasty ruled Bell for two generations.

The record of Charles Sise's term at the head of Bell Telephone would not be complete without mentioning two figures. In 1880 Bell had 2,165 subscribers. In 1914 it had 237,068 subscribers.[20]

THE MARITIMES AND NEWFOUNDLAND

New Brunswick: A Famous Independent

Beyond Bell's territory, the telephone industry was growing out of its infancy, but, unlike the telegraph industry, it had not yet reached maturity. With a few limited exceptions, the "great policy" advocated by Theodore Vail in the United States had not yet crossed the border.

In the Maritimes, there had been a series of major manoeuvres that resulted in throwing control of all phone companies into new hands. But this was simply a movement towards financial rationalization and not an overall shake-up of the industry. In contrast to the rest of the country, the Maritimes saw no grand confrontation between the independents and the dominant companies. In fact, the existence of dominant local companies in touch with the population kept things from becoming politicized.

New Brunswick was the only province in Canada where the phone business was under the control of an independent company (which *ipso facto* ceased to be one). This brilliant move was the work of Central Telephone, founded in 1904 to operate a line between Fredericton, Saint John, and Sussex, a small town halfway between Saint John and Moncton.

Everything started innocuously enough. Howard Perley Robinson, a young Sussex industrialist imbued with civic spirit, agreed to take on various management positions in the public utility. When the company, with just one dead-end line, was on the verge of bankruptcy, Robinson took his greatest risk and became general manager. Naturally, he wanted to sell Central Telephone off to New Brunswick Telephone, but the latter refused to buy. Furious and determined, he paid off the debts of "his" company and began a battle against New Brunswick Telephone, taking extreme financial risks (such as doubling some of the lines in his competitor's network) and pushing his legal options to the limit. Finally, New Brunswick Telephone had to agree in principle to merge with Central Telephone on a share-for-share basis.

But Robinson was not content with this half-victory and made special preparations for the shareholders' meeting of June 1907, at which the merger was to take place. Bell Telephone held one-third of the voting shares in New Brunswick Telephone. Robinson had learned that NB Tel was purchasing telephone equipment in the United States and not from Northern Electric, to Bell's great displeasure. He obtained Sise's assurance that Bell's representatives would not vote. Thus, during the merger meeting, with the support of half the shareholders and the abstention of another third, Robinson was elected general manager and chief executive officer NB Tel, a position he held until his death.[21]

This merger aroused a public that was always ready to denounce private monopolies. When New Brunswick Telephone asked to increase its capital from $600,000 to $2 million, Saint John brought a lawsuit, arguing an attempt to disguise revenues, and the newly created Union of New Brunswick Municipalities demanded that the company be nationalized. Neither of these actions came to anything.

Howard Robinson, however, was not content with managing wisely. Under his watch, the new company's style was distinguished by a voraciousness that swallowed up the last independents in just a few years. The Conservative premier, John Douglas Hazen, calmed things down in the end by creating the Public Utilities Commission in 1910.[22]

Nova Scotia: Regulation Takes Hold

The situation in Nova Scotia seems to have been very different. Nova Scotia Telephone thrived on good relations with the independents; it even agreed to sell its installations to new companies in exchange for stock and interconnected their local systems to its long-distance lines. In fact, the NS Tel archives show that the independents were in the habit of offering shares to the company whenever they needed money to expand or modernize their systems.

There was only one blight on this idyllic picture: the Chambers affair in Truro. Interference between the electrical and telephone systems led NS Tel to seek an injunction against the Chambers Electric Light Company in September 1891. After two years of constant squabbling, the owner of Chambers created a telephone company and orchestrated an advertising campaign to offer telephones for free to everyone who took electricity service.

The affair was deemed serious enough for Charles Sise to be consulted. He had been on the board of directors of NS Tel since Bell Telephone's peaceful retreat in 1888. Sise dispatched a specialist in fighting independents to lend a hand. The expert's solution was, of course, the strategy of "turn-around is fair play": Bell should create an electricity company. A charter was even filed to this effect at Nova Scotia's legislative assembly. Chambers Electric Light backed off and the war between electricity and the telephone never took place. NS Tel had to follow other telephone companies and convert its Truro network to metallic circuits.[23]

In 1903 Nova Scotia was the first province to regulate the telephone industry: any request for a rate hike had to be addressed to Cabinet. However, the provincial ministers complained so loudly about their part in regulation that a public utilities commission was created in May 1909 on the model of Wisconsin's regulatory agency.[24] In 1913 the law was amended to make the commission a court of final appeal. Telephone service was thus defined as a public utility.

Unification of Nova Scotia and Prince Edward Island

In April 1910 a new company was formed in Halifax: Maritime Telegraph and Telephone. The grand ambition behind this initiative was to unify all telephone companies in the Maritimes in order to improve the long-distance network among the provinces on the Atlantic coast. MT&T's first priority was Telephone of Prince Edward Island, which had been having financial difficulties for some years, as it had been unable to recover from the modernization effort it had had to undertake when urban electrification had forced it to replace its wire-with-ground system with metallic circuits. New Brunswick Telephone also had designs on Telephone of PEI.

Prince Edward Island had wanted a telephone link with the mainland since the beginning of the century, but it was opposed by Anglo-American Telegraph, which had controlled telegraphy on the island since Frederick Gisborne's first efforts to establish the first transatlantic cable. Anglo-American claimed that it had a monopoly on all off-island electrical communications (as we have seen, this company had taken the same position when Bell wanted to introduce telephone service to Newfoundland some twenty years earlier). Telephone of PEI claimed, however, that the monopoly applied only to international links and not to interprovincial ones.

After several years of talks that led nowhere, the situation was resolved when MT&T laid a rudimentary cable under the Northumberland Strait between Pictou, Nova Scotia, and Wood Island, PEI. The first telephone conversations with the mainland took place in January 1911. All Anglo-American could do at this point was accept the *fait accompli*.

PEI's rural network was in such poor condition, however, that the Wood Island–Charlottetown section had to be modernized immediately. Even before the undersea cable was inaugurated, Telephone of PEI had to ask MT&T for financial assistance, and the response was an offer to purchase in September 1910. After laborious discussions, MT&T purchased the majority of Telephone of PEI's stock with voting rights in December 1911. At the time, the acquired company had 1,009 telephones.

Nova Scotia: MT&T Triumphs

Even in Nova Scotia, MT&T's presence was at first unobtrusive. In October 1910 it purchased small local companies in Antigonish and Sherbrooke. In January 1911 it offered to lease NS Tel's installations. Unlike Telephone of PEI, NS Tel had no financial problems: it regularly paid a 6 percent dividend to its shareholders and coexisted peacefully with the independent companies.

At first, NS Tel refused the offer to lease; MT&T responded with an irresistible offer to purchase: an exchange at par of NS Tel shares at 6 percent for those of MT&T at 7 percent and payment of a premium of one MT&T share of three NS Tel shares. The shareholders offered no resistance to this deal; Bell was a little reluctant, but it fell into line when the entrepreneurs made a pilgrimage to Montreal and guaranteed that Bell's two seats on NS Tel's board of directors would be maintained on MT&T's. The sale was concluded in December 1911. During this time, all independents in Nova Scotia had been purchased and integrated into the new company, for a total of 12,908 telephones. In spite of this promising début, MT&T could go no farther in its dream of unifying the telephone industry in the Atlantic provinces.

Newfoundland: Persistent Difficulties

In Newfoundland, the telephone industry developed slowly because of the difficult geography and, no doubt, because of Anglo-American's lack of interest, oriented as it was towards telegraphy and the transatlantic cable. Telephone service was confined to the capital, St John's, and a number of pulp-and-paper, mining, and public utilities companies had built their own telephone systems.

One of these networks, United Towns Electric, expanded noticeably. The company was incorporated in 1902 with its head office in the Carbonear region, on the Avalon Peninsula. Soon after, a man named J.J. Murphy opened a number of exchanges in the Avalon and Burin peninsulas in order to link his establishments with each other and with the outside world. Without knowing about it, Murphy had created the ancestor of Newfoundland Telephone.[25]

THE PRAIRIES

Manitoba: Difficult Times after Nationalization

As soon as nationalization was completed, Manitoba had to face the same economic problems as Bell. The first completely publicly owned telephone network in North America was having start-up difficulties. Bell had left nothing behind in terms of organization, since all of its management activities had been centralized in Montreal. On top of that, access to the American Bell System, a valuable asset, was cut off from one day to the next.

An administration had to be built from scratch, a job assigned to a telephone commission composed of three ex-Bell managers that had no legal autonomy from the government; expenses had to be authorized by the Department of Telephones and Telegraphs. There was a lack of qualified personnel, so people had to be recruited from independent companies in the United States. One month after his government took control, Premier Rodmond Roblin said, "We did not know anything about telephones and we admit it."[26] If only he had listened to Charles Sise, who had wanted to create a company jointly owned by Bell Telephone and the provincial company.

The managers of Manitoba Government Telephones did their best to preserve the professionalism within their system. They kept the government's promise to lower rates as far as was reasonable. One month after nationalization, in February 1908, the rural basic rate dropped from twenty-four to twenty dollars per year. Overall, depending on population density, rates fell between 16 and 28 percent – far from the 50 percent reduction announced by the government during the municipal referendums. One year later, party lines made their début in Manitoba. Finally, all of the independent companies were interconnected to the long-distance network. But MGT did not avoid the pitfall of all nationalized companies on the Prairies: no budget line was created to account for depreciation of equipment, and so the profits posted in the annual financial statements were misleading.

From the beginning, the Roblin government had used the telephone as a political tool. Rate reductions had been imposed by political power, over the opposition of the three MGT commissioners. During the 1910 provincial election, rural lines were frenetically thrown up: "The greatest drawback has been the unavailable supply of experienced men. If it had not been for this much more could have been accomplished. As it is, a number of rural and long-distance lines planned to be built had to abandoned and left until the season of 1911. It was impossible to comlete [*sic*] them ... A force of over 650 men were employed on the rural, long-distance and exchange work throughout the province. We could have used another 400 men, but they were not available."[27] Things were utterly out of control. After just one year, the losses were so huge that MGT had to declare a deficit and proposed the introduction of local measured service (LMS), whereby each local call would be charged, instead of having a uniform basic rate. In this instance, the press used the same arguments against MGT that it had used against Bell in previous years.

Under popular pressure, Roblin named a commission of inquiry in January 1912, which proceeded with a comparative study of the Manitoba network and the neighbouring American network. It concluded that MGT had done excellent work on the technical level but "that the system has generally been administered extravagantly and that very large savings could be made by economical management."[28] The commission of inquiry cancelled the rate hikes and cleared the government of accusations made by MGT management that it had mishandled MGT's affairs. Reassured, Roblin summarily fired the "independent" commissioners who were managing MGT – and found himself with the same problem as before: how to manage the public system? The fact was that the three fired commissioners hadn't caused MGT's problems; on the contrary, they had always fought to preserve MGT's integrity in the face of political influences, and they ended up being the government's scapegoats.

Telephone service had become such a burden for Roblin that he made over-tures to AT&T and even to Bell to resell the Manitoba network. Such a move would have been politically suicidal, so Roblin finally opted for creation of a regulatory screen between MGT management and the government and set up the Public Services Commission for this purpose. It was certainly the best de-cision the Roblin government made with regard to telephone policy.

The commission's first move, in July 1912, was to raise rates in Winnipeg by 20 percent. One month later, rates across the province rose by 10 percent. The rural rate was now twenty-two dollars a year, and on top of that the farmers were responsible for connecting their residences to the line on the road, work that had been done by the telephone company in the past. On the other hand, LMS disappeared with the new rates.[29]

Four years of folly thus came to an end. It must be said in defence of the inexperienced MGT administrators that the number of telephones had more than doubled, from seventeen thousand in 1908 to forty thousand in 1912. Manitobans had wanted telephones quickly and cheaply. They obtained them quickly but not as cheaply as planned.

Most important, after the 1912 crisis the principle of amortization began to be used. In 1913 and 1914, the MGT books included a reserve fund and an average-depreciation grid for various types of equipment, making Manitoba the first province in the Prairies to adopt this accounting principle. The am-ortization rate at the time hovered around 5 percent, which was comparable to Bell's.[30]

Also in 1912, when Rupert's Land was annexed by Manitoba, more than tripling the province's area, MGT had to be reorganized. It then adopted a principle that would mark the history of the telephone industry from then on – not just in Manitoba but throughout Canada: "All parts of the province that are accessible are to be supplied with service, regardless of the fact that sup-ply of telephone service to rural and distant areas will, in most cases, be so supplied at a loss, but that other areas and services of the system will change such rates as will enable Manitoba Government Telephones to avoid financial losses."[31] This principle, called cross-subsidization, is essential for any public utility and has been the basis of Canadian rate policy up to the present day.

Alberta: A Mixture of Enthusiasm and Amateurism

Enthusiasm for the telephone must be given credit for the major expansion program undertaken by Alberta Provincial Telephone. Alberta lacked skilled technicians, so they were hired from England. In December 1909 the Alberta network was connected to thirty-five towns in British Columbia's Kootenay Valley, but the Rockies were too big an obstacle to connect most of BC Tel to APT. The idea of linking up the three Prairie provinces would not be realized until 1923.

The Alberta government had refrained from creating a government administration on the model of Manitoba's Telephone Commission, and APT was part of a "normal" government structure – not even a department, but simply a division integrated into the Department of Public Works. Like any departmental division, APT's, accounting system consisted of posting the sums collected and paid, any positive difference being considered a "surplus." Like MGT's, APT's managers forgot one detail – that their telephone equipment was wearing out – and there was no "amortization" posting in their expense column. This accounting heresy went unnoticed during the first years because the equipment was new. But, unlike in Manitoba, no one protested and no corrective measure was adopted. What's more, promises of lower rates were kept with far more determination than in Manitoba: the basic rate in Calgary, the showcase of the provincial network, immediately plummeted from forty-eight to twenty-four dollars a year. This rate was far from a break-even level and would lead straight to bankruptcy.[32]

The construction program for the telephone network soon surpassed the capacities of the province's Department of Public Works. Thanks to the enthusiasm of APT's managers, the number of telephones tripled in the three years following Bell's departure. The provincial Department of Railways and Telephones was created in December 1911. This was also when the name "Alberta Government Telephone" began to appear in official documents, and from then on it was the name of the company. Work on the network pushed ahead and the Alberta network was connected to Saskatchewan's in 1914. An interprovincial service was inaugurated on the eve of the First World War, linking Edmonton to Saskatoon and Calgary to Regina.

The rapid expansion of the network did not go totally smoothly. AGT ran into the same problems that had been faced by Bell's managers: mobility of the population and the tendency of people to disappear without paying their bills: "The fact must also be taken into consideration that this province is still in its development stage; the rural as well as the urban population are more or less transient, there is more trading, bartering and moving about than in the older provinces, where the "Old Homestead" has a significant meaning unknown in the newer western provinces. As a result we find that each year there are a number of subscribers lost to the system which leaves us with an ever increasing portion of idle plant, upon which the fixed charges have to be met."[33]

The telephone boom ended as the First World War began. Calgary was hit right away: two thousand subscribers returned their telephones. The Conservative opposition suddenly began to question the management of the system and make accusations of corruption. The government replied by creating the Board of Public Utilities Commissioners in 1915, with authority over the public network but not the municipal systems.[34]

All systems in the Alberta network belonged to AGT or to rural cooperatives supported technically and financially by AGT. The Edmonton municipal network, jealously guarding its prerogatives as an independent, was an exception. In the early years of public telephone service in Alberta, a dispute had arisen with the city's telephone department. The first skirmish went back to the creation of AGT, when Edmonton was getting rid of the rural installations it owned on its periphery, going so far as to cut service to Fort Saskatchewan. AGT had to take up the less-profitable sections. The affair got nastier when the neighbouring towns of Strathcona and Edmonton were merged in 1912 and AGT sold its Strathcona installations at a price that Edmonton thought was exorbitant.[35]

The crisis came to a head in 1914, when Edmonton's Telephone Department realized that AGT was sharing long-distance revenues with other phone companies but not with it, even though a good proportion of the long-distance calls made between Alberta and the other provinces came from Edmonton. AGT replied that, on the contrary, Edmonton should be paying royalties to AGT which was "improving" the municipal service by giving it access to the long-distance lines.

It is not clear whether AGT was acting in bad faith. Long-distance service at the beginning of the century was far from being the sure thing that it became later, and it was difficult for AGT, which was investing at a loss in a costly long-distance network and in an even more costly rural service, to think of making payments to a wealthy municipal network. This dispute was not settled until the purchase of Edmonton Tel by AGT, then Telus, in 1995.

Saskatchewan: The Triumph of the Cooperative

Like Alberta, Saskatchewan entrusted administration of the telephone business to a single department administered by a commissioner, who was none other than the minister. However, the Saskatchewan Department of Telephones refrained from any demagogic act such as a wholesale rate reduction. The day after nationalization, long-distance rates continued to be aligned with Bell's. As well, in the first annual report, the department emphasized the uniqueness of telephone operations, particularly in the financial domain. In 1912 the books for the telephone system were separated from the rest of the government accounting and the expenses were applied directly against revenues. It was the first step towards autonomy in management. In spite of this commendable attempt at structural clarification, chaos reigned within the accounting of the enterprise itself.

In general, Saskatchewan applied Francis Dagger's hypotheses with great fidelity. The provincial government concentrated its efforts on the long-distance network, which required a large investment, leaving it to the municipalities and cooperatives to develop local service in towns and the countryside. One of the explicit reasons for the government to stay away from local service was its fear of being forced to establish uniform rates, so that some towns were defraying the deficits of others. Unlike Manitoba's government, Saskatchewan's considered interfinancing unfair.

The Dagger system failed in urban areas and succeeded in the countryside. The municipalities proved to have great reservations about the government policy, and only five of them took advantage of their right to create their own telephone network. Given this obvious failure, the government withdrew the 1908 laws regarding municipal telephone service. Municipal ownership was limited to the five cities that already had telephone systems.

On the other hand, the number of rural telephones grew rapidly, with the government donating telephone poles, technical aid, and tax breaks. In 1912 the government network comprised fifteen thousand subscribers, with an additional nine thousand in the rural cooperatives, for a percapita penetration double that of Manitoba and Alberta.

That this result was obtained without interfinancing was unusual, because all telephone companies and regulatory agencies in North America considered this the cornerstone of their social policy and one of the justifications for a monopoly. It seems that Saskatchewan's success was due to extreme decentralization, which reduced costs and mobilized the population. This was a triumph of the telephone militancy of the West over the cold rigour of eastern economic laws. But such a success could not be reproduced elsewhere, and, above all, it was linked to a lucky conjuncture of circumstances.[36]

This spectacular expansion was not without its problems; the cooperatives were handicapped by a lack of capital, the perpetual problem of young telephone companies. Indeed, they had to raise the funds to create their system in order to be eligible for state funding. To solve this problem, the provincial government passed a law in 1913 that permitted cooperatives to issue debentures by means of a tax on the land crossed by the telephone lines: "[This guarantee was] intended to remove the timidity with which capital the world over hitherto seemed to have viewed investment in rural telephone extension work. Its repayment was not to be left to chance or caprice or personal convenience but was assured so long as land continued to have value and that time will not be ended so long as homes are required for the people."[37]

The Saskatchewan Department of Telephones had just had its most ingenious idea. It had broken through the wall of mistrust and freed up an almost unlimited float of capital. On the other hand, the government suspended its donations of telephone poles and required cooperatives to conform to provincial telephone standards if they wanted to be connected to the long-distance network, though it maintained the technical assistance that it had provided since the beginning. This policy linked Bell's technological rigour to the West's vaunted social dimension, providing Saskatchewan with the most dynamic telephone network on the Prairies.[38]

Unfortunately, on the afternoon of 30 June 1912 Saskatchewan was hit with the worst tornado of the century in Canada. The wind speed in the eye of the cyclone was estimated at eight hundred kilometres per hour. In downtown Regina, the official count was 28 dead, 200 injured, and twenty-five hundred homeless. The roof of the central government telephone exchange was torn off, a wall collapsed, and a fifteen-tonne standard on the third floor fell, crashing through two floors into the basement. Miraculously, none of the eleven Department of Telephones employees on the premises were killed. They had to work all night to re-establish a line between Regina and Moose Jaw. Two days later, the long-distance line was up and running, but telephone service didn't return to normal until a month later. Since the old equipment had been completely destroyed, the ministry decided to replace it with automatic equipment, which placed the city in the vauguard of the Canadian telephone industry.[39]

Overall, were the results of telephone nationalization in the three Prairie provinces on the eve of the First World War positive or negative? Harold Innis answered this question by enumerating the flaws in the new government-run companies: defective accounting; politicization of management; loss of access to the Bell System, which was the largest reservoir of telephone expertise in the world. However, his overall judgment was mixed: "It would be difficult to say whether [nationalization] fulfilled the direct purpose of production more economically, but it is not difficult to say that it fulfilled it more quickly."[40]

The greatest justification for nationalizing was that a public company could raise capital more rapidly than private enterprise could. Thus, the Prairies filled the gap left by the federal government in telephone service. Moreover, Saskatchewan mobilized farmers and spread the real estate guarantee to accelerate further investment in telephone service in the Prairies.

In a North American anti-government context, nationalization of the telephone industry by provinces with no experience in the business was a risky venture. But as soon as Bell failed in its mission and the federal government was no longer interested in telephony, nationalization was the only possibility for the Prairies.

BRITISH COLUMBIA: BC TEL'S BONES OF CONTENTION

On the far side of the Rockies, thanks to energetic work by James Matthew Lefevre and his acolyte, William Farrell, BC Tel had unified telephone service in British Columbia. However, like Bell at the other end of the country,

BC Tel was incapable of supplying telephones quickly enough outside of the main centres, and on the eve of the First World War, independent companies began to spring up: in addition to those in the Kootenay and Okanagan valleys and the towns of Mission and Chilliwack, there were telephone cooperatives formed by farmers and a municipal company in Prince Rupert.

Even in Vancouver, BC Tel had a hard time responding to the demand of a population that had exploded from 27,000 to 147,000 between 1901 and 1911. It was hardly surprising that relations with the city were stormy. In June 1905 the mayor went so far as to have six BC Tel employees, including the company's general manager, thrown in jail because they had dug a hole in a street without permission. The Supreme Court had the employees released, and the mayor had them sent back to jail the next day. This might have gone on *ad nauseum* had the court not quickly rendered a decision confirming the precedence of the company's provincial charter over municipal regulations.

Almost all towns in the region around Vancouver studied the possibility of setting up municipal companies at one time or another. Did this mean that telephone service in British Columbia was worse than in other provinces? The figures show, on the contrary, that British Columbia had the highest rate of telephone penetration in Canada (11.2 percent versus 7.5 percent in the country as a whole).

The turn of the century was marked in British Columbia, as elsewhere, by civic populism and its strident, demanding spokespeople. The politicization of the debate was underscored by various attempts by municipalities to convince the province's Conservative government to nationalize telephone service. The Conservative party even adopted a resolution to that effect at its 1910 convention. But the premier, Richard McBride, wasn't listening; he felt that the role of the provincial government was to encourage investors to build British Columbia and not to take their place or frighten them away.

Between 1905 and 1910, BC Tel had to fight off attempts by Automatic Telephone, a Chicago company, to take over the Vancouver market. The company, founded by the inventor of the step-by-step switch, Almon Strowger, had the support of Vancouver City Hall. However, after the Supreme Court's decision, BC Tel signed an agreement with the City, and relations between the two parties grew less chilly. The debate turned to the respective merits of manual and automatic exchanges. BC Tel purchased all of its equipment from Northern Electric in Montreal. It appealed to Bell's experts, who declared that telephone operators were superior to machines. A municipal referendum was organized, and the majority voted in favour of machines. Nevertheless, the American offensives were rebuffed, and British Columbia kept its operators.

In the countryside, BC Tel, like Bell, was accused of not serving sparsely populated areas and refusing to connect independents to its long-distance lines. On the other hand, this was the only territory in Canada in which the federal government maintained a telephone network to serve isolated communities. It was the old Overland telegraph network built by Perry Collins, taken over by the British Columbian colonial administration after the project was abandoned, and purchased by the federal government when BC entered Confederation. In the late 1880s, the government converted the network to telephony; now called Dominion Telephone System, it served some fifty communities. Like Dominion Telegraph Service, the network was administered by the federal Department of Public Works. The existence of Dominion Telephone no doubt helped to alleviate some of the pressure on BC Tel. In any case, it was used by Premier McBride to justify his refusal to have his government involved in the telephone industry.

Yet the pressure was intense. From 1905 to 1910, the number of telephones in use went from six thousand to twenty thousand. BC Tel was functioning at the absolute limit of its capacity. One might surmise that Farrell and Lefevre were reluctant to appeal to foreign capital for fear of losing control of the company for good. It would have been difficult for them to perform the 1898 coup twice in a row. Their inaction could also be explained by the threat that nationalization posed to investors during these years. At any rate, in 1911 BC Tel was in solid enough shape to issue shares to the public, which improved its situation considerably.[41]

INDEPENDENT COMPANIES
IN QUEBEC AND ONTARIO

In Quebec and Ontario, regulation of Bell by the Railways Commission meant a clear improvement in the situation for independent companies, which gained connection to Bell's system. But the federal agency did not regulate the independents. In Quebec, the Liberal government of Jean-Lomer Gouin created the Public Utility Commission in 1909.

In 1910 the Ontario government adopted a law on the telephone that created a provincial regulatory agency called the Ontario Railway and Municipal Board, to bring some order to the independents (there were 460 in the province). Having played an active role in the drafting of this new legislation, Francis Dagger was named the commission's supervisor of telephone systems.

Everywhere, competition was yielding to regulated monopoly, with very few exceptions. Indeed, the main Quebec independent company, Demers's Compagnie de Téléphone de Bellechasse, became the Compagnie de Téléphone Nationale in March 1907, and under this name it began another battle with Bell: the battle of Quebec. Demers had the support of Rodolphe Forget, the giant of Quebec finance who owned Montreal Tramways Company and Montreal Light, Heat and Power Company. Forget purchased shares in Téléphone Nationale, and a first exchange was built in Quebec City in 1913. What's more, the new network was completely underground, which pleased the municipality. Demers did not take his commitments lightly, and on the eve of the First World War, the outcome of the battle between Téléphone Nationale and Bell was not a foregone conclusion.[42]

7 The Birth of Northern Electric and Technological Advances

If Alexander Graham Bell invented the telephone, Theodore Vail "invented" the telephone company. The Canadian telephone industry was completely in the American orbit, not only because of the financial and institutional links that united Bell Telephone and AT&T but, in a more general way, because of easy access to American suppliers: both the newly nationalized companies in the Prairies and the private companies in British Columbia and the Maritimes were also dependent on American manufacturers. The history of Canadian telephone technology thus parallels that in the United States, with one exception: the brief but fiery adventure of the Lorimer brothers in switching.

MAKING TELEPHONES IN CANADA

To understand the importance of the connections between Bell Telephone and American Bell, National Bell, and AT&T, one must look at technological exchanges, not just investments. As we have seen, American Bell quickly lost its majority share in Bell Telephone and had to be content with a minority share.

On the other hand, as Bell's territory was being carved out in Canada, all of the technological innovations were coming from the United States. This technological umbilical cord had been created in the general agreement of November 1880, in which the American parent company ceded to the Canadian company all of its own patents on the telephone, those that it had acquired from Western Union, and all those that it "may hereafter acquire," in exchange for its share of the company. But the American company was not obliged to file its own patents in Canada or to keep Bell apprised of new applications for, or purchases of, patents. This accord was confirmed in October 1882, when Bell Telephone wound down the activities of Canadian Bell, which had held the rights up to then. The modalities of transferring future patents had to be negotiated case by case.[1]

When Charles Sise decided to launch a manufacturing division in Canada in July 1882, the first instrument to be mass produced was the magneto, or Blake, telephone. However, some of the parts still had to be imported from Western Electric in the United States. As we have seen, it was because Bell hadn't managed completely to Canadianize production that its patents were cancelled at the beginning of 1885.

Sise reacted to the cancellation by pushing production: one year later, he had to move the factory to larger quarters; two years later, there were fifty employees. From then on, telephone equipment used by Bell and in Canada in general was, for the most part, Canadian, and it has remained so up to the present. Although Sise wasn't convinced of the virtues of Prime Minister Macdonald's National Policy, he saw in control of manufacturing a way to confirm his personal independence from the American holding company. From 1885 on, he favoured a voluntarist production policy and did not back down from confrontations with Western Electric.

Western Electric had been founded in Chicago in 1869 by Elisha Gray, Alexander Graham Bell's ill-fated competitor. In 1879 Western Electric was making the Edison telephones that Western Union sold in the United States and Canada. When Western Union had to withdraw from the telephone market, American Bell signed a long-term contract with Western Electric, which immediately converted its production to Bell telephones. In November 1881 American Bell purchased Charles Williams's patents and bought a majority share in Western Electric. The two companies merged under the latter's name in February, 1882, and Western Electric became the official supplier for AT&T, a position it held until the 1996 breakup that led to the creation of Lucent Technologies, an independent telecom manufacturer.[2]

Although it was part of the same financial group as Bell Telephone, Western Electric continued to export telephones and network and switching equipment to the companies in the Maritimes at prices lower than Bell's, forcing the latter to lower its prices. But Bell Telephone was itself a customer of Western Electric's, mainly for cables. Sise reacted to the competition from Western Electric by multiplying his calls for public tender and encouraging European manufacturers to do the same. He went so far as to send samples of Western Electric's cables to a German company so that it could make high-quality copies. For its part, Western Electric refused to allow Bell Telephone access to its patents, which Bell wanted for free.

It took years to reach a truce. In the accord of February 1892,[3] arbitrated by American Bell, Western Electric agreed no longer to supply Bell Telephone's competitors, while Bell agreed to purchase Western Electric's Canadian rights. For twenty-five dollars per patent or a lump sum of one thousand dollars per year, Bell Telephone gained access not only to technical innovations but to the training necessary to apply them. In addition, Western Electric was obliged to warn Bell in advance if it applied for or purchased patents. Bell immediately began to order its network and switching equipment from Western once again.

On the financial side, the accord was advantageous to Bell Telephone. It systematized the transfers of technology provided for since 1880 and was the forerunner of the service contract that would come into effect in 1923. On the technological level, it kept the Canadian company within the American sphere of influence. Western Electric's goal seemed complementary to its parent company's and, overall, more imperialistic. Where American Bell had intended to control the Canadian market with a minimum investment, Western Electric wanted to impose its technology at any price, including grabbing a majority share. In both cases, however, the goal was the same: to eliminate any competition in the northern reaches of the company's empire. The 1892 agreement left a number of blurry areas in terms of how this was to be accomplished.

NORTHERN ELECTRIC: BORN TWICE

During this time, Bell Telephone's manufacturing division was expanding. In addition to telephones, it produced alarm systems and various hardware items, electrical and otherwise. Without such a sideline, the plant would have been forced to close for several months of the year because of the small size of the Canadian market. Sise's revenue-diversification operation had borne fruit, and Bell was attacking several markets at once. Its charter forbade it, however, from making anything but telephone equipment. Sise had the charter modified, but to avoid any question of the legality of this type of activity, he decided in 1895 to transfer Bell's manufacturing division into a subsidiary called Northern Electric. Bell owned 93 percent of the new company, which was federally incorporated, and the rest of the stock was held by the seven members of Northern Electric's board of directors.

In the same year, a man named Graham Barrie began to manufacture telephone cable in a Montreal workshop with three young assistants. It was obvious that his small firm was producing cable superior to what was on the Canadian market at the time. In 1899 Sise purchased the enterprise in his own name and found himself in a classic conflict-of-interest predicament: the company of which he was president was purchasing the product of the company he owned. The Bell Archives are discreetly silent on this questionable move in Sise's career, but he must have received a sound scolding from AT&T management. In any case, he resold Barrie's company to Bell at cost and obtained a Quebec charter in December 1899 under the name Wire and Cable Company. Naturally, he appointed himself president of the new subsidiary, and his son, Edward Fleetford Sise, general manager; Barrie stayed on as superintendent. Manufacturing activities quickly expanded to include telephone cords and electrical bobbins. Wire and Cable's head office, like Northern Electric's, was in Montreal.

With this feverish expansion, Bell Telephone's purchases from Western Electric soon dropped off and conflict between the two companies flared up again. In 1901 Sise wanted to increase Wire and Cable's production capacity. He needed American expertise and asked Western Electric to send an engineer. After stalling for a while, Western Electric demanded shares amounting to 40 to 50 percent of both Northern Electric and Wire and Cable in exchange for its cooperation with Bell Telephone. Sise agreed to cede only 15 to 20 percent of Northern's capital shares. Things were at an impasse once again, and Sise went back to war for "his" company against Western Electric and even against AT&T, which had a hard time confining itself to its role of arbiter – the parent company tended to favour its direct subsidiary rather than its interests in Canada, which comprised a minority holding.

Western Electric used the carrot and the stick. It extolled the advantages that its international network of plants and subsidiaries could bring to the small Canadian company in terms of knowledge of markets and patents, and it also reminded Bell Telephone of the constant stream of orders it received from Canadian customers, "almost without solicitation." In 1901 Sise yielded 40 percent of the shares in Wire and Cable to Western Electric, which permitted him to triple the capitalization of the company, bringing it up to $300,000, with each of the two partners investing *pro rata* to its share.[4]

A few weeks after the compromise over Cable and Wire, yet another dispute arose with regard to Northern Electric. Sise wanted to send Bell Telephone engineers to Western Electric's plants in Chicago so that they could study the latest innovations; again, Western demanded more stock in exchange for a transfer of technology.

Things came to a head when New Brunswick Telephone put out a call for tenders for a telephone switchboard and Northern Electric lost the contract to Kellog, a large American competitor. Western Electric seized upon this otherwise insignificant incident, implying that Northern was incapable of serving the Canadian market properly and that the 1892 accord had ended up giving the green light to competitors outside of the Bell group. Sise jealously guarded his prerogatives: "I do not assent to the Western Electric Company coming in here in competition with Northern Electric ... We do not ask to be allowed to compete with them in the United States."[5]

Sise stubbornly resisted until the beginning of 1906, when the situation became untenable. The Canadian government was beginning to regulate Bell Telephone and the Prairie provinces were preparing to nationalize their telephone services. New telephone companies, being neither competitors nor subsidiaries of Bell's, could purchase supplies from Western Electric without breaking the 1892 agreement.

Forced to capitulate, Sise nevertheless did so with some sleight of hand. He issued new stock for up to 40 percent of Northern Electric's capital shares and sold it for double its value to Western Electric. The Chicago giant had to invest a total of $400,000 to obtain 40 percent of Northern's shares, while Bell Telephone, which had invested $300,000 over eleven years, remained majority shareholder with 55.8 percent.[6] Western thus acquired an interest in Northern's profits, but this strategic advantage had been dearly bought. The deal bore the personal stamp of Sise, for whom a cent was a cent, even in defeat – especially in defeat – as the province of Manitoba was soon to discover.

In 1911 Wire and Cable changed its name to Imperial Wire and Cable and doubled its capital to bring it up to $2 million. Sise chose not to purchase all the stock to which he had a right, letting his share slide to 50 percent; he did the same with Northern Electric. Western Electric increased its share in Imperial Wire and Cable to 42 percent and in Northern Electric to 45.2 percent. Sise was certain to retain his majority of shares since the balance belonged to members of the boards of directors of the two subsidiaries, and he chaired both boards.

In January 1914 Imperial Wire and Cable and Northern Electric merged on a parity basis: since the former was larger than the latter, they had to dilute the value of the shares to arrive at equity.[7] However, the new federally chartered company retained the name of the smaller company, Northern Electric. A new plant on Shearer Street in Montreal opened its doors in January 1915. (The centre of all of Northern Electric's activities until the 1950s, it grew into a true industrial enclave within the city of Montreal, employing ten thousand workers, and became the symbol of the Canadian telephone industry.) Edward Sise became president of the consolidated Northern Electric, and his brother, Paul, was named vice-president and general manager.

In the telephone industry, there was no conflict between an American (north-south) axis and a British (east-west) axis on the way to Canadianization. Clear domination by the north-south axis faced a single adversary: Sise and the interests of "his" company. It was nevertheless around this fragile bulwark that a Canadian telephone industry was built that was distinct from the one in the United States, even before it gained autonomy with regulation of Bell Telephone by the federal government in 1906 and nationalization of networks in the Prairie provinces.

THE TELEPHONE COMES OF AGE

American control or not, Northern Electric still made equipment under American licence, and research and development remained concentrated in

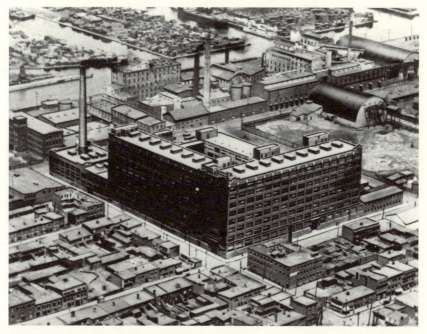

The plant on Shearer Street, opened in 1915, was the centre of all of Northern Electric's activities until the 1950s. This "city within a city" in an industrial district of Montreal became the symbol of telephony in Canada. Courtesy Bell Canada.

the United States. And the telephone had evolved in leaps and bounds since Alexander Graham Bell's first experiments.

Bell had invented a telephonic instrument, which is now known as a set or terminal. Over the years, local systems grew into tentacles of a network that would soon spread throughout the continent. Very quickly, the network stole the technological show from the individual telephone: as the focus of the telephone companies, it underwent the most profound metamorphosis.

Every telecommunications network involves both transmission – routing calls from one point to another – and switching – selecting where the calls are to be transmitted to. The switching took place in telephone exchanges. At first, this essential function was fulfilled by switchboard operators. The great advance in this budding technology was automation of the telephone exchange. This is precisely the point at which the Lorimer brothers made a major technological breakthrough, which has since fallen into obscurity.

Curiously, at least in North America, automation of exchanges was completely blocked by the major telephone companies. Bell Telephone, following AT&T's policy, refused for a long time to automate its exchanges.

The multiple switchboard allowed incoming calls to be distributed among all operators in the exchange, opening the door to industrial operation of the telephone network. Shown, the Montreal exchange between 1886 and 1897. Courtesy Bell Canada.

In the 1880s, Almon B. Strowger, an undertaker in Kansas City, Missouri, began to wonder why his main competitor was getting more telephone orders than he was. He figured that the operators were telling potential customers that his line was always busy when it wasn't, so he designed an apparatus that would do away with operators once and for all. He carefully studied the Kansas City exchange and, according to a no doubt apocryphal account, built a contraption out of pencils, pins, and a series of starched shirt collars. History doesn't tell us whether this machine worked; one source claims that it was meant simply to illustrate the principle of automatic switching. At any rate, in 1889 Strowger filed a first patent. The fundamental distinction of this switching equipment was its double movement – first vertical, then horizontal – which reproduced, "step by step," the operators' actions.[8]

The ingenious undertaker shopped his invention around to telephone companies, but he had no takers. In 1891, after Western Electric turned him down, he and a nephew set up their own company, Strowger Automatic Exchange, later called Automatic Electric. The Strowgers assembled a team of top-notch researchers, and the first step-by-step switch was put into service in 1892 in LaPorte, near Chicago, at an independent telephone company. In 1896 two engineers at the Strowgers' firm invented the numbered dial, heralding the birth of the modern telephone.

This pattern was duplicated in Canada, where telephone exchanges were also automated first by independent companies. An attempt was made to manufacture Strowger switches in Montreal: in 1893 a company with the

The telephone takes its modern shape. The dial telephone was invented in Chicago in 1896 by two engineers at Automatic Electric Corporation, Almon Strowger's firm. Courtesy Bell Canada.

ambitious name of Automatic Telephone and Electric Company of Canada issued a financial prospectus for future shareholders, bearing a no-less ambitious title: "Survival of the Fittest."[9]

The first permanent automatic exchange in Canada was in Whitehorse, Yukon Territory. The gold rush had drawn an ex-employee from the Strowgers' firm in Chicago, John Wyley, to the Canadian Far North. When

The step-by-step switch was invented by American undertaker Almon Strowger in 1889 and is still in operation all over the world. The last Canadian step-by-step switches were retired from service in the late twentieth century, when they were replaced by more efficient digital switches. Shown, the Lévis exchange in the 1950s. Courtesy Bell Canada.

he didn't find gold, Wyley founded the Yukon Electric Company in 1901, using his ex-employer's technology. With this success under his belt, he went to Saskatoon and started another company.[10] But it wasn't until telephone service was nationalized on the Prairies in 1908–09 that automatic switching reached maturity in Canada. The provincial administrations immediately opted for automatic systems and, with more funds at their disposal than independent companies had, standardized their use – although with some hesitation regarding the technology to be used.

THE TRUE BRANTFORD INVENTORS

Strowger wasn't the only maker of automatic switching systems. Many competitors sprang up in the United States, but the main alternative to the step-by-step system came from Canada. The earliest of the Canadian competitors, Romaine Callender, had been in Graham Bell's entourage. A Brantford music professor and organ maker, he was known for inventing an automatic organ player. He founded the Callender Telephone Exchange Company, a small firm employing a maximum of fourteen people in its heyday, among them two young brothers, George William and James Hoyt Lorimer, from the neighbouring town of St George.

Romaine Callender developed the first Canadian switch in Brantford between 1892 and 1896. Callender's principles were used in the Panel and Rotary switches, which were used throughout the world. Courtesy Bell Canada.

Callender filed three series of patents between 1892 and 1896. The first successful experiments with a wooden model known as the Brantford, or Callender, Exchange took place in January 1895 in New York. It is not known exactly how or why Callender abandoned his work; it seems that he left Canada for Great Britain. His research was continued by the two Lorimer brothers. They produced the first commercial models of the Callender Exchange, which were put into service in Troy and Piqua, Ohio, by independent companies in 1897.

The switchboards required constant maintenance, so the Lorimer brothers set up a repair workshop in Piqua, where the larger of the two switchboards (with a capacity of five hundred telephones) was located. After major financial difficulties, the Lorimer brothers opened a plant in Peterborough, Ontario, in March 1897, named Canadian Machine Telephone. Egbert, the youngest Lorimer brother, then joined the company. Soon after, they named the workshop in Piqua American Machine Telephone, a name they used internationally. At the same time, they kept up their research, ending up with a machine that was completely different from Callender's invention. The Lorimer brothers marketed their switchboards very quickly – too quickly, perhaps, because the switchboards never quite worked satisfactorily.[11]

James Hoyt Lorimer, the "mechanical genius" of the Lorimer family, developed the first Canadian commercial switch. He died in 1901, aged twenty-five. Courtesy Bell Canada.

The "mechanical genius" of the Lorimer family, as the newspapers of the time liked to label him, was Hoyt. He died in November 1901, aged twenty-five, of exhaustion. Then it seemed that the creative spark was no longer there and the technology evolved no further. However, the Lorimer brothers' marketing appears to have been effective. They demonstrated a switchboard for several hundred lines in Ottawa for two months. Francis Dagger wrote a report for the city of Toronto recommending that the Lorimer equipment be tested. It seems that Canadian Machine Telephone even opened a plant in Toronto in the hope of receiving a municipal contract. Toronto dropped its project & creating a telephone service, however, and the Lorimer brothers had to fall back on the small independent companies in Ontario. In 1905 Lorimer switchboards were installed in Peterborough, where the Lorimers' plant was, and in Brantford, where the old Callender plant was. Three years later, they installed switchboards in Burford, St George, and Lindsay, Ontario, but they made no sales in the United States. Edmonton Telephones had placed an order with the Lorimer brothers in 1906, but it refused delivery since the firm was incapable of delivering a model adapted to its needs. After two years of waiting, the municipality turned to the Strowgers. It took Automatic Electric only about two months to install the step-by-step switchboard. Compared to the Strowger team, the Lorimer brothers looked like

Above: The Lorimer workshop, probably in Peterborough.
Below: The Lorimer workshop with a series of switch panels. Egbert is at the far right.
Courtesy Sandra Lorimer Connell.

rank amateurs. Nonetheless, it is interesting to note the reasons for Edmonton's original choice: the purchase price for a Lorimer line was thirty-four dollars, compared to forty dollars for a Strowger line; the Lorimer switchboard took up 50 percent less space; and its centralized technique seemed simpler and more elegant than that of its competitor, even though its connection time was slightly longer.[12]

The Canadian technology did see some success overseas. The European rights were ceded to a group of Frenchmen, who, in May 1908, filed incorporation papers for Société Internationale de l'Autocommutateur Lorimer, with its head office at Galerie Vivienne in Paris. France and Great Britain each purchased two switchboards, and Italy bought one. The system was not widely used by these public utilities, however, because it was never reliable enough – it lacked the "polish" that only a powerful research team, like the one supporting the Strowger system, could have given it. Moreover, delivery dates were respected whimsically at best by Canadian Machine Telephone, as the Edmonton example showed. The company finally went bankrupt in 1923, and its assets were placed in trust. Bell acquired it two years later.[13]

If the Lorimer brothers have a place in the history of telecommunications, it is for their theoretical contribution. The principles they developed influenced all subsequent development of electromechanical switching. When AT&T converted to automatic switching, it adopted a technique derived from the Lorimer system. An international expert on switching, Robert J. Chapuis, enthusiastically described this hidden heritage: "Onto the vigorous and resilient sapling planted by the Lorimer brothers, Western Electric engineers proceeded to graft the shoot that would turn it into a healthy and productive fruit tree."[14]

The most important innovation in the Lorimer switch – the principle of preselection – in fact came from Romaine Callender. Instead of having as many connection mechanisms as there were subscriber lines, the Callender system started from the principle that not all subscribers used their telephone at the same time, and the number of connection mechanisms was only a fraction of the number of lines, resulting in a huge savings in equipment. The great challenge of preselection at the beginning of automation was speed, since the subscriber could not be kept waiting for the dial tone for too long. Introduced in 1893 by Callender, this innovation was picked up by all switchboard manufacturers, including Strowger.[15]

The Callender-Lorimer system was based on the action of a constantly spinning wheel. The contacts were established by stopping the wheel at the height corresponding to each number dialed. Therefore, just one movement was needed to make a contact, instead of two as in the step-by-step system.[16]

The Lorimer telephone apparatus had four levers, rather than the dial on the Strowger telephones. Each lever was moved to a position beside a digit in order to compose a four-digit number, a procedure inspired by the signalling

The Lorimer telephone had levers rather than a dial. To make a connection, the user placed each lever beside a number. This procedure was borrowed from the signalling commands used in the railways. Courtesy Bell Canada.

system used by the railways. One of the main criticisms of automation of the exchanges was precisely the complexity of operations required of subscribers. This was why it was important to design a telephone that was as simple as possible. The Lorimer telephones seemed to meet this requirement.[17]

The Lorimer technology was the latest thing in automatic switching, and observers were quick to recognize this Canadian contribution. Kempster B. Miller wrote in 1914: "These young men, with no prior training and – so they say – without ever having seen the inside of a telephone exchange, invented and developed the system in question and put it into operation. Knowing something of their struggles and efforts to achieve their purpose,

The Lorimer switch was based on the principle of a constantly turning wheel. At the beginning of the twentieth century, the Lorimer technology was the latest thing in automatic switching. Courtesy Bell Canada.

we find their creation one of the most remarkable we have ever seen whatever the value of the system."[18] And in 1925 Professor Friz Lubberger, a German expert in telephone switching, wrote: "In Canada towards 1900, the Lorimer brothers invented a system which, although it has not been introduced anywhere, is of such a richly inventive design that even today any specialist in automatic systems would benefit from studying it in detail."[19]

But the surest proof of the value of the Lorimer brothers' research came from AT&T, when in 1903 it purchased the Lorimers' patent and decided to transform their laboratory experiments into a marketable product. Not just one but two teams of researchers were set to the task. And out of the research emerged two of the most popular systems in the history of electromechanical switching: the panel and the rotary. Both took from the Lorimer system the principle of selection in a single movement. The rotary also borrowed the principle of a motor in constant rotation (from which it took its name). The panel system was chosen by AT&T and panel switches served most large cities in the United States until the 1950s, when they began to be replaced by the crossbar system. It was never used outside the United States. The rotary, on the other hand, was chosen by a number of European public utilities, notably France.

Thus, the legacy of the Lorimer system was international as the principles they established were adopted in the United States and in Europe. This double lineage was rediscovered only recently, however, by Robert Chapuis.[20] One may wonder how the work of the Lorimer brothers fell into such utter obscurity, even in their own country – even among telephone engineers. Chapuis's answer to this question is interesting: "The reason for its posthumous eclipse is that most of the works published on switching describe former or existing systems of telecommunication manufacturing companies that were well established, which ... was soon no longer the position of the Lorimer brothers' company."[21]

In any case, it is interesting that Canada's only technological contribution to telecommunications development before the contemporary period involved electromechanical switching. Three quarters of a century later, Northern Telecom would repeat the exploit with digital switching, this time with better luck on the commercial front. Switching is the brain of the system, the key to its evolution. It was as if Canadians, overwhelmed by the disproportionate geography of their country, realized that survival as a nation depended on mastering telecommunications. And what better instrument than the telephone switch to ensure this mastery?

At the turn of the century, however, foresight was not much in evidence within the telephone industry establishment. Bell Telephone in Canada, following AT&T in the United States, installed no automatic switchboards, Lorimer or otherwise, before 1924. Bell's mistrust was well expressed in a letter from Charles Sise, dated October 1892, at the time the step-by-step switch was launched at LaPorte: "Both experience and observation have united to show us that an operation so complex as is that of uniting two telephone subscribers' lines, and bringing the two sub-stations at the outer end of the lines into communication, can never efficiently or satisfactorily be performed by automatic apparatus, dependent upon the volition and intelligent action of the subscriber."[22]

Two months later, Sise sent his special agent, William C. Scott, to LaPorte. Posing as a journalist, Scott visited the LaPorte telephone company's exchange. He wrote a detailed report of what he had seen, right down to the colour of the telephone poles. Of course, in his view, the automatic system was full of flaws: it had too many wires, it was expensive to maintain, and it took away the human element that telephone service entailed.[23]

Bell's rejection of automatic telephony was the rejection of technological progress by a supremely self-confident monopoly. All arguments, even the most far-fetched, were made by Sise and Scott, who seemed determined to personify the reasons for which critics of the monopoly had always claimed that this economic regime slowed technological innovation. It is true that Bell had acted erratically during the pioneering era of telephony, favouring innovation when it served its purposes (long distance) and suppressing it when it didn't (automation). But these incidents would not survive the end of unbridled capitalism, and Bell in Canada, like AT&T in the United States, later acquired a reputation for leading-edge technology.

So why did AT&T and Bell reject automation at the time? They later claimed that the increase in number of technicians needed to maintain the automatic exchanges cancelled out the decrease in number of operators, but that was only partly true. Another argument, advanced by Lewis McFarlane, was more revealing: "Party lines cannot be cared for with the Automatic system; and as these occur in the outlying sections of all exhanges, a manual board with operators has to be provided."[24] The manual telephone was thus linked with party lines. This takes us back to the North American concept that a cheap technology with poorly paid labour seemed to be the best way of bringing service within everyone's grasp.

DISTANCE IS CONQUERED:
BELL AND WATSON'S SUCCESSFUL MAKEOVER

The advent of electronics at the beginning of the twentieth century enabled voice transmission to conquer distance. In 1906 an ex-Western Electric employee named Lee De Forest succeeded in controlling the flow of electrons in a vacuum tube called a triode.

At the time, the absolute priority at AT&T's new research division was long-distance transmission. Its laboratory, headed by chief engineer John J. Carty, soon discovered the amplification potential of the triode: electrons, stripped of their inertia, amplified the electrical current – hence telephone signals – with no distortion. The firm purchased De Forest's rights to the patent in 1912 and threw all of its resources into manufacturing triodes. In 1913 the first triode repeaters were installed on the New York–Baltimore line; in July 1914 they were incorporated into construction of the first transcontinental line, then in progress. Three electronic repeaters were placed between New York and San Francisco, and the first transcontinental phone conversation took place in January 1915.

In a nostalgic repetition of the very first telephone conversation in 1876, a white-haired Alexander Graham Bell repeated, in the AT&T offices in New York, the famous sentence that had inaugurated the telephone era: "Mr Watson, come here, I need you." To which the renowned electrician replied, from San Francisco, "This time, I'm afraid it will take a little more time, Mr Bell."[25]

The spectacular progress that was made in the area of transmission demonstrated, if there was need to, the power of the AT&T research machine when it set itself the task of solving a problem. As much as the company had dodged the question of automatic switching, it threw itself into electronic long-distance telephony. The result was explosive progress. This first call be-

The first call from Montreal to Vancouver took place in February, 1916, routed through the American network. In Montreal, a huge portrait of Alexander Graham Bell presided over a pre-Stalinist décor (with Charles Sise and Lewis McFarlane at the head table). Courtesy Bell Canada.

tween New York and San Francisco placed the telephone in the long-distance realm. From then on, the division of roles between a telegraph system devoted to long distance and a telephone system limited to local service was a thing of the past: telephone technology had come into its own.

Canada's first transcontinental conversation, between Montreal and San Francisco, took place in May of 1915, but it was not until 14 February 1916 that the first Montreal–Vancouver conversation took place. More than 4,500 people witnessed this first call, which connected Bell Telephone executives in Montreal's Ritz-Carlton Hotel with top BC Tel brass in Vancouver's Globe Theatre. It was one of the last public appearances of Charles Sise, who had retired the year before. The transmission was routed via Toronto, Buffalo, Omaha, Salt Lake City, and Portland – at more than 6,700 kilometres, it was a long-distance record. It wasn't until the 1930s, when the Transcanadian Telephone System was built, that the long-distance infrastructure became completely Canadian.

The first Canadian electronic repeaters were installed in Kingston in 1917. Previously, telephone conversations could not be transmitted beyond 2,500 kilometres; now, the range of the human voice was practically unlimited.

Nevertheless, long-distance communications remained a rarity for a long time. Although the technological barrier had been removed, a structural one remained: a collection of exchanges, even linked to each other, did not constitute a network. To make a long-distance call, one had to go through many exchanges, and therefore many operators, which made communications lengthy and problematic. The telephone had been designed as a local technology, and it would be many years yet before the administrative structures of companies adjusted to its new potential. In the paired evolution of technology and administrative structure, technology always preceded structure, which had to wait for managers to understand the nature of the changes.[26]

8 Unbridled Capitalism and Language Clashes

Not only was the business environment changing but within the telephone industry employee conditions and labour relations were also evolving; this is the least-studied facet of the history of telecommunications. The firms themselves were changing in nature, moving from craftsmanship and folklore to big business. In Bell Telephone's case this problem was compounded by the language issue, since Bell served both of Canada's linguistic communities.

FEMINIZING THE JOB OF OPERATOR

Most of the first telephone operators were men. The telegraph companies had launched telephone service in Canada, and the trade of telegraph operator was reserved for men and highly respected because of the technical expertise it required. Aside from knowing Morse code, the telegraph operator had to be capable of repairing the electrical equipment, and he was therefore regarded as a specialized electrician. No one would have thought of hiring a woman for such a job.

Bell Telephone hired the telegraph operators from Montreal Telegraph's and Dominion Telegraph. At the other end of the country, Victoria and Esquimalt Telephone had been founded by an ex-employee of British Columbia's Government Telegraph Service, Robert McMicking. At first, there weren't enough subscribers to keep the operators busy full time, so they served as jacks-of-all-trades: they repaired and maintained equipment, took care of accounts receivable, delivered messages, and even cleaned the offices – at least, in the large cities, where exchanges had their own premises.

The first telephone operators were men. Often former telegraph operators, these young men were generalists: they repaired and maintained equipment, took care of accounts receivable, delivered messages, and even cleaned the offices. Courtesy Bell Canada.

In small towns, the exchanges were located in shops or even in private homes, and the merchant and his family took on the job of operator. During the first years of telephone service, the Vancouver exchange was located in the Tilley bookstore. The son of the owner, Charlie Tilley, was responsible for the switchboard. Since there were no calls at night, he came up with the idea of giving concerts by telephone. The bookstore became a place where young Vancouverites gathered to entertain subscribers; respectful of tradition, they always ended the concerts at 10:30 PM with a rendition of "God Save the Queen"

Under these conditions, there was no control of labour. On the one hand, the operators were more like the independent craftsmen of the pre-industrial era than office employees. On the other hand, job security and social benefits were unheard of. There were no pension funds (but who would have thought of retiring from a trade in which everyone was young?), no paid holidays, no regular working hours, and no disability insurance.

In 1881 the work week for telephone operators was supposed to be forty-eight hours, and the maximum monthly salary was twenty dollars.[1] In most exchanges, the telephone service was closed at night, but in certain locations, municipal authorities required night service for emergencies, and someone

In small towns, telephone exchanges were installed in businesses, and even in private homes. Here, the Bowmanville, Ontario, exchange in 1899, with the exchange manager (top centre), the lone operator, and the manager's brother. Courtesy Bell Canada.

slept beside the switchboard. In 1882 a telephone technician earned a maximum of thirty-five dollars a month and worked sixty-hour weeks.[2] There were no social benefits of any sort for operators or technicians. Of course, telephone company management might come to the rescue of deserving employees when they had a problem. Work-related accidents were sometimes compensated, but management always stated in black and white that it was on a voluntary basis.[3]

The first labour conflict in the Canadian telephone industry took place on 28 May 1884. A group of Toronto linemen walked out to support a salary demand. Bell management sacked them. In a recent labour-relation document the company explained that the firing, "however deplorable today, was perfectly consistent with the business philosophy of the time."[4]

The freedom of the craftsman-employee to manage his work in his own way corresponded to the employer's freedom to be arbitrary. There was a semblance of equality. But an operator who answered, "Hold the line, I've got my hands in the dough" because he was also a baker did not make a very good impression. It was a holdover from the pre-industrial era that the industrialized nineteenth century would soon eliminate.

The first step towards scientific management of personnel was to replace male operators with females – and, if possible, teenagers. The oft-repeated reason in the official documentation was that feminine qualities were better suited to the job of telephone operator than masculine qualities: "Why are women chosen as operators instead of men? There are several good and sufficient reasons. In the first place, the clear feminine quality of voice suits best the delicate instrument. Then girls are usually more alert than boys, and always more patient. Women are more sensitive, more amenable to discipline, far gentler and more forbearing than men."[5]

The reality was less genteel. Unlike telegraphy, telephony required no technical skill on the part of the operators, so the companies fired the technician-operators and hired young apprentices. But because of rising unionism, this solution did not hold. A strike organized by the Knights of Labour in 1883 in the neighbouring telegraph industry seems, while unmentioned, to have been the main impetus behind feminization of the job of telephone operator. In fact, until the 1920s when the first automatic exchanges were introduced, operators held the entire network in their hands. A work stoppage by operators would instantly paralyze the system, while the impact of a strike by technicians would take longer to make itself felt.

After the telegraph strike, the telephone companies rationalized the telephone operator job: the function of maintaining installations was separated from the operator function. To be a telephone operator was simply to operate the switchboard, which allowed apprentices to be replaced with non-unionized women. The unions accused the women of "stealing jobs" from men and refused to organize them. Very quickly, the telephone companies abandoned male labour, both teenage and adult, and opted for women's work.

The pioneering character of the first years of telephony did not vanish the moment women began to work in the exchanges. There continued to be a direct relationship between operators and subscribers, as evidenced by the gifts the operators received each year, which could mount up to fifty dollars, much higher than the highest salary, and the many calls to order in the telephone company archives show that operators continued to have private conversations with subscribers. Operators had their favourites, to whom they transmitted messages and gave out weather information, the time, and hockey scores. Like the erstwhile telegraph operators, they sometimes ran errands and carried messages for subscribers, which earned them a five-cent tip. In on everything that went on in the influential community of subscribers, the operators played a role that was disproportionate to their social status.[6]

Under Taylorism, male operators were replaced by female ones, often teenagers. They quickly became symbols of the telephone companies. Courtesy Bell Canada.

But the industrial model finally made its mark on the job of operator. With the increase in number of subscribers, isolated switchboards in houses and businesses disappeared and discipline was enforced in the exchanges. At the turn of the century, Frederick W. Taylor came up with the "scientific" organization of work, based on three principles: planning of tasks and radical separation between the design of the work and its execution; a systematic search for economy of movement to arrive at "one best way"; and maximal use of machinery. Taylorism hit operators fast and furiously. As unskilled employees, they lost all control over their work. They were seated elbow to elbow before rows of switchboards, on the model of the assembly lines that were springing up at the same time. Supervisors, standing behind the operators, oversaw their work. At Bell, it wasn't until the 1970s that operators obtained the right to go to the bathroom without raising their hand to ask permission from the supervisor. Even during slow periods, it was forbidden to read on the job (an interdiction that still holds today). Laughing was also forbidden, since the Victorian conception of work forbade pleasure. One story has it that an employee was fired for having danced a tango at a staff party.[7]

The theory of "one best way" was taught in the 1910s in training centres for operators set up by the telephone companies. The operators learned the stock phrases with which they were to answer subscribers; they weren't allowed to answer spontaneously, since this would cost time. "One best way" also meant that the microphone in the switchboard was replaced by one integrated into the operator's headset. This new and heavier apparatus, the Gilliland headset, weighed three kilograms and resembled an ox yoke, but it eliminated several superfluous gestures. Finally, when traffic was slow, an operator could take care of two neighbouring switchboards (the physical size of an applicant for the position of operator became a criterion for hiring her).

At the turn of the century, most telephone operators were young women aged seventeen to twenty-five who were working until they got married. Their temporary status explained the low regard in which they were held by their employers – and by the unions. The salary was clearly insufficient for the almost half of all operators who had to pay their own way. For them, the only way to survive was to work overtime, take on a second job, or find a "sugar daddy." This last solution was a great concern at the end of the Victorian era, when some people regarded the practice as prostitution: the telephone companies required of aspiring operators three certificates attesting to their clean living and high morals, including one signed by a clergyman.

In spite of all of this, the job of operator was very much sought after since, for young working-class women, it offered a rare opportunity to elevate their social status. The first operators benefited from their association with the prestige of the still slightly mysterious telephone technology, and the public treated them with as much deference as it did teachers, with whom they shared a dress code. The telephone companies thus had no problem finding and replacing operators.[8]

Under such conditions, it is not surprising that labour demands in the telephone industry came from linemen and not operators, even though the former were earning twice as much as the latter. In 1897 the National Brotherhood of Electrical Workers, an American union renamed "International" for the occasion, began to organize telephone technicians in Hamilton, Quebec City, Montreal, Toronto, Winnipeg, and Vancouver. In Vancouver, operators joined the International Brotherhood of Electrical Workers as a local auxiliary women's section, where they received only half a vote per person.[9]

EARLY LABOUR PROBLEMS AND THE BATTLE FOR ALLEGIANCE

Social issues exploded in the telephone industry at the very moment when populist agitation was at its greatest. The first organized strike in the Canadian telephone industry took place in Vancouver, when the New Westminster and Burrard Inlet Telephone Company required linemen to purchase their own work tools. In September 1902 the linemen walked out. Against all expectations, management gave in to their demands, agreeing to pay for their work tools and recognizing the union.

Encouraged by this first victory, the Vancouver local of the IBEW established a long list of demands, including elimination of unremunerated labour by apprentices, paid sick leave, and wage parity with electricians in Seattle. Management refused to negotiate and a new strike was called for 26 November 1902. This time the operators joined the action and telephone service was paralyzed. Two days later the New Westminster operators also walked out, followed by the linemen in Victoria, even though they were employed by a different company (Victoria and Esquimalt).

The operators' involvement in the conflict gave it a dimension that the first strike hadn't had (service stoppage) and began a sympathy movement among the general public, especially among the two groups of professionals who used the telephone the most: journalists and businessmen. It must be said that, on the crest of a wave of populism, the entire city was up in arms against New Westminster and Burrard Inlet Telephone. Newspapers took up the cause of the operators, and the businessmen organized an arbitration committee. The union agreed to return to work under the supervision of the businessmen's committee; the businessmen were eager to be involved because they were thinking of opening their own, competing telephone company. Management at New Westminster and Burrard Inlet Telephone refused arbitration and hired strikebreakers.

Telephone service became even more chaotic when bad weather was added to the mix. A storm blew down most of the above-ground lines,

On the telephone poles, the craftsman-
employee worked at his own pace. Telephone
technicians were members of the worker aristocracy. Courtesy Bell Canada.

leaving only the downtown underground system in service. The company
had to yield, and on 15 December it signed a collective agreement that satisfied the strikers on most points, in particular the essential questions of paid
sick leave, remuneration of apprentices, and institution of the closed shop.
The strike had lasted two weeks and it had been won thanks to the telephone
operators. Not only had they paralyzed the telephone company but the fact
that they were young women, many of them teenagers, had turned the population in their favour.

But the closed shop was objectionable to New Westminster and Burrard
Inlet Telephone, as it was to all other telephone companies. What it did was
to force companies to recruit technicians only from among IBEW members,
and the technicians were to be trained by the union. The result was that technicians identified with their union and not with the company where they
worked, so they maintained the telegraph "boomer" tradition of switching
companies and cities as they liked, while keeping their job security.

When, in 1904, BC Tel brought together most telephone companies in the
province under its umbrella, it made very sure that the closed-shop clause
did not extend beyond Vancouver, where the collective agreement had been
signed. What is more, the clause was never applied to operators, even in Van-

couver. To be effective, however, the closed-shop system had to be universal, and so a strike was called over the issue in February 1906. This time, management was ready. Operators from every town in British Columbia converged on Vancouver on the first day of the strike to maintain service. It was the first "dirty" strike in the Canadian telephone industry.

There was a case of sabotage in the third week of the strike, when it became obvious that the company was not going to give in. Management accused the strikers, since only "the telephone technicians know where the important cables are," and offered a thousand-dollar reward for clues to the identity of the guilty parties. The union replied by stating that the acts of sabotage invariably consisted of cuts that were easy to repair; the technicians could easily have destroyed the entire system by wetting the inside of one of the cut cables, which, according to the union, would have caused major short-circuits. The issue of vandalism was not settled during the 1906 strike, and it has continued to plague all labour conflicts within the telephone industry until today.

In any case, the violence scared business circles, which allied themselves this time with the telephone company, and the employees had to break ranks and go back to work in November without having made headway. The strike had lasted more than eight months.[10]

After the failure of the 1906 strike, the union took some time to reorganize. By 1910 all BC Tel technicians had rejoined the IBEW. Conflict soon arose over wages and over non-unionized apprentices, which the company used in large numbers.

In fact, the union and BC Tel management were battling over a much more basic issue: the workers' allegiance. The IBEW had no category for "telephone technicians." There were only "electricians," who were working at a telephone company today but might work at an electricity company tomorrow and at a streetcar repair yard the day after that. Tension continued to mount between the company and the union, and a new strike was called in March 1913.

This strike was the first to mobilize all of the technicians in a Canadian telephone company. After a ten-day walk-out, the company gave in to the demands of the three hundred technicians. But it also marked the beginning of a new strategy by BC Tel. The company now understood that to win the battle for allegiance, it would have to abandon its strong-arm management style and adopt a series of social measures, the most tangible of which was the creation of a medical-insurance fund in 1914. It was the start of implementation of social policy by telephone companies – or, in the unions' view, the triumph of paternalism.[11]

In general, conditions for Bell Telephone employees were not very different from those for BC Tel employees. The first social benefit was instituted in 1894 by the operators themselves, not by the company, in the form of a

Telephone technicians celebrate completion of a new line laid. Linemen had to have very specific technical knowledge, which meant that they could not be easily replaced by strike-breakers. They were the first strikers in the telephone industry. Courtesy Bell Canada.

"Lady Operator Benefit Association" that provided medical insurance. But, as at BC Tel, Bell Telephone employees had to pay for a replacement out of their own pocket if they fell ill. It is not known exactly when the first paid holidays were granted – maybe in the early 1890s. All that we know for sure is that in 1897 operators obtained two weeks of paid holiday (they had the right to hire a replacement if they wanted a longer holiday).[12]

The first retirement pension was paid by the company in 1896 to Hugh Neilson, a high-ranking manager. There is no mention of pensions for low-level employees until the company started a special pension fund in 1911.[13] In fact, however, it was the length of the work day that gave rise to the first Bell Telephone labour conflict, which took place in Toronto. It was on a different order of magnitude from that in Vancouver, if only because of the numbers of employees involved. The 1906 strike in Vancouver involved twenty technicians and thirty-four operators, while the 1907 strike in Toronto involved four hundred operators.

At the turn of the century, the work day at Bell Telephone was ten hours long for technicians and eight hours long for operators, and the work week was six days long. Most of the technicians were organized in the IBEW; although the company didn't recognize their union, they had some security if they were dismissed. Thus, it was among the operators that dissatisfaction ran the highest.

Bell Telephone tried an experiment in Toronto in 1903, when repairs reduced the workland and led to a temporary surplus of labour at the main exchange. The work day for operators was decreased to five hours, with no pay cut, but the fifteen-minute breaks every two hours were eliminated. The idea, according to the company, was to make the operators work more intensely over a shorter period of time.

In fact, things turned out differently. Although the operators' salaries weren't lowered, they were frozen at twenty-five dollars a month, while salaries everywhere else in the company were rising. Since it was impossible to live on twenty-five dollars a month, operators did two successive five-hour shifts.[14] The local Bell manager, Kenneth J. Dunstan, was absolutely opposed to introducing a five-hour day, as his correspondence with Charles Sise and the rest of the top Montreal management reveals. He did nothing to encourage the success of the reduction in work time; on the contrary. His hard line was all the more paradoxical given that, as a young man, he had been the first Canadian telephone operator when he worked for Hamilton Telephone. He was a self-taught man who had slowly risen in the company, starting what would become the tradition of "telephone men," but he was inflexible. The labour atmosphere in Toronto degenerated rapidly and stayed poor.

Two AT&T experts were consulted in 1906, but neither could find a solution. One of them, however, pointed to an unexpected possibility: Wasn't

Dunstan, who was overworking his troops, the source of some of these problems? Why not transfer Dunstan to Montreal for a while?[15] Bell paid no attention to the two reports, which suggested further studies.

Instead, seeming to throw oil on the fire for no reason, the company decided to re-establish, starting on 1 February 1907, the eight-hour day and to increase salaries by proportionately less than the increase in hours. These changes were announced without explanation in a drily worded memo that was posted on the afternoon of Friday, 25 January 1907, so most operators learned the news the next day. They were worried: they could barely earn a living with a five-hour day that allowed them to do a second shift on overtime; with an eight-hour day, a second shift would be impossible.

Meetings were held at operators' homes on Sunday, 27 January, and the operators decided to consult J. Walter Curry, a lawyer known for his position in favour of municipalization of private monopolies. The next day, Curry called Dunstan, who declared with ill humour that Bell's decision was final and that it was "a matter between the company and its own employees, and not one in which a third party should interfere."[16] The impasse was complete. The operators had no choice but to give in or go out on strike. They decided without hesitation to call their strike for Friday, 1 February, the day the schedule changed. In spite of the decision, the long-distance operators tried a last-minute negotiation with Dunstan – in vain.

Another group of operators had the idea of contacting the mayor of Toronto, Emerson Coatsworth. They couldn't have found a better person. Like Curry, Coatsworth was a proponent of civic populism and therefore a sworn enemy of monopolies. After lending a sympathetic ear to the operators, he dashed off a telegraph to the federal deputy minister of Labour, William Lyon Mackenzie King, to ask him to intervene in the public interest.

There was little social legislation in Canada at the time, and what there was fell under provincial jurisdiction. In 1900, however, Wilfrid Laurier's Liberal government had adopted the Conciliation Act, on the British model. A junior labour minister had even been appointed to oversee its application, but federal intervention in the social domain remained the exception.

King took a train to Toronto right away. He arrived on the morning of Thursday, 31 January. The strike was planned for the next day, so every hour counted. He sent a message to Dunstan to offer his good offices, including organizing a meeting between the company and its employees, while "the operators should continue in the company's service under the present schedule of rates and hours."[17]

Meanwhile, operators from Peterborough, Kingston, Ottawa, and Montreal were flocking to Toronto. Working hard to break the strike, Dunstan didn't take time out to respond to King. He waited until Friday afternoon, when the strike had begun, to reply, and he then refused the good offices of the government with the excuse that it was too late.[18] That evening, a meeting of four hundred operators at the labour exchange asked for the government to call a commission of inquiry.

Straight away, the operators had the press and public opinion on their side. Most Torontonians had been against private monopolies, and the crisis in the electricity industry looked likely to be won by the proponents of civic populism.[19] Just as important, the operators had King with them. It is not known how Dunstan's epistolary slap in the face influenced him, but a commission of inquiry was set up the very next day – a Saturday – with King chairing it.

The hearings proceeded full tilt: they started on Monday, 4 February and ended that Friday. A total of seventy witnesses were heard, generating 1,700 pages of transcript, not counting exhibits. The operators went back to work on 4 February, in conformity with the terms of their demand for an inquiry. Sise had specified that "under no conditions can we allow these operators access to our building," but he hadn't counted on the strength of public opinion or on King. After three days out on strike, the operators returned to work under the old conditions.[20]

The commission of inquiry was highly critical of Bell Telephone management. It provided evidence of the contempt in which the company held the health of its operators: to a question about fatigue among operators, the Bell representative responded that they arrived at work "already exhausted" from roller-skating at night.[21] The commission denounced the operators' salaries, which were insufficient to make ends meet, even according to the company managers. This excerpt from the interrogation of the Bell representative by the operators' lawyer is typical of the state of mind that ruled in the business world at the time:

Q. Where you saw that a number of operators were not receiving a sufficient wage to enable them to gain a livelihood, and to be self-supporting, would you consider in looking to a readjustment of that wage, would you think of picking up a balance sheet of the company, and finding out to what extent the profits might admit of an increase?

A. No.

Q. That would never enter your mind?

A. That would never enter my mind.

Q. ... What is your reason for saying you think the matter of profits should not be considered?

A. For the reason that all that is required is to pay the market price for your goods.[22]

Bell management seemed eager to make the conflict exemplary to the point of equating operators with goods – an unconscious reference to Karl Marx's thesis on reification of human relations.

The commissioners had no choice but to condemn the way the company had informed – or rather, failed to inform – the operators of the change in schedule and salaries; they emphasized the arrogance with which the company had rejected the good offices of the government and was refusing to be bound by the recommendations of the commission of inquiry itself. It found that the operators' strike was justified given the company's attitude. The recommendations included a six-hour work day with one break; an end to hiring girls under eighteen; a medical inquiry into the health of telephone operators; and a forum for dialogue between the employer and the operators in the form of a permanent conciliation board.

Was the Toronto operators' strike a success or a failure? In the context of the Canadian labour movement, it was a failure. The operators had to return to work with no guarantee other than a government commitment to proceed with an inquiry. However, the government's speedy intervention in the work conflict marked a decisive turning-point in Canadian social history. For the first time, the government had indicated that a strike was justified and laid responsibility at the feet of the employer. A strong signal was sent to employers that they could not approach labour relations on the basis of economics alone but must also consider other social criteria, such as health, and that if they did not keep up their end of the bargain, the government would force them to do so.

The King report hit Bell Telephone like a bomb. Sise had told the press that the strikers were no longer with the company and had refused to be bound by the commission of inquiry. In fact, however, he found himself scrounging for a last-minute compromise in the middle of the commission's hearings, in an attempt to soften the blow of the anticipated condemnation, and he agreed to rehire all striking operators. As for the dispute itself, the daily work schedule would be seven hours with breaks every two hours, and the salary scale would be raised a little.

But the real meaning of the 1907 strike went farther: it constituted a turning point in the evolution of labour relations at Bell Telephone, and it had repercussions throughout the Canadian telephone industry. Starting in 1907, a series of social benefits were developed by the company (first among them retirement and medical-insurance plans). Operators' working conditions, which had deteriorated continually since Taylorism was introduced, slowly began to improve. However, it wasn't until Sise's death and the First World War that the rhythm of social progress accelerated.

Nevertheless, the lesson had been learned, as suggested by the self-criticism in the official account of Bell's labour relations written by the company itself in 1963. Management's responsibility in the operators' strike is presented plainly. This is rare in the industrial world; not many North American companies have been around long enough to "remember" facts more than a half-century old.[23]

On the political level, the 1907 strike at Bell served as a testing ground for King's social thinking. At the very moment when the commission of inquiry's hearings were taking place, the Industrial Disputes Investigation Act was in second reading in the House of Commons. This law, also called the Lemieux Act after the Labour minister of the time, had been provoked by the unions' closed-shop clause and, among others, the 1906 BC Tel strike. It made provision for a government inquiry in the case of labour conflict in a public utility and for a thirty-day suspension of the right to strike, but it had a major sohrtcoming: its recommendations were not enforceable. The government had intervened in the Bell strike in exactly this way even before the law came into effect.

After the First World War, labour relations at Bell Telephone changed completely, as the company conformed with the letter and the spirit of the labour legislation that King was passing. The Lemieux Act had been the antecedent of a series of social measures enacted during the First World War to provide the opposing parties with a permanent forum for discussion moderated by the government. After face-to-face confrontations between employer and employee came tripartite dialogue between employees, employers, and government. It was precisely this "class collaboration" policy that was put into practice by Bell during the 1920s.

The tone of the 1907 commission's recommendations, which were probably written by King himself, has been much criticized.[24] It is true that repeated allusions to the fragility of the young girls and their future role as mothers seem anachronistic today, but King was all the more touched by the Toronto strikers because they were young women: "We agree entirely with the view expressed by the local manager that *it is the pace that kills*, and the working of women at high pressure at work of this kind should be made a crime at law as it is a crime against Nature herself."[25] King's emotion every time he evoked the Bell operators was sincere; all of his private correspondence, his journal, and his later work bear this out.[26]

At the turn of the century, men's attitudes towards women usually ranged from neglect of their problems to downright exploitation. IBEW's refusals to unionize operators was an example of this, likewise its relegating of Vancouver operators to an auxiliary local with only a half-vote per person. If the very prejudices paradoxically aroused compassion for women, they must be seen as a positive factor in the power relations of the time. King's own position with regard to the striking operators helped to improve their working conditions, whatever the old-fashioned tone of his writings. His attitude was not so unusual – in many countries, outrage over women's and children's labour served as an impetus to pass social legislation. This was one of the priorities of the International Labour Organization when it was formed in 1919.

MASSIFICATION OF THE WORKFORCE
AND THE OPPOSITION BETWEEN THE MALE
AND FEMALE SITUATIONS

On the eve of the First World War, both standardization and feminization of labour forces marked the Canadian telephone industry. For instance, Bell Telephone had 150 employees when it was founded in 1880 and almost eight thousand in 1915. A firm with 150 employees is a medium-sized business in which it is possible to know everyone by name and solve work-related problems case by case. A company with eight thousand employees, however, is definitely large: anonymity reigns and collective controls are required.

As explained above, the advent of Taylorism seemed to offer a miracle solution to the problems posed by controlling large masses of people. From one day to the next, the company's sorcerers' apprentices underook to survey and measure workers – in short, to bring them into line. "Scientific management" was applied by a management group that had no appropriate special training: at the time, Bell did not have a human resources department and internal communications were non-existent (as we have seen, the 1907 strike was triggered as much by the way management announced the new working conditions as by the working conditions themselves). Bell didn't even have an internal newsletter (the first organ of this type, the *Telephone Gazette*, was created in 1909 for management) and, significantly, top management never spoke to the rank-and-file workers.[27] In this regard, the managers' personalities were of paramount importance: at Bell Telephone, obviously, Sise's total indifference to human problems aggravated an already difficult situation, just as Dunstan's did in Toronto. At BC Tel, the change in ownership and multiple mergers created an even more explosive situation, although the much smaller size of the company should have enabled it to grow smoothly.

The proportion of female employees at Bell Telephone, which had been insignificant during its first decade of existence,[28] rose to about 60 percent by the eve of World War I, and other telephone companies in Canada followed the same path. This rapid increase corresponded to the growing popularity of the telephone among increasingly broad segments of the public.

Because it either could not or did not want to automate its exchanges, Bell was forced to hire more and more operators. The operator position was, in the service sector, the equivalent of the assembly line in the manufacturing sector: detailed and repetitive movements led to complete dehumanization of the job. Thus, it is not surprising that the most significant strikes were held by the operators, or that their success was always uncertain because of the precarious position of women in the job market and the refusal of unions to organize them.

The technicians, for their part, retained certain of the advantages of the pre-industrial craftsman, and retain them to this day. Always on the road to build new lines or install telephones for new subscribers, they were far from

Always on the road, far from the administrative hierarchy, telephone technicians retained some prerogatives of the pre-industrial artisan. In many respects, they still have this freedom. Courtesy Bell Canada.

the administrative hierarchy and bureaucratic ambience. What's more, their trade required very specific technical knowledge, which meant that they could not be replaced on a moment's notice by strikebreakers.

When the telephone companies, under the combined pressure of unions and government, agreed to improve working conditions, restore salary levels, and introduce social benefits, the technicians were the first beneficiaries. For instance, retirement plans instituted just before the First World War had no impact on the operators, who didn't stay on the labour market long enough to profit from them. At Bell Telephone, the first pension plan, set up in 1911 on a non-universal basis, was "for the purpose of providing funds for Pension and Gratuities to worthy employees, who, through length of faithful service or other reasons are entitled to such consideration."[29] The feminization of Bell's labour force thus had the effect of creating a long-lasting schism in the labour force between male technicians – permanent and qualified, "unionizable," and to a certain degree respected by the companies – and female operators – temporary and untrained, ideal guinea-pigs for "scientific management."

Labour relations in the telephone industry followed the same curve as that for introduction of any new technology. First, there was a prolonged period during which the individual initiative of a few pioneers took the place of an organized market. Then, improvised industrialization resulted in a terrible degradation of working conditions. It was a time of child and teenage labour and one-way industrial relations. Once the technology and the enterprises in charge of implementing it reached a certain level of maturity, labour relations slowly improved.

In the telephone industry, the movement that restored dignity to workers had hardly begun on the eve of the First World War. There is an effective way to determine when an industry as a whole becomes an element of liberation for the people who work in it: this is the moment the workers themselves begin to subscribe. It is axiomatic that until workers can benefit from the services that they labour to provide, they cannot talk of liberation. In 1915 that time had not yet come for telephone workers.

LINGUISTIC SKIRMISHES

In Quebec City and Montreal, the social issue was further complicated by the linguistic issue. Bell Telephone was run by anglophones for anglophones. Since the business was based on oral communications, language had a special significance.

Sometime in 1884 or 1885, the *Montreal Daily Witness* stated that there were as many francophone as anglophone subscribers in the city,[30] but this had no effect on company policy. The first reference to linguistic questions in Bell's archives appears in the 1960s. Before then, the "French fact" simply did not exist for Bell Telephone.

Bell's policy was for its operators systematically to answer in English. Since there were subscribers who insisted on speaking French, the company taught unilingual anglophone operators the numbers one to nine in French and two or three stock phrases (even in English, the operators had to use stock phrases). The company also assigned one or two bilingual operators to each exchange to answer unilingual francophones who needed help.

As we have seen, Bell's insensitivity to its sociocultural environment had provoked members of the Quebec petit bourgeoisie to create the Merchants Telephone Company, which provided hopelessly outpowered but obstinate competition for Bell between 1894 and 1913. For years, the Quebec press railed against Bell: "One cannot obtain an answer in French on the telephone, one cannot send a telegram in French without having the spelling and the meaning garbled, one cannot present onself to the courier service without being insulted by some rude, loudmouthed clerk."[31] What had been difficult to take from a private company was intolerable from a public utility. In the Quebec legislative assembly, a member named Armand Lavergne began a campaign to impose bilingualism on the public utilities: streetcars, railways, water, natural gas, electricity, and, of course, telephone service.

For things to change, however, Henri Bourassa's nationalist movement had to throw its weight into the battle. The issue of bilingualism was raised alongside anti-militarism in mass demonstrations. Lavergne tabled a bill on bilingualism in the assembly, but the debates were stormy and the commission charged with studying the bill introduced so many amendments that Lavergne withdrew it in April 1909. He returned to the issue the next year,

The Telephone speaks FRENCH

The telephone is bilingual. French is as much its native tongue as English. It serves all French-Canada.

In the Province of Quebec alone, there are 266,300 Bell telephones. There are also 37,700 connecting telephones: a total of 304,000, predominantly French.

These telephones are a potent factor in the life of this great community. They are part and parcel of its daily routine. They serve it in emergency; they facilitate business and social communication throughout the vast area they serve.

To reach the Quebec market, effectively and economically, use the telephone. We shall be glad to help you in preparing your sales plans.

The telephone speaks French: in the mainly English-speaking business world, no one thought that Francophones might also want to use the telephone. Bell undertook advertising campaigns to say that French could be used on the telephone! But it was not until the 1930s that such advertising appeared; in the 1910s, Bell spoke English only. Courtesy Bell Canada.

and finally, on 4 June 1910, the legislative assembly passed Bill 151, imposing partial bilingualism on public utilities. But the law had very few teeth; the maximum fine for infractions was twenty dollars.

Yet it was too much for most large Anglo-Saxon firms in Montreal, including Bell. They formed a common front against Bill 151, hiring a lawyer, R.C. Smith, who had made representations against the bill in the assembly, and prepared to fight it in the courts. It took interventions by Senator Raoul Dandurand and the president of CP, Thomas Shaughnessy, to convince Grand Trunk, Montreal Light, Heat and Power, Bell, and the other companies concerned to adopt bilingualism.[32]

Bell refused to respect the agreement, however, and threw oil on the fire through sheer thoughtlessness. During a public hearing of the Railways Commission devoted to complaints about deficiencies in telephone service in Montreal, the Bell representative invoked bilingualism as a source of error: "It is very difficult," he said, "to find young women who speak both French and English, and there is often confusion in the numbers because of this."[33]

The company thus continued to communicate with its subscribers in English only, as many articles published in the press of the day complained. This journalistic guerrilla war did not involve in-depth exposés, however, just little repeated items, buried carefully in the middle pages of the papers.[34]

On 13 March 1914, Henri Bourassa wrote an editorial in Le Devoir. (Strangely, for a French daily, it was written in English.) Summarizing Lavergne's legislative battles and the many pressures exerted by the press, Bourassa sounded the battle cry against the Anglo-Saxon arrogance of the large private companies that ran public utilities: "A few years ago, a movement was started in this province to induce, and if need be, to force the various companies doing public service *in the province*, to make use of the two official languages of Canada, or, in other words, to put the language of the vast majority of the people of this province on a *footing of equality* with the language of the minority" (italics in original). Bell was singled out: "It has taken years to bring the Bell Telephone Co. of Montreal to send accounts and notices in both languages. It is but the other day that a new exchange having been opened in the northern part of Montreal, all notices sent to the French subscribers were printed in English alone." And Bourassa pointed out: "It is but in the last two or three years that the French in Ottawa, who are one third of the population, have succeeded in getting telephone operators to understand and speak French. On that basis, all telephone communications in Montreal, where all English-speaking people form between one-fourth and one-third of the population, should have been given in French alone. What would the 'big business men' of Montreal think of it?" Faced with the growing discontent of the French Canadian élite, Bell reacted ... by doing nothing.

Like the telegraph industry, the telephone industry in Canada was born in an anglophone cultural milieu. As long as telephone service remained limited to élites, the question of language remained under the surface. But by the 1910s, democratization of the telephone was raising the problem of Bell's language, and especially its culture, in Quebec.

9 The Telephone Industry in Canada and the International Scene

As the telephone industry attained maturity, its development was marked by disparities between Canadian provinces, between Western nations, and between different parts of the world.

Within Canada, the great debate on nationalization of the telephone industry, already under way at the beginning of the century, had ended with a compromise solution: federal regulation of the main actor, Bell, and provincial regulation of the other enterprises – or their nationalization, in the Prairie provinces. Did the forms of development chosen for the telephone industry have an impact on its performance? The best way of evaluating the success of the telephone industry is to look at the number of instruments per hundred inhabitants (see Table 5).

Would Manitoba have performed as well under Bell management? Probably not. The excellent results of 1910 were manifestly the fruit of nationalization and do not correspond to the relative wealth of this province compared to Ontario. In the early years of the telephone industry, businesses' actions played a decisive role and planning could somewhat correct for economic and geographical disparities.

On the other hand, the imbalance between Quebec and Ontario honestly reflects an economic and perhaps a linguistic discrepancy (francophones did not subscribe to telephone service as easily as did anglophones), matters of policy aside. Both provinces were dominated by one company, Bell, which excluded any major operational differences. Furthermore, since the company's head office was in Quebec, that province should have been favoured. Manifestly, this was not the case.

Table 5
Telephone Penetration Rate in Canada by Province, 1901, 1910, 1915

	Number of telephones per 100 Canadian residents		
Year	1901[a]	1910[b]	1915[c]
British Columbia	1.6	5.0	11.2
Ontario	1.1	5.4	9.8
Quebec	1.0	2.6	4.7
New Brunswick	0.8	2.9	5.1
Nova Scotia	0.8	2.6	4.6
Manitoba	–	7.4	10.4
Alberta	–	2.7	9.4
Saskatchewan	–	1.6	6.5
Prince Edward Island	–	1.1	2.5
Newfoundland	–	–	–
Canada, average	1.2	3.9	7.6

Notes: a. Dawson, *Proceedings*, 1:8.
b. Estimated from Bell Canada data (BCA no. 34 564) and Statistics Canada. Since Bell did not make public the distribution of telephones between Quebec and Ontario, I applied the ratio for the years 1915 and 1920 to the year 1910. The data concerning independent companies were added in.
c. L.B. McFarlane, *Telephone Stations, Dominion of Canada, December 31, 1915* (BCA, no. 34 564).

CANADA: NUMBER TWO WORLDWIDE

Just as development of the telephone industry did not take place evenly across Canada, there were discrepancies on the international scale. In general, there were two groups of countries: the United States, Canada, and Sweden were clearly ahead of France, Belgium, Great Britain, and Germany, with Switzerland somewhere in between (see Table 6). This division does not appear to correspond to the division of wealth among these countries at the time. Some discrepancies were caused by arbitrary political choices – for instance, Great Britain and France favoured mail and telegraphy over telephony.[1]

In fact, it would be tempting to posit a political equation such that public administration corresponds to delays in technology and private enterprise corresponds to technological progress. In 1914 the telephone industry in the United States, Canada, and Sweden was dominated by private enterprise.

Table 6
Number of Telephones and Subscribers in the World, 1914

Country	Number of subscribers	Number of telephones per 100 residents
United States	9,542,000	9.7
Germany	1,428,000	2.1
Great Britain	781,000	1.7
Canada	500,000	6.5
France	330,000	0.8
Sweden	233,000	4.1
Japan	220,000	0.4
Switzerland	97,000	2.5
Belgium	65,000	0.9

Source: AT&T, *Telephone and Telegraph Statistics of the World.*

There was no doubt an overexpansion of the public sector in Europe, and government always has a tendency to favour the written over the spoken, control over freedom. But on the eve of the First World War, no precedence had been definitively established. Serious planning by the German and Swiss PTTs presaged future growth.

On the world scale, imbalances were even greater. The telephone was a technology of industrialized countries. Unlike telegraphy, which rapidly penetrated the colonized Third World as a means of domination, telephony was more or less completely confined to North America and Europe (see Table 7).

On the eve of the First World War, it was obvious that economic factors were playing an important role in development of the telephone industry. But this role was not determinant; operating formats and political choices also helped to shape the contours of world telephony.

AMERICAN LEADERSHIP

The world leader in all categories was, obviously, the United States – a success due not to chance or to the "laws of the market," but to a single man: Theodore Newton Vail. When he hired on with the new company in June, 1878, Vail was thirty-two and had been director general of the United States Post Office, where he had reorganized routing of mail by railway and had the reputation of being a management magician. He seemed destined for his new position: he was the cousin of Alfred Vail, who had helped Morse develop the telegraph.

Table 7
Distribution of Telephones in the World, 1914

Continent	Number of subscribers	Distribution (%)
North America	10,121,000	68.0
Europe	4,013,000	27.0
Asia	306,000	2.0
Oceania	217,000	1.5
South America	166,000	1.1
Africa	65,000	0.4
Total	14,889,000	100.0

Source: AT&T, Telephone and Telegraph Statistics of the World.

After the founding team of the Bell Telephone Company (Hubbard, Sanders, and Bell) was pushed aside, everything changed. The lawsuit with Western Union was settled out of court. Western Union withdrew from the telephone market, while the Bell Telephone Company, renamed National Bell Telephone Company, agreed to stay out of the telegraphy market. National Bell's future was secure. Its shares, worth fifty dollars each in early 1879, leapt to nearly one thousand dollars following the agreement with Western Union.

National Bell, which soon changed its name to American Bell, did not meddle in daily management of the operating companies that were its subsidiaries, but it imposed very strict technical and financial standards on them. It also provided itself with production capacity: in 1882, it bought up most of the firms that made equipment for exchanges, including the largest ones, Western Electric and Charles Williams's factory.

The first long-distance line between two major cities, Boston and New York, was established in March, 1884. This was the launching point for Vail's grand design: long-distance would provide American Bell with a strategic advantage over future competitors who might try their luck by offering service limited to a single local area. In February, 1885, development of long-distance lines was handed over to a new subsidiary, American Telephone & Telegraph. Vail left his position as director-general of American Bell to become president of AT&T and make his vision a reality.

Unfortunately, a disagreement soon arose between Vail and Forbes. Vail wanted to reinvest as much of the revenues as possible in expansion of long-distance lines in anticipation of the competition that would surely strengthen after Bell's patents expired in 1893 and 1894, while Forbes wanted to take advantage of this respite to increase profits to the maximum. Forbes prevailed, and Vail resigned in 1887.

Theodore Newton Vail was an empire builder. In his terms as president of AT&T, from 1885 to 1887 and 1907 to 1919, he infused the company with the culture of public service. Courtesy AT&T.

The post-Vail period was marked by slowed expansion and increasing discontent among the public, which felt that rates were too high. The advent of competition after 1894 severely penalized this short-sighted policy. The monthly rate of creation of telephone companies broke record after record: seventy-one in January, 1907; a hundred and ten in February. Ultimately, there were some twelve thousand independent companies in the United States.

When the 1907 recession arrived, AT&T found itself in an untenable position: a $100 million loan issue floated in 1906 had found no takers. Financial institutions that had agreed to sell shares to the public inherited, involuntarily, a large portion of AT&T's assets. Heading up the bankers' group was John Pierpont Morgan. The famous magnate was not one to support lost causes. He forced the president of AT&T to resign and turned to Theodore Vail, who, in the meantime, had made a fortune in Argentina building dams, starting an electric company in Cordoba, and reorganizing the tramway system in Buenos Aires.

In May, 1907, after twenty years of self-imposed exile, Vail, aged sixty-two, became the president of the new AT&T. It was the beginning of a telecommunications policy that relaunched the company toward the heights, making it the largest company the world had ever known, in any category. To mark the change in leadership, the company's head office was moved from Boston to New York.

Vail met no resistance as he redressed the financial situation and reformed accounting practices in the operating subsidiaries. He was assisted by his banker friends, many of whom he had placed on AT&T's board of directors. Although he had the confidence of the financiers, Vail did not take his marching orders from them. No one had the slightest doubt who was running the "Bell System"; Vail was the only master on board, and his supporters guaranteed a stable source of funding through loans and stock issues. In return, Vail made profits. Although they were not Forbes's fabulous but fleeting profits, they were something better: stable profits.

One of Vail's first moves when he returned to the telephone business was to transfer AT&T's small research laboratory from Boston to New York and appoint his right-hand man, John J. Carty, to run it. Carty infused pure and applied research with a degree of energy unprecedented in the history of the technology. It was the initial kernel that gave rise, in the 1920s, to the famous Bell Laboratories, which produced eight Nobel Prize winners.

However, Vail's main contribution to AT&T during his second term was an intellectual one. He defined his vision of telecommunications in a formula that became instantly famous: "One policy, one system, universal service." Everything he did as president of AT&T between 1907 and 1919 flowed from this vision.

The telecommunications historian Louis-Joseph Libois summarizes Vail's thoughts and actions in four premises. According to Libois, Vail's first premise concerned shareholders, who had to be as numerous as possible. His second premise dealt with subscribers and with the public in general: the drops in rates that accompanied his return to the business reversed public opinion, which had been hostile to AT&T's monopoly. Vail felt that over the long term, "The public must be served," as he liked to say. This was a prerequisite for public relations and advertising.

Vail's third premise concerned company employees, whom he wanted to treat as well as possible. Retirement funds and health and death insurance were instituted (the Benefit Funds were created in 1913); to crown his social policy, he created an employee stock-purchase programme in 1915. There were great coherence to Vail's actions. Improvements in working conditions for operators and Vail's policy of operator courtesy led to one of the most effective slogans ever used in public relations: "The voice with a smile." Killing not two, but three birds with one stone, as was his habit, Vail launched another slogan, this one aimed at the subscribers, urging them to "Return the smile." Each of these actions consolidated the others and fed on their success, creating an irresistible synergy.

Vail's last premise dealt with regulation and government relations. When he returned to the business, he understood the need for state regulation. The Canadian model, in which the price paid for establishment of the monopoly had been the institution in 1906 of government regulation, played a decisive role. Until 1910, with the Mann-Elkins Act, the American government classified telephony (and telegraphy) as public services. As such, telephony fell under the authority of the Interstate Commerce Commission, which up to then had been concerned with regulation of transportation and electricity. Vail freely cooperated with the public authorities, but he always drew a line between regulation and management: the government might regulate, but the private sector must manage.

Vail was certainly a man of vision, and he is still held up as the American paradigm of a corporate leader. But his generosity was "targeted" to employees, subscribers, journalists, and even independent companies. When he returned to the business, he agreed that if AT&T was in a weak position in a district, it would withdraw and interconnect its long-distance network with the independent company's system. On the other hand, if AT&T was in a position of strength, Vail was both devious and merciless. He developed a "clean" way of eliminating competition. Using the resources of his financier friends, he dried up all sources of credit for competitors: independent companies that thought that they could circumvent his influence by going to London to raise funds saw the doors of British banks slam shut in their faces as if by magic. This meant, however, that subscribers were spared the inconveniences of "telephone wars" in which skirmishes featured telephone poles being chopped down with axes. Peace in the telephone sector was thus acquired at the price of sharing a bitterly disputed market. Vail was no choirboy but, indeed, a captain of industry.

Vail improved working conditions for operators and invented one of the most effective slogans in public relations, "The Voice with a Smile," which was used in Canada in both English and French. Courtesy Bell Canada.

All of Vail's genius did not keep his grand policy from running into obstacles. With the help of his friend J.P. Morgan, he acquired Western Union in 1909 and put his former opponent back on its feet in three years by prescribing the usual remedy: raising salaries by 50 percent and achieving spectacular gains in productivity. This union of the two American telecommunications giants was the consummation of Vail's dream of a single, universal service. But it was also the last straw for the always hair-trigger North American anti-monopolists. The US Department of Justice began anti-trust lawsuits in July, 1913. In November, the Post Office published a report advocating nationalization of the Bell System.

Vail got the message and dispatched his senior vice-president, Nathan Kingsbury, to sign a compromise. The Kingsbury agreement stipulated the abandonment of Western Union, extended the policy of interconnection with networks to independent companies that had been excluded up to then, and forced AT&T to ask for permission from the Department of Justice before purchasing an independent. Although Vail was abandoning his dream of creating a single system to provide universal service, however, he gained the endorsement from federal authorities to build a dominant national network. The federal government gave a nod to a monopoly on telephone service in exchange for the setting of strict limits and regulation.[2]

AT&T was nationalized anyway during the First World War. In spite of his opposition to this measure, Vail cooperated with the government to minimize its effects. He also deployed all of his considerable energy in an attempt to reverse the decision, since he felt that a government-run administration would always be submitted to the vagaries of political games; even if it managed to be efficient, it would never be economical.[3]

Vail's arguments finally triumphed, and the US government returned AT&T to its former status in July 1919. But Vail had left the company the previous month to take the honorary position of chair of the board of directors. He died one year later on his property in Vermont. This giant of a man – he was over six feet tall and weighed three hundred pounds – left no fortune behind. He had spent all his money on his Vermont farm, giving elaborate parties, handing out money to those around him, and investing in anything that seemed to him to be scientifically valuable or innovative, without regard to its profitability. To his successor at the head of AT&T, he declared, rather impishly, "Thayer, I never saved a dollar in my life."[4]

Vail's life's work did not die with him. The Kingsbury agreement was the cornerstone of relations between AT&T and the American government throughout the interwar period. AT&T itself, modelled in his image, continued to operate according to his principles. Although Alexander Graham Bell invented the telephone, Vail founded the Bell empire and a model of the operating company that would be copied in Canada, of course, and also in

Europe – his sense of public service and propensity to plan for the long term could not help attract the highly state-controlled PTTS.

In the United States, the structure thought up by Vail was perpetuated until the Bell System was dismantled in January 1984. Today, while his work has been cut into pieces, the specialized press has taken a new interest in the "Vail years." His visionary side fascinates a North America devoted to the laws of the market and short-term profits.[5]

Britain: The Hesitation Waltz

The beginnings of the British telephone industry are especially pertinent to a Canadian history of telecommunications because of the influence exerted by England: the Canadian anti-Bell Telephone movements sprang up more or less along the lines of the British model demanding nationalization.

In Great Britain, things did not begin well for the telephone industry. From 1874 to 1876, Alexander Graham Bell, a British citizen in good standing, tried to interest the British Postmaster in his invention, but his efforts were rejected in a letter of typically bureaucratic arrogance, and Great Britain lost the opportunity to repatriate its prodigal son. Thereafter, the Postmaster always had a mistrustful attitude toward the telephone. Telegraph service had been nationalized in 1871 and placed under the aegis of the Post Office, which showed its hostility when the two first private telephone companies were created in 1878 to operate Bell's and Edison's patents.

The first British telephone system was set up by the National Telephone Company, which built a long-distance network, including a London-to-Paris link. NTC, however, was never popular with the public; the Post Office refused to allow the private company to use its poles, causing construction delays and long-standing discontent.

Finally, in 1892 the Post Office decided to purchase NTC's long-distance lines. NTC was thereafter confined to local service, but at least it obtained some rights of way along railway lines and canals. This did not mollify proponents of nationalization, however, and they continued to press for nationalization of the entire telephone network. The Post Office felt that this would be too expensive, and it limited itself to constructing a phone network in London. In 1899 a law was passed encouraging municipal authorities to create public systems.

This policy, however, immediately raised its own funding problem. In Great Britain, as in Canada, there was a chronic lack of capital. Among the 1,344 municipalities covered by the 1899 law, only 13 asked for permission to build phone systems, and only 6 of them actually did so. Of these, five went bankrupt and had to sell their installations to NTC or the Post Office. Only the Hull municipal system survived; it prospered so well that it still exists today.

The policy of competition – nationalization of long distance along with municipalization of local service – definitively failed on all fronts. It didn't take long for the British government to realize its error; in 1905 (the year the Select Committee was set up in Ottawa), it stated that monopoly was the most practical and economical business form. However, the Postmaster at the time felt that telephone service was too important to remain in the private sector and decided that NTC's licence would not be renewed when it expired in 1911. In that year, the Post Office purchased NTC's 1,565 exchanges. At the time, the company had 560,000 subscribers, which were then added to the Post Office's 77,000 subscribers in London.

The slow start of British telephony compared to that in Germany and, especially, the United States, cannot be imputed to socioeconomic weakness. At the time, Great Britain was the most powerful country in the world, its companies the wealthiest, its population the most cultivated. Rather, it was no doubt due to a hesitant political and administrative class that wanted to be involved in the telephone industry but didn't know how to deal with the new technology.[6]

Francis Dagger's attempts in Canada to nationalize long-distance service and municipalize local service, following the British model, were even more obviously a lost cause when, in 1905, the failure of that model was patent – a fact that Bell representatives were only too happy to point out.

The Rest of Europe

In Europe, two groups of countries with very different development models were vying for top spot in the telephone industry: Scandinavia and the Switzerland–Germany group. Scandinavia had allowed the industry to develop in a context of sometimes very violent competition, while the Switzerland–Germany group had immediately conferred management of the telephone industry to their respective post and telegraph administrations.

Stockholm offered a unique example of prolonged competition between networks owned by three companies, two private and one public. In the 1880s a company called Ericsson began to export its equipment, acquiring a long-standing reputation for excellence. Nationalization came much later, in 1918, when Sweden had the third-highest rate of telephone penetration, after the United States and Canada.

In Norway, which was still united with Sweden, telephone development was mostly the work of agricultural cooperatives and municipalities. An 1892 Royal Commission noted that telephone service was not a luxury reserved for a business élite but had a social vocation. This was a first in Europe, and perhaps in the world.

In Germany, the post and telegraph administrations immediately adopted the telephone, not only installing it in homes but also using it instead of the telegraph in long-distance communications. This unique inception explains why the long-distance network was always a priority in Germany, and it quickly became the most extensive in Europe. The first conclusive experiment with automatic switching took place in Berlin in 1900. Although it was only a private switchboard for government buildings, with this success under its belt, Germany put into effect a bold plan to automate telephone service before the rest of Europe and the United States. At the time, only independent companies in the United States and the Canadian Prairies were doing the same thing.

At the turn of the century, Germany had an urban penetration rate equal to that of the United States. On the other hand, rural service in Germany lagged seriously behind that in the US.

Some International Cooperation

Transformation of the telephone into a public utility led to the beginnings of cooperation within the International Telegraph Union. At the Berlin Conference in 1885, the telephone had been considered simply an extension of the telegraph. A new chapter was added to the telegraph regulations: it contained just five articles, leaving national governments free to negotiate bilateral agreements.

Real regulation of the telephone began in 1903 at the London Telegraph Administrative Conference. The French delegate proposed fifteen articles addressing all aspects of telephony that were distinct from telegraphy, while the agreements made at the founding convention of the International Telegraph Union continued to apply. The name of the organization itself was not changed to incorporate the word "telephone" because it wasn't a diplomatic conference, nor was the issue of distribution of revenues from international communications to the countries transited through addressed. Unlike telegraphy, the entry of telephony onto the international stage was quiet.[7]

THE REST OF THE WORLD

In all countries, telephone service was first an urban phenomenon. The asymmetry between the cities and the countryside was epitomized in France in 1902, when Paris had forty-four percent of all telephones in the country. At the turn of the century, the only countries in which use of the telephone was beginning to be widespread in rural regions were the United States, Canada, and the Scandinavian nations. As we have seen, rural service could raise problems, as it had in Canada, provoking a series of political crises in Ottawa and on the Prairies.[8]

As a rule, competition was what encouraged penetration of telephone service. In Canada, the United States, and, to a point, Scandinavia, the initiative of many small-business entrepreneurs took the telephone outside the limited circle of the upper classes. The difference between private and public monopolies was indeed the faculty of the latter to nip this diversified, spontaneous grass-roots initiative in the bud. Governments suppressed local attempts much more effectively than did private monopolies, since they could quickly adopt laws; when laws were insufficient, regulatory power provided them with a panoply of strategies likely to dissuade those who might otherwise dare to get into the telephone business.

In particular, governments had the moral legitimacy lacking in private monopolies. They could righteously invoke the national interest or social policy to smother an entrepreneur. AT&T in the United States and Bell Telephone in Canada were atypical outgrowths of the market economy. At the outset, their monopoly situation made them suspect. It was precisely due to the ambiguous status of private monopolies that competition could spring up and favour telephone penetration.

However, competition was found to be untenable in the long term; it was a transitory state that got rid of ineffective structures. All experiments with competition ended up with the creation of a new monopoly or a geographic distribution of markets between a number of monopolies. This rule applied without exception: no country, large or small, maintained a number of competing networks for a long period – not even Sweden, which was host to the longest confrontation between the public and private sectors.

At the beginning of the telephone industry, competition, wherever it was manifested, was an indispensable corrective to the industry's blockages, allowing an instrument of power reserved for the use of the upper classes to become a public utility. But as soon as the telephone rose to the rank of public utility, it became or returned to a monopoly. At the beginning of the twentieth century, the telephone industry, like all other industries based on operation of a network, was a natural monopoly.

Radio

Radio was born as an outgrowth of telegraphy, from which it took its first name: wireless telegraphy. It was the Canadian inventor Reginal Fessenden who first "made radio speak." Following his invention, radiotelephony and radio broadcasting developed in parallel. The success of this constantly changing technology was rapid and spectacular.

It should be mentioned that this book deals with radio from the point of view of wireless telegraphy and radiotelephony exclusively. The history of radio broadcasting belongs to another universe where content is dominant.

10 Radio: A Spectacular Success

Telegraphy and telephony are both based technologically on electrical waves carried on metallic wires. Throughout the nineteenth century, a series of inventions demonstrated that electricity could also move through the atmosphere in the same way as light. If this was the case, surely it would be possible to transmit Morse code signals, or even voices, without the use of wires?

THE MARCONI PHENOMENON

Radio is a technology with a direct pedigree in scientific theory. The Scottish physicist James Clerk Maxwell demonstrated that it was possible to transmit information by electromagnetic waves in 1864. In 1887 the German physician Heinrich Hertz measured electromagnetic waves between an oscillator and an "electric eye" – the ancestors of the transmitter and receiver. All of these principles were known, and the only leap from them to radio was the notion of using electromagnetic waves to transmit intelligible signals over a longer distance (the oscillator had a very limited range).

It was a young Italian electricity buff, Guglielmo Marconi, who discovered in 1895 that an antenna would extend the range of the oscillator – a discovery that seemed trivial but was utterly decisive for the future of telecommunications. Radio was simply Hertz's oscillator plus an antenna.[1] Yet again, as for the telegraph and the telephone, one might be tempted to object that Marconi's invention was hardly worthy of the word, such was its utter simplicity. The Italian Government even refused Marconi's offer of the patent rights. Harold I. Sharlin, a historian of electrical technologies, wrote, "Why hadn't anyone else thought of such an antenna? Marconi's invention was not based on any known principle. The mystery is not why others did not think of it sooner, but why Marconi had thought of a grounded antenna."[2]

Marconi moved to Great Britain and founded the Wireless and Signal Company in 1897. It was an immediate success, sending telegraphs across the British Channel by 1899. At first, radio evolved along the model of telegraphy. Marconi intended to set up a network of point-to-point links, as no one was thinking of mass broadcasting at the time. In fact, the radio was not yet "talking."

The first market that opened up for radio was maritime communications. The Marconi International Marine Communication Company was launched in 1900 to exploit this niche with a wireless network. Marconi intended to follow the example of AT&T: he rented equipment to maritime companies and supplied radio operators. A series of coastal stations enabled ships to stay in constant communication with shore.

It should be noted that in September 1899, Alexander Graham Bell had suggested to the Canadian minister of oceans and fisheries to "make a thorough examination of the Marconi System of wireless telegraphy before deciding to lay a cable to Sable Island."[3] The government did nothing, but radio came to Canada anyway.

CONQUERING THE ATLANTIC

Indeed, Marconi's ambition was to set up a worldwide network. To reach this goal, he had to be able to cover the Atlantic Ocean by wireless. In 1901 he undertook to link the giant station he had just built in Cornwall, England, to Cape Cod in the United States. The scientific community loudly objected that this was impossible; the curvature of the earth permitted transmission only over limited distances; the radio waves, which went in a straight line, would be lost in space. Marconi was certain that, on the contrary, the waves would follow the curvature of the earth. Both theories were wrong: radio waves did indeed travel in a straight line, but they were brought back to earth by the "reflective layers" of the ionosphere.

Whatever the theoretical gaps, Marconi's star was on the rise. He travelled to Cape Cod with a technical team, but the antenna they erected was broken twice by storms a few weeks before the planned date of the trials. Would fate get the better of him?

Marconi then decided to take the shortest possible route. Newfoundland was the part of North America closest to England. He would have to act quickly to be the first to link Europe to America by radio waves. In December 1901 he and his team made their way to St John's, the capital of Newfoundland. They installed their equipment on Signal Hill, which stood between the city and the Atlantic, and began to transmit.

In December, 1901, from a makeshift installation in Newfoundland, Guglielmo Marconi made the first wireless telegraph transmission between North America and Europe. Marconi, a brilliant inventor, was greedy and showed his competitors no mercy. In just a few years, he built a worldwide empire. Courtesy National Archives (C 5945).

Unfortunately, one of Newfoundland's frequent storms swept away the balloon to which Marconi's antenna was attached. The next morning, the antenna was hooked up to a kite. In another stroke of bad luck, freezing rain grounded the kite. Marconi launched another kite. This time, luck was with him. At 12:30 PM, he heard the three characteristic short clicks signifying "s" in Morse alphabet, coming from Cornwall. It was Wednesday, 12 December 1901. Wireless was now transatlantic.

The following Monday, while the entire world was talking of nothing but Marconi's exploit, he received a letter from Anglo-American Telegraph's laywers informing him that the company had a monopoly on communications between Newfoundland and the rest of the world and ordering him to stop his experiments or be subject to a lawsuit. It must be said, in defence of the venerable cable company, that it dropped its legal action when the first transatlantic transmission by wireless was announced. Anglo-American was still headed by Alexander Mackay, who had rejected Bell Telephone of Canada sixteen years earlier (see chapter 4).

Graham Bell invited Marconi to construct a permanent coastal station in Nova Scotia, offering the use of some land on his property at Baddeck. The site turned out to be too far from the coast and was rejected, but Marconi retained the idea of locating in Nova Scotia, where the provincial government put a special train at his disposal so that he could tour likely sites. Finally, Marconi setttled on Glace Bay, near Sidney, on Cape Breton Island, which thus became the bridgehead for wireless in North America.

The federal government quickly became interested in radio. Marconi was greeted in Ottawa by Prime Minister Wilfred Laurier, Conservative leader Robert Borden, and telegraphy pioneer Sandford Fleming, with all the pomp reserved for heads of state. He received not only the land he wanted at Glace Bay but also a grant of seventy-five thousand dollars – more than telephone technology had ever received in Canada. The Marconi Wireless Telegraph Company, predecessor of Marconi Canada, was created right away and soon managed to establish a complete monopoly on wireless transmission in Canada.

In December 1902, a year after the historic experiment in Newfoundland, a commercial transatlantic service began operation, but from Nova Scotia. Three months later, the London *Times* published the first transatlantic dispatch transmitted to it from the Glace Bay station. But all of this made little difference to the general public, for it was not possible to send private messages to Europe by wireless until October 1907.

FOUNDATION OF THE INTERNATIONAL RADIOTELEGRAPH UNION

By 1903 the Marconi company had forty-five coastal stations covering the entire planet and three high-power stations in Great Britan, the United States, and Canada (the Glace Bay station), not including military installations. Marconi was a British company and its installations were added to Britain's underwater cable network. As noted earlier, the British telegraph network was the most extensive in the world. Was Marconi going to try for a stranglehold on international telegraph communications? The power of the Marconi empire began to frighten governments in a number of countries.

Recall that Marconi's policy was to establish a marine monopoly as powerful as AT&T's telephone monopoly in the United States. Not only did it rent equipment and supply operators to ships but it also forbade all radio posts using its equipment to communicate with the posts of its competitors. According to Marconi, it was impossible to ensure the quality of transmissions between equipment of different brands. This policy raised not a little anxiety about the safety of ships on the high seas.

Germany led the counteroffensive with regard to international cooperation. It convened a preliminary radiotelegraph conference in Berlin in August, 1903. Nine countries participated in the discussions on the central issue of the obligation to respond to radio calls without discrimination by brand.[4] The United States, which had been notably absent from the International Telegraph Union, immediately came out in favour of cooperation with regard to radio.

The Berlin Preliminary Radio Conference put "Marconi-ism" on trial. Germany, along with all countries except Great Britain, which supported the interests of its standard-bearer company, and Italy, which stood up for its "neglected" son, wanted intercommunication between all marine radio systems to be obligatory, regardless of brand. Over the protests of the two dissidents, the final protocol stipulated that coastal stations and ship stations had to exchange messages "without distinction as to the systems of radio used by the latter."

Marconi contested the authority of this administrative conference, which had no plenipotentiary power, and continued to follow its own rules. The US government reacted in 1904 by retracting the concession for the coastal station at Nantucket.

In October-November 1906, a conference was held in Berlin bringing together radiotelegraph plenipotentiaries from twenty-nine countries.[5] It was at this conference that "SOS" (in Morse code, three dots, three dashes, three dots) was chosen for distress signals in place of the old signal, CQD. The United States demanded that intercommunication be obligatory not only between coastal stations and ship radios but also between ship stations.

Issuing from the conference was the 1906 Radiotelegraph Convention (also known as the Berlin Convention), confirming the principle of freedom of communication, including tacit recognition of the notion of public service on the international level. The Marconi empire had had its day and now was reduced to the status of a company like all the others. Canada ratified the convention in 1907 under the so-called "colonial vote" of Great Britain, thereby timidly entering the international telecommunications stage. A basic structure was set up under the name International Radiotelegraph Union to apply the terms of the Berlin Convention, in particular those dealing with rates. Such details as interference and distribution of the electromagnetic spectrum were addressed by the International Telegraph Union staff in Berne.[6]

The International Radiotelegraph Union merged with the International Telegraph Union in 1932. Some claim that Canada had been a member of the IRU since 1907, which is false. To achieve such seniority, Canada would have had to be part of the International Telegraph Union, which, as we have seen, it was not. Great Britain, which had had only one additional vote in St Petersburg in 1875 (India), now had five votes, one each for Australia, Canada, India, New Zealand, and the South African Union. Although it was a signatory in name, Canada in fact had to be content with a back-seat role.

A CANADIAN MAKES THE RADIO TALK ...
IN THE UNITED STATES

Guglielmo Marconi had invented wireless radio, but he had not made it talk. That distinction would belong to Reginald Aubrey Fessenden. Born in 1866 into a Loyalist family in East Milton,[7] Quebec, Fessenden was initiated into the mysteries of electricity and chemistry in Thomas Edison's laboratory in New York. After Edison went bankrupt, Fessenden worked briefly at Westinghouse before heading for the University of Pittsburgh in 1893, where a professor's position in electrical engineering had been created for him.

As had Alexander Graham Bell's, Fessenden's career as a researcher took off in the supercharged American scientific culture. In Pittsburgh, Fessenden began to work on transmission of speech through electromagnetic waves. At the time, Marconi's work in Great Britain was making headlines around the world; every experiment represented new progress.

Fessenden wasn't discouraged by this. In fact, he felt that Marconi was on the wrong track when he compared radio transmission to a whiplash effect: by this model, electromagnetic waves were produced by electrical sparks strong enough to propagate waves in the air, but between the sparks there was nothing. Fessenden believed, on the contrary, that radio waves were continuous, like light projected from a candle flame. If this was the case, it would be possible to transmit speech, he thought; all one had to do was increase its frequency. In his Pittsburgh laboratory, he was able to modulate radio waves several thousand times a second. But the university wasn't supporting his work, and the rector even accused him of having purchased electrical equipment without permission.

However, the secretary of Agriculture invited him to conduct wireless experiments at the weather station on Cobb Island, a small island on the Potomac River. His task was to transmit weather forecasts to Washington, ninety kilometres away. On the evening of 23 December 1900, Fessenden conducted his historic experiment: in a small wood cabin lit by an oil lamp and topped with a fifteen-metre antenna, he started the generator that was to run his transmitter; 1.6 kilometres away, his assistant, Alfred Thiessen, did the same thing. "One ... two ... three ... four. Can you hear me, Mr Thiessen?" Fessenden asked. Thiessen replied in Morse code, "Your voice comes through sounding like the flapping wings of a flock of birds. I can make no sense of it."[8]

Discouraged, Fessenden recalculated and checked his settings: everything seemed to be in order, but nothing was happening. Two hours later, when it was pitch-dark out, he started the experiment again. He spoke very slowly into his microphone, pronouncing each syllable distinctly: "One, two, three, four ... Is it snowing where you are, Mr Thiessen? If it is, telegraph back, and let me know."[9] Thiessen heard him. The first wireless voice transmission had taken place.

Reginald Fessenden invented the talking radio just before Christmas, 1900, and invented the variety-programme format. Marconi called the Canadian inventor "a preposterous upstart." Courtesy National Archives (PA 93 160).

Fessenden had been right, and everyone else had been wrong. However, his equipment was as primitive as Marconi's: a spark transmitter and a receiver directly descended from the Branley coherer.[10] The coherer, which transmitted and received waves by surges, was well suited to transmission in Morse code but was very erratic in speech transmission. After the Cobb Island experiment, Fessenden improved the receiver by replacing the metal filings with sulphuric acid. This electrolytic detector received continuous waves with relative fidelity.

Meanwhile, Marconi was also developing an improved receiver: he replaced the filings with a needle that became magnetized under the impact of radio waves. The result was a compact and reliable magnetic detector, but it was no more suitable for speech transmission than was the coherer. In any case, Marconi didn't believe in wireless telephony.

Encouraged by his success, Fessenden left the Department of Agriculture in 1903 to set up his own company, National Electric Signalling, with two Pittsburgh financial backers. His research on the electrolytic detector had led him to a solution to the transmitter problem. He had found that his detector was so precise that he could recognize the characteristic sounds of each transmitter. Knowing that telephony used the principle of current variations to transmit the voice, he decided to do the same thing with radio.

The frequency of the human voice was too weak to produce easily trasmissible electromagnetic waves. For the syllable "ah," for instance, the frequency is about 800 cycles per second (800 Hertz). The length of an electromagnetic wave for a sound of 800 Hz is about 370 kilometres. Transmission of this length of wave would require an antenna thirty to forty kilometres high.

Fessenden decided to use variations in an electrical current as a function of the voice and to modulate the radio waves by these variations. To do this, all he had to do was plug a telephone into the antenna. The result would be a radio frequency the amplitude of which would vary according to an audio frequency. Fessenden had invented amplitude modulation (AM).[11]

Once again, in hindsight, the simplicity of the reasoning seems striking; yet the entire scientific community greeted Fessenden's idea with scepticism. Marconi called the young inventor a "preposterous upstart"; Edison, less nasty, nevertheless commented to his former student, "Fezzy, what would you say are man's chances of jumping over the moon? I figure one is about as likely as the other."[12]

A practical problem remained to be settled: how could the audio frequency of the alternator be increased sufficiently to amplify the human voice enough to cross the Atlantic? An ordinary alternator functions at sixty Hertz – that is, it oscillates 120 times per second. Fessenden needed to raise the frequency to tens of thousands, or even hundreds of thousands, Hertz to transmit telephone signals – that is, much more than he had managed to do so far. He borrowed from the physicist Nikola Tesla the principle of the high-frequency alternator: 100,000 Hz (or one hundred kilohertz), which is in the low frequencies of today's radio.

The First Public Radio Broadcast

In 1906, after three years' work, everything was ready. One station had been built at Brant Rock, near Duxbury on the Massachusetts coast, and another at Machribanish, Scotland. Each facility was equipped with a rooftop 130-metre antenna, topped with a sort of parasol. The electrolytic detector was two thousand times more sensitive than Marconi's, and Fessenden sent wireless signals across the Atlantic in both directions, while Marconi had always had to

be content with unidirectional signals. But the goal, which was voice transmission, was reached almost by accident.

In November the operator of the Scottish station was surprised to receive
an experimental radiotelephone transmission between Brant Rock and a
smaller station eleven kilometres away on the New England coast. The first
words to cross the Atlantic by radio were not at all historic: Fessenden's assistant was explaining to the operator of the subsidiary station how to make a
dynamo work.

Fessenden invited the press to a public demonstration, but a few days before the planned date the Scottish station's antenna was destroyed by a storm
and the station itself was so badly damaged that it could not be rebuilt. Fessenden, however, was not terribly discouraged. National Electric Signalling
had equipped all the ships in the fleet of the then powerful company United
Fruits with ultrasensitive electrolytic detectors, which were normally used
for radiotelegraphy but could also be used for radiotelephony.

On Christmas Eve 1906, from the Brant Rock station, Fessenden inaugurated radio broadcasts as we know them today. American historian Alvin
Harlow described it this way:

Early that evening, wireless operators on ships within a radius of several hundred
miles sprang to attention as they caught the call, "CQ, CQ" in the Morse code. Was it a
ship in distress? They listened eagerly, and to their amazement, heard a human voice
coming from their instruments – some one speaking! Then a woman's voice rose in
song. It was uncanny! Many of them called to officers to come and ilsten; soon the
wireless rooms were crowded. Next some one was heard reading a poem. Then there
was a violin solo; then a man made a speech, and they could catch most of the words.
Finally, every one who heard the program was asked to write to R.A. Fessenden at
Brant Rock, Massachusetts – and many of the operators did.[13]

It should be no surprise how quickly Fessenden grasped what was to become the main application of radio technology: broadcasting of variety programs. (He even invented a rating system.) For good measure, he repeated
the feat on New Year's Eve. But this impressive series of technological firsts
and experiments went unpublicized, for Fessenden, like many inventors, did
not have the marketing abilities of an Edison or a Marconi. After all, where
would Graham Bell have been without Hubbard or Sanders?

Fessenden's career as a businessman came to an end, paradoxically, in
Canada. He had obtained from the British Post Office a licence to create a radio service between Canada and Great Britain. In 1909 he suggested to his
two American partners that they set up a subsidiary of National Electric Signalling, under the name Fessenden Wireless Telegraph Company of Canada,
to compete with Marconi in transatlantic transmission. Some Montreal
investors had agreed to provide funds for this venture.

But Fessenden, whose career had been built in the United States, then committed a cardinal error: unexpectedly, he demanded that control of the new subsidiary remain in Canadian hands (those of the Montreal business-men and his own – he had always maintained his Canadian citizenship). The Americans, who had already invested two million dollars at a total loss in National Electric Signalling, refused outright and tried to squeeze him out of the company. Fessenden turned this dispute into a Canadian national issue – or, more precisely, an issue of honour for British imperialism. He sued his ex-partners and won his suit in 1912, but by then it was too late. Marconi had locked up the Canadian market.

Fessenden still had a long career as an inventor before him. During the First World War, he became interested in submarines, even participating in a submarine battle in the English Channel. Two of his inventions were used by the Allies: an early version of sonar, for detecting enemy submarines, and a radio system for communicating between land and submarines. He filed a to-tal of more than five hundred patents, but his prickly character got him into trouble with all of his partners, and none of his industrial projects came to fruition.

As radio advanced by spectacular leaps in the 1920s, Fessenden's patents were shamelessly pirated and he lost much precious time in court. In 1928, after seventeen years of trials, he ended up settling out of court with Radio Trust of America, the latest pirate user, for $2.5 million, instead of the $60 million he had demanded. But by this time, he was ill, and he never got to enjoy this windfall. He died in July 1932 of a heart attack.[14]

PART TWO

Creating Universal Service, 1915–56

After the pioneering era came the search for universal service: a telephone in every home, an objective that seemed to be within grasp. All of the major technological principles were known, and telephone companies had finally acquired the respectability that gave them access to capital. On the other hand, recovery from the First World War turned out to be more difficult than foreseen, and when prosperity finally returned, the 1929 stockmarket crash and then the Second World War delayed realization of universal telephone service.

Telephony stole the show from telegraphy when its network began to include long-distance and overseas lines. Telegraph companies were forced to diversify their activities to include voice transmission for radio programs, but this did not keep the telephone from dominating the telecommunications sector. In 1909 telephone revenues surpassed telegraph revenues for the first time.[1] During the inter-war years, this domination was complete. The division into three industries – telegraph, telephone, radio – that described the first years was no longer possible; telecommunications was unified under the reign of the telephone companies.

In 1956 a combination of events interrupted the previous balance. During the postwar period, the long search for universal service had come to an end: telephone companies had fulfilled their mandate, and now they had to find a new reason for being or lose their monopoly. This progressive transformation extended into the late 1950s and early 1960s. But 1956 was the symbolic year in which the Consent Decree between the United States Justice Department and AT&T was signed. This agreement ended the influx of research and development into the Canadian industry that had kept it in a state of subservience. At this point, the Canadian industry became committed to technological independence.

The year 1956 also marked the advent of the first transatlantic telephone cable. It was the beginning of a technological connection between North America and Europe that foreshadowed the "Global Village."

11 Creation of a National Industry

At the end of the First World War, the new technology of radio took off. When it burst into the self-contained telecommunications sector, it provoked a confrontation between telephone and telegraph companies.

In December 1919 Marconi launched the first commercial radio station in Canada, and perhaps in the world, with a regular broadcast schedule; it was in Montreal and its call letters were XWA (later CFCF). It took another three years for commercial radio really to light a spark with the public, and then its popularity exploded. Everyone, it seemed, was buying a radio set – or making one, as they had made telephone sets in the first years of telephony. Newspapers of the time made much of this fad, which lured husbands away from their wives to tinker endlessly with vacuum tubes.

In the United States, AT&T considered that it had a monopoly on all voice communications and put forward a grandiose scheme to set up radio stations in all American cities and link them up through its long-distance network. The protest was so loud that in 1924 AT&T gave up on this plan and forsook the radio market.

The Canadian situation was entirely different because of the early involvement of Canadian National, the newly created state-owned railway company, in radio. CN was a federal Crown corporation born of the merger of two railway companies on the verge of bankruptcy, Grand Trunk and Canadian Northern (which had absorbed Great Northwestern), whose telegraph subsidiaries were integrated at the time. Canadian National Telegraphs was formed on 1 January 1921, with a mission to supply telegraph service to the new railway and serve the general public within the territories it crossed. Bell and CN made perfect partners: CN purchased all its radio equipment from Northern, and when it wanted to do network broadcasting it rented Bell's long-distance lines between Ottawa and Montreal.

The success of this collaboration gave Bell the idea of imposing the CN model on the entire young radio industry, as spelled out in a draft agreement with Northern in which Bell would refuse access to long-distance phone lines to radio stations that had not purchased their equipment from Northern, while Northern would not sell equipment to radio stations in Bell territory that did not rent Bell's lines. Needless to say, this highly collusive plan never saw the light of day.

A different strategy was hatched in the Prairies. In April 1922, the Manitoba Telephone System created its own radio station in Winnipeg, with the call letters CKY. It was the first state-owned commercial station in Canada. The government was well placed to put pressure on competitors, which it did: when their licences expired, they didn't ask for renewals. It should be noted that although operating licences were given out by the federal government, it was always difficult for a private company to operate against the will of the province. Thus, MTS began with a monopoly on radio broadcasting, but it did not hold on to it for long. In 1928 private competitors gained a foothold on its territory, although it did launch a second station in Brandon in 1930, with the call letters CKX.

MTS kept its two stations until after the Second World War. In 1945 the premier of Quebec, Maurice Duplessis, had a law passed creating Radio-Québec. The federal government reacted by forbidding provinces to enter the field of radio broadcasting and revoking MTS's licence. In 1948 the CBC picked up CKY, while CKX went to private interests, putting an end to the longest intrusion by a telephone company into the radio broadcasting sector.[2]

To return to the birth of commercial radio broadcasting, an agreement was concluded in 1923 between Bell and Marconi, on the one hand, and between Northern Electric, Western Electric, General Electric, and Canadian Westinghouse, on the other. The "services" part of this ten-year agreement gave Bell the right to use all of Marconi's radio patents in the telephone sector and Marconi the right to use all of Bell's radio patents in the wireless sector, while the "manufacturing" part pooled all patents held by the participants. This agreement eliminated all causes of friction between Bell and Marconi and began a fruitful collaboration with regard to radiotelephony that would lead to the creation of the transatlantic service explicitly planned for in the 1923 document.[3]

Having protected its base of operations, Bell turned towards its plan for a radio network. For some reason, it was Northern Electric, not Bell, that opened a station in Montreal in January 1923, with the call letters CHYC. The studios were in the Shearer Street plant and much of the programming comprised religious services for the employees. Other stations were opened in Toronto (CHIC) and Halifax (CHNS). Programming was more or less regular up to 1929, and possibly until the beginning of the 1930s.[4]

CHYC had an unexpected customer: CN. In 1925 and 1929 the federal government took control of Western Union's networks on Vancouver Island and in the Maritimes. Soon after, CN Telegraphs' mandate was broadened to cover all of Canada with a complete range of services, similar to CP Telegraphs' mandate. CN's new president, Sir Henry Thornton, had discovered the commercial potential of radio broadcasting. In June 1923 he created a radio division with two objectives: improving the quality of service to train passengers and improving CN's image with employees and the general public.

To attain his first objective, Thornton had radios installed in the trains. In July 1923 CN's first transcontinental train fitted with radios left Montreal to the sound of music. The programming was produced and directed by CN's radio division but broadcast by Northern Electric's station. As the train rolled westward, it received other radio signals, most of them American. Over time, radios were installed in all of CN's express trains.

The other part of Thornton's radio policy was to broadcast to CN employees and the general public. To counter the railways' negative image – the bankruptcies and scandals that had led to their nationalization were still fresh in people's minds – Thornton had the radio station CKCH (today CBO) built in Ottawa, where CN's head office was, so that he could address his various audiences directly. Starting in 1924, the fiery president of CN held forth in a series of public affairs programs designed to rehabilitate the company's image.

Under Thornton's guidance, CN's radio division expanded rapidly. For reception of radio transmissions in the trains, technicians were trained to operate small sound studios housed in special cars. Their mission was to maintain reception quality, and so they had to know the geography of radio waves in North America. Naturally, Thornton thought that these rolling studios and experienced technicians were underutilized; why not offer telephone service to passengers? In April 1930 radiotelephone service was inaugurated on Montreal–Toronto trains. This first experiment with mobile radiotelephony might have led to a conflict with the telephone companies, but in November 1931 the economic crisis put an end to radio in the trains.

On the broadcasting front, things were evolving as well. The public was concerned less with performance than with quality. It became essential to use network broadcasting – the term at the time was "simultaneous broadcasting." In December 1923, before its own radio station opened, CN began to rent Bell's long-distance circuits for network broadcasting between Ottawa and Montreal. For a company that had its own trans-Canada long-distance system, it was vexing to have to make use of another company's installations, but CN's lines could not provide the necessary quality to carry voices;

in any case, it was overloaded at the time. The solution was technical: the entire network would have to be overhauled.

It was Alexander Graham Bell's research on "harmonic telegraphy" that had led him to discover the telephone. He had found that if a musical tone was sent through a wire, it acted only upon the mechanism set to receive that same tone, or frequency, leaving free the mechanisms set to receive other frequencies. A new tone had to be transmitted to make each one function. This type of multiplexing had finally been implemented by the Bell Labs in New Jersey, which called the technology carrier currents. Upon CN's request, the Bell Labs sent a talented engineer, J.C. Burkholder, to Canada to supervise installation of a carrier-current system between Montreal and Toronto. In April 1927 it suddenly became possible to send thirteen signals – twelve telegraphic and one telephonic – on an existing line.

A few months later, the Jubilee celebrations sealed the return of the telegraph companies to voice communications. Bell now realized that it had been playing with fire by allowing its parent company, AT&T, to help CN acquire a telephone technology; now, CN was well established in the network-broadcasting sector. After the success of the Jubilee, the carrier-current system was quickly expanded to cover all of Canada, and by December 1928 CN's telegraph department had the means to broacast regular programming on a national scale for the first time in Canada. Most of the lines used belonged to CN, although many sections were still rented from telephone companies. CN thus confirmed its mastery over voice transmission.

One apocryphal story has it that during this first broadcast a train was caught in a blizzard in the middle of the Prairies. The anxious technician scaled a telegraph pole and attached his emergency telephone to the line. Surprised to hear Christmas carols on the line instead of the usual Morse code clicks, he let loose a few choice swear words before asking where the Winnipeg station manager was. This unexpected interruption was heard from Winnipeg to Vancouver by thousands of amused listeners.

By December 1929 CN's carrier-current network stretched from Vancouver to Halifax. The telephone companies had been beaten on their own turf: voice transmission. This transcontinental long-distance network served thirteen English-language and four French-language radio stations, most of them privately owned. At the height of its involvement, CN itself owned three radio stations, in Ottawa, Moncton, and Vancouver, and two studios, in Montreal and Halifax. Notably, most programming on both public and private stations was local, while national broadcasts were rare: one hour three days a week on the English-language network, and one hour a week on the French-language network.

CP neglected radio broadcasting at first, but as it saw the growing success of the new medium and the expanding commitment of its nationalized competitor, it began to reconsider. It was the threat of government intervention in the sector that pushed CP to act. In September 1929 the Royal Commission on Broadcasting – also called the Aird Commission, after its chairman – had recommended that radio be nationalized. As the largest private company in the country, CP tried to steal a march and prevent this government intrusion. With Bell's help, it improved the speed of its network by adopting, in its turn, the carrier-current system in 1930. CP did not own radio stations, but it had music to sell, for there were a number of orchestras in its hotel chain. It sold musical programs – and network broadcasting – to a number of Canadian and American stations.

COMPETITION BETWEEN TELEPHONY AND TELEGRAPHY BEGINS

The telephone companies could not sit still for this resurgence of the telegraph companies through voice transmission. They developed their own transcontinental network in 1931 and in their turn broached the strategic stector of very-long-distance communications. As long as there had been a market for very-long-distance text transmission, the market for very-long-distance voice transmission was a moot point. The technology (electronic repeaters) was relatively new, the rates high, and the market uncertain, with the depression in full swing. Only the radio sector offered good prospects for the future. Not that radio stations were exempt from the economic crisis – on the contrary, many smaller staions had to shut down – but the imminence of federal legislation in this sector encouraged both telephone, railway, and telegraph companies to position themselves in the market for very-long-distance voice transmission.

Faced with the counteroffensive by telephone companies, CN and CP formed a four-person committee at the beginning of 1932, with representatives from the radio and telegraph divisions of both companies, to study what measures to take. They decided to maintain their respective network-broadcasting infrastructures as long as the federal government had not legislated on the future of the industry. On the other hand, operational coordination was instituted to ensure greater flexibility in network management and a supply of back-up circuits in case of breakdown. All revenues from radio service were pooled, and this marked the beginning of the CN/CP consortium.

Everything was in place for the telegraph-telephone war. The first round took place under the reign of the Canadian Radio Broadcasting Commission, which had been created in May 1932. The CRBC was a hybrid institution in the sense that it was both referee and player. As referee, it handed out licences for radio stations and regulated the industry. As player, it produced its own programs and operated a national network. In March 1933 it purchased

CN's three stations and an independent one in Quebec City. Its initial team was composed of ex-CN employees, who were naturally inclined to side with the telegraph companies.

The CRBC was headed by three commissioners, two of whom had no idea what radio was. Its chairman, Hector Charlesworth, had been a music critic and editor-in-chief of the most prestigious magazine in Canada, *Saturday Night*. One vice-chair, Thomas Maher, was a forestry engineer from Quebec City who had been appointed because he was a member of the Conservative party, then in power in Ottawa. The other vice-chair, William Arthur Steel, was the only industry expert. An excellent engineer from the Canadian army's Signal Corps known for his work on cathode radiogoniometry, a precursor technology to radar, Steel was a poor administrator. As we shall see, a less well matched trio would have been hard to find.[5]

In this unfavourable context, Bell made a spectacular *faux pas*. The British Post Office and the BBC had decided to organize an empire-wide program to celebrate Christmas 1931, during which King George V was to address his subjects all over the world for the first time. The major radio networks of the Commonwealth countries were to participate in the event, and CN was representing Canada. The Canadian land-based circuits were to be supplied jointly by the CN and CP networks, and transatlantic transmissions between Great Britain and Canada would normally have been provided by Marconi Canada. Unfortunately, Marconi could not guarantee the audio quality at the chosen time and AT&T had to be called upon as the only company that measured up to the task. At first, AT&T agreed to take on the project, but on 16 December it changed its mind and peremptorily notified the British Post Office that all voice communication between the United States and Canada was the responsibility of the telephone companies, not the telegraph companies. AT&T refused to connect its network to CN/CP's and the BBC had to cancel the transmission.

It's easy to imagine the hue and cry in the press: Bell and AT&T received a sound verbal thrashing. They tried to save face by offering to carry the program free of charge, but it was too late; the BBC's cancellation decision was final. The affair ended up in front of the parliamentary Radio Commission, where Bell was severely chastised. The following year, the BBC reiterated its proposal to organize an empirewide program. This time, it was the CRBC that represented Canada, and of course Bell and AT&T committed themselves enthusiastically to broadcasting the event free of charge. The radio historian Austin Weir commented dryly, "That pledge was carried out energetically and in complete harmony."[6]

It was against this background that, in April 1933, the CRBC awarded an operating contract for network broadcasting to the CN/CP consortium. The contract was for a seemingly low $275,000, but it corresponded to the minimal number of network broadcast hours. In any case, the telegraph companies had won the first round, and they used the opportunity to complete the merger begun the previous year; in September 1933 they merged the operations of their radio facilities on a parity basis.

In 1935, when the contract was renewed for five years, the annual fee rose to $375,000. It rose again in 1936, when the Canadian Broadcasting Corporation began operating, but remained at a modest level. The CN/CP consortium retained its monopoly on network broadcasting until 1962, relegating the telephone companies to a few marginal subcontracts.

When television began network broadcasting in 1952, Bell finally got its revenge. It persuaded the other telephone companies to launch construction of a transcontinental microwave network and grabbed the CBC's contract. Later, when a private television network was created, its broadcast contract also went to the telephone companies. Finally, radio, in its turn, fell into the hands of the telephone companies; the CBC contracts alone were worth ten million dollars per year. For the railway/telegraph companies, the coup of the 1930s was turning into an albatross. They began to plead – a bit late – for grouped contracts dividing the markets between all applicants.

CN AND CP: CLOSER TIES

Two dates mark the peaks of telegraphy in Canada: 1929 and 1956 (see Table 8). Until 1929 the industry grew vigorously, in spite of growing competition from the telephone. The depression brought growth in both industries to a screeching halt. It took the Second World War for telegraphy to recover from the crisis, and it then reached a second summit in 1956, when it began the long, slow slide towards its current state of decline. The telegraph companies thus had to adapt to survive. They did this in two ways: uniting for joint projects and diversifying their activities towards radio.

In the late 1930s, the inauguration of the first commercial national airline, TCA, obliged the Department of Transport to set up complete meteorological and air traffic control systems. In September 1937 CP received a mandate to supply a teletype service between Winnipeg and Vancouver; CN received a similar contract for the eastern part of Canada. The first national system for gathering and disseminating meteorological information was set up in 1939 thanks to a cooperative effort by CP and CN. The meteorological teletype network was complemented in 1953 by dissemination of maps by facsimile to all Department of Transport weather stations.

In 1942 CN and CP set up the same kind of system for air traffic control at Canadian airports, using a private telephone network. The involvement of telegraph companies with voice transmission, which had started with contracts for retransmission of radio programs, had now come full circle. Their strategy had proved effective: with their own technology in jeopardy, they united to penetrate neighbouring sectors such as the telephone market.

Table 8
Evolution of the Telegraph Industry in Canada, 1915–56

	Number of telegrams sent (000)	Revenue ($000)	Number of Employees
1915	10,929	5,536	6,243
1920	16,752	11,337	7,508
1930	17,580	14,265	7,331
1933	11,710	9,268	5,263
1940	14,389	10,923	6,588
1945	19,859	18,016	8,230
1950	22,166	23,922	9,757
1956	22,811	40,720	10,833

Source: Dominion Bureau of Statistics, Statistics on Telegraphs and Cables (cat. no. S564–522–126).

Both telegraph companies, however, had their own expansion projects. During the Second World War, Bell and AGT had built a telephone line for the United States government crossing Canada from north to south to link Alaska to the lower forty-eight states. The Canadian Department of Transport inherited this line in 1946; the following year, it turned it over to CN to operate. After the war, CN was thus both a national telegraph company and a regional telephone company.

This marked the beginning of CN's activities in the North. Starting with a "network" that in 1947 barely merited the term – five isolated independent companies with no link to the outside world – CN created a complete telephone system to serve residents of the Northwest Territories and the Yukon; by the early 1970s the northern network comprised 4,800 kilometres of line and seven thousand subscribers. Meanwhile, the federal government had handed over its northern British Columbia telegraph system to CN in 1954.[7]

CN's telephone activities received a further boost when on 31 March 1949, Newfoundland became the tenth Canadian province. The federal government asked CN Telegraphs to operate the new province's telecommunications service, previously under the jurisdiction of Newfoundland Post and Telegraphs. These services included telegraphy, of course, but also telephone service covering most of the island, as well as specialized systems for the military bases established by the United States during the Second World War. Western Union's installations in St John's and the federal government's line between Newfoundland and the mainland were integrated into CN's network in March 1950 and April 1951.[8]

Immediately after the Second World War, Canada was the only country in the world with two national telegraph companies. In the United States, Western Union had purchased its only competitor in 1943; everywhere else, the rule "one territory, one network, one company" applied. The Canadian "duopoly" had to make a choice between confrontation – as in the time of Montreal Telegraph versus Dominion Telegraph, then Great North Western versus CP Telegraphs – and cooperation. The latter worked so well that CN's and CP's telegraph divisions were gradually merged, finally forming a monopoly.

In August 1947 CP Telegraphs and CN Telegraphs pooled their direct telegraph and telephone lines (except for those of a few major users) and merged their direct-lines sales, facilities, and maintenance divisions. (However, they maintained their competitive activities in all other sectors, including what was still their principal market: public telegraph service.[9]) In 1956 CN/CP adopted the Telex International Standard, thus ensuring its supremacy in the field of written transmissions, and five cities (Montreal, Toronto, Winnipeg, Vancouver, and Ottawa) were linked to the world network. The following year, telex was extended to the rest of the Canadian market, and in 1958 to Western Union's American network.

AT&T, on the other hand, developed a more sophisticated standard called TWX, which was incompatible with the rest of the world. Canadian telephone companies adopted TWX and therefore became isolated in a technological ghetto. CN/CP took advantage of this miscalculation and took over the text transmission market of all corporations that had international relations. The number of subscribers to the Telex service of CN/CP reached a peak of 20,000 in 1970. The success of CN/CP's Telex venture represents the advantage of an international standard over a more-advanced but local technology.

At the same time, a major joint project was realized: the trans-Canada microwave network. Microwaves had first been used in telecommunications at the end of the 1940s to link Prince Edward Island to Nova Scotia. In 1954 Bell linked Montreal, Toronto, and Ottawa, and CN/CP launched two networks (Montreal-Quebec City and Toronto-London-Windsor) to carry the CBC's television programs. It was the beginning of the microwave era.

Faced with the inexorable decline of telegraphy, the slowly forming CN/CP consortium had no choice but to diversify, but it found the telephone companies blocking its way. Would it join forces with the telephone companies or compete with them? By the mid-1950s, in spite of the skirmishes of the 1930s, nothing was settled yet.

UNIFYING CANADIAN TELEPHONE COMPANIES

At the end of the First World War, there was no Canadian telephone network, and no company strong enough to build one. Bell's monopoly had been broken up by the withdrawal of the Maritimes in 1885–89 and the series of nationalizations in the Prairies in 1908–09. The Canadian telephone industry was fragmented, uncoordinated, and very fragile.

In addition, the country was divided by geographical obstacles. The Rockies formed a mountainous wall between BC Tel and the three Prairies networks. A rugged stretch of more than fifteen hundred kilometres from Winnipeg to Sudbury separated the Prairies from central Canada. Low population density in the regions where Quebec bordered New Brunswick discouraged any expansion of networks in that direction. Finally, there was the special case of Prince Edward Island. On the other hand, all of the provinces – except Prince Edward Island – were connected to the United States.

In the Maritimes, it took one of the federal government's rare direct interventions in the telephone industry to link Prince Edward Island to the mainland. It should be recalled that Nova Scotia's telephone company, Maritime Telephone & Telegraph, had broken Anglo-American Telegraph's monopoly in 1911 by hastily laying an underwater cable between Nova Scotia and Prince Edward Island. But the cable that was used did not meet telephone-quality criteria, and it had to be abandoned after a few months. The Prince Edward Island government was insistently invoking the constitutional agreement that had made the province's entry into Confederation conditional upon establishment of communications links with the mainland. In 1918 the federal government finally agreed to construct not one but two telephone links: one with New Brunswick (Cap Tourmentin-Cap Travers), and one with Nova Scotia (Caribou-Wood Island).

The western telephone companies had made the first steps toward a Canadian national policy. The Manitoba minister of Telephone and Telegraph services invited the directors of the four main regional companies – Alberta Government Telephone, BC Tel, Manitoba's Department of Telephones and Telegraphs, and the Saskatchewan Telephone Department – to a meeting in Winnipeg. (Bell representatives attended the meeting as observers.) The resulting association, Western Canadian Telephone, formed in October 1920, "invented" cooperation between public-sector (the three Prairie companies) and private-sector (BC Tel) telephone companies in what was to become Canada's hallmark in the telephone industry. As well, this historic meeting marked the reconciliation between Bell and the Prairies. Sise had died, as had most of the directors who had organized the secession, and a new generation with a more pragmatic attitude had risen to power.

The top item on the agenda was construction of a telephone network for the West. Quickly, the debate broadened to a national scale. The Winnipeg meeting resulted in the creation of the Telephone Association of Canada, comprising seven companies: to the first four were added Bell Telephone, MT&T, and NB Telephone. Each company covered one province, with the exception of Bell, which covered two (Quebec and Ontario), and MT&T,

which also covered two (Nova Scotia and Prince Edward Island). This embryonic telephone administration assigned itself a mainly technical mandate.

The first meeting of the TAC took place in Vancouver in August 1921. Again, the discussions went beyond the planned framework: the representatives had come to talk about standards and accounting, but they ended up remaking the telephone map of Canada. The vice-president of BC Tel, George H. Halse, proposed to create a trans-Canada telephone line, and it seemed that he was speaking for all four western provinces. It was in the West that the telephone network was the most dependent on the United States. Halse cited the example of certain telephone districts in the interior of British Columbia, on the eastern slopes of the Rockies, whose communications with the rest of the province had to go through Washington State; similar inconvenient situations existed at other locations on the Prairies. It was John E. Lowry, the brand-new commissioner of Manitoba Government Telephones, who pushed to a vote the motion proposing that a transcontinental link be studied.

Tellingly, at this first meeting, the telephone companies compared building a trans-Canada telephone line in the twentieth century to building the railway in the nineteenth century. Realizing the importance of creating an inspirational myth around their technology, they tried to resuscitate the link between national unity and the railway and adapt it to their purposes. At first, the trans-Canada project was dubbed "All Red," like Sandford Fleming's old around-the-world telegraph project.[10]

In spite of the hopes raised by this first meeting, the trans-Canada network project barely passed the stage of good intentions. A number of feasibility studies were conducted, all of which demonstrated that low-population-density areas would not be profitable. One solution would have been to rent the railway companies' telegraph circuits, but it was felt that such a proposal would be turned down. The telephone companies lacked the common will to rise above self-interested calculations and an absence of long-term vision.

Throughout this period, Manitoba played an essential role in keeping the unification flame alive, at least in the West. In the 1920s some long-distance lines were crossing provincial borders: all three Prairie provinces were linked up in August 1923. Lowry persuaded the telegraph companies to play a role in this patchwork service. When one recalls the deep hostility with which the West regarded CP, one can truly appreciate Lowry's skill as a negotiator. In May 1926, the Manitoba Telephone System used CP Telegraphs' Winnipeg-Fort William line to link Manitoba to Ontario. Significantly, this was the first time since 1880 that a telegraph company had been involved in voice transmission. The time was right for cooperation between the telegraph and telephone industries, even though the situation contained the seeds of future competition.

In August 1928, CP Telegraphs allowed MTS and Bell to install their wires on its poles to create the pivotal Winnipeg-Fort William line. There was even cooperation in the Rockies, which were crossed using CP Telegraphs' poles in November 1928, enabling all of British Columbia to be connected to Alberta (the part east of the Rockies had been connected since 1908). In this case, however, completely new lines had to be built.[11]

These interprovincial lines were of uneven quality. Worse, they functioned like a milk-run train, being switched in each telephone exchange. A long-distance call was treated as the sum total of calls from one local exchange to the next local exchange. This was not so bad for intercity calls, but it became very awkward when someone wanted to communicate between two noncontiguous provinces – especially since, south of the border, AT&T had set up a transcontinental line that worked like an express train, with a specialized long-distance exchange that allowed a caller to get on the line and be connected directly. Thus, even between the nominally interconnected Western provinces, communications continued to transit through the faster, more efficient American network.

This not only had a negative impact on Canadian nationalism, it also led to substantial losses of revenue. The telephone companies shared long-distance revenues according to the number of kilometres that their respective networks traversed. Since most of the Canadian network was located within a few kilometres of the United States border, most of the revenues from long-distance calls had to be paid to American companies. And as the volume of calls mounted, so did revenues lost to the US.

There was also the danger of internal competition. As we have seen, CN, in its infatuation with radio broadcasting, was in the process of equipping its telegraph lines with carrier-current systems, turning its network into a true telecommunications system capable of transmitting both text and voice. It didn't take the telephone companies long to realize the size of the threat.

THE CONFEDERATION JUBILEE: COOPERATION RETURNS

The federal government had decided that radio would be the medium to carry the jubilee celebrating the sixtieth anniversary of Confederation. Dominion Day, or 1 July, 1927 would provide Prime Minister King with an opportunity for a media bonanza, and he decided to address the Canadian public from sea to sea to see from Parliament Hill in Ottawa. For the occasion, all radio stations across the country would have to be connected to the government offices in Ottawa and the nine provincial capitals. To be sure to reach the largest possible audience, loudspeakers were to be hooked up in parks, schools, hospitals, and every other public place in the cities of Canada.

Which network would be used? By April, CN had a transcontinental carrier-current line capable of carrying voice transmissions. Telephone lines still provided better transmission quality, but the telephone infrastructure was incomplete and the Canadian government wanted to avoid transiting through the American network, as was the case for ordinary calls. Therefore, a hybrid network would be used.

The telegraph companies enthusiastically agreed to put their networks at the government's disposal for transmission of the official speeches – it was an opportunity for them to break back into the field of voice transmission. A total of 16,000 kilometres of telephone lines and 14,300 kilometres of telegraph lines would be connected up; not only would the telegraph and telephone lines have to be made compatible, but so would the telephone lines of different companies. This delicate operation would be executed throughout the subcontinent by engineers from Bell, acting as the technological mediator. All main arteries were doubled to prevent interruption of the rebroadcast. For security, a back-up route through the United States between Windsor and Winnipeg was made available.

On "D" Day, forty thousand people gathered at the Parliament Buildings in Ottawa to join Prime Minister King and the elite of the political and artistic worlds. The last living telegraph and telephone pioneer, Thomas Aheard, was present, and he later received as a gift the gold-plated microphone used for the ceremony. Twenty-three radio stations and an untold number of loudspeakers transmitted this first Canadian media high mass. The operation was a technical success, but it starkly highlighted the incompleteness of the telephone system.[12]

At the next meeting of the TAC, held in Minaki, Ontario, in August 1928, Bell was asked to conduct a feasibility study for a Halifax-Vancouver line, dubbed the Trans-Canada Telephone System, including an evaluation of the costs of improving existing sections and construction of others to fill the gaps. Just one more study? No, this time, the telephone companies decided to proceed without taking into account the immediate profitability of the operation – or rather, they decided to gamble that there was elasticity of demand: if they made an effective infrastructure available to the public, the number of calls would grow. There wasn't a minute to lose, since the government was getting ready to name a commission on the future of radio, the Aird Commission. The telephone companies had to be ready for reorganization of this market.

Bell's chief engineer and the chairman of the engineering committee of the TAC was Robert Vernon Macaulay. This First World War veteran, hero of the battles of Passchendaele and the Northern Channel, was the first in a series of great Bell engineers, with a national conception of the network. If Bell couldn't have a monopoly on all of Canada, in his opinion, it would have to form a "federation" of monopolies. The following year, he presented a report to the TAC meeting, which adopted it on the spot. On 31 August 1929, he was put in charge of coordinating construction of the trans-Canada

The first trans-Canada radio transmission took place in an atmosphere of general jubilance, and with the participation of American aviation hero Charles Lindbergh. The event took place in July 1927, on the sixtieth anniversary of the founding of Canada, with the telephone and telegraph companies cooperating by connecting their networks. Courtesy Bell Canada.

network. Less than two months later, the stock market collapsed and the world economy entered the worst depression since the Industrial Revolution, but construction of the trans-Canada network never stopped. Nothing would stop it.

CREATION OF THE TRANS-CANADA TELEPHONE SYSTEM

Each of the TAC's seven member companies assumed responsibility and costs for the work undertaken on its territory. Even before construction was completed, it was obvious that the association's structures were insufficient. On 1 March 1931 the seven companies signed an agreement that brought into existence the Trans-Canada Telephone System, a juridical form unique to Canada. It wasn't a company or a cooperative, but an organization not constituted as a legal entity, which meant that it had no capital, property, or staff. The TCTS functioned with people borrowed from and paid for by its

Erecting a telephone pole: half the men pulled the pole, while the other half pushed. The trans-Canada line from Halifax to Vancouver had 185,000 poles. Courtesy Bell Canada.

members. Of course, Bell played an essential role, with all presidents of the TCTS until 1971 coming from the ranks of its top executives.

One among many examples of Bell's leadership was its decision to lend a strong hand to the Manitoba Telephone System with construction of the terrible Sudbury-Winnipeg section, the bane of all trans-Canada infrastructure builders, an uninhabited region with no economic value. Bell itself built the line, using CP's telegraph poles, and maintained it for twenty years ... all in the interest of a transcontinental network. The beginning of Bell's Canadianization can be dated from this time. Beyond the American hold on its capital and technology, Bell had discovered a Canadian vocation and was associating its own development with national unity.

The mandate of the TCTS was both broad and specific. Of course, it created a national telephone service, which Canada had not had until then, and plans were afoot to provide international service. It represented the seven Canadian companies in negotiations with foreign governments, including that of the United States. For the Canadian government, it was a unified structure that could finally respond to the country's growing communications needs and would work closely with the armed forces to ensure the national defence.

But the TCTS was above all a management organization, where the distribution of revenues from long-distance calls and the technical and administrative standards of the long-distance network were decided. There-fore, it was not surprising that the new association's first move, in May 1931, was to create a clearing-house for trans-Canada business. An initial agreement, signed in August, stipulated that revenues be divided in propor-tion to the length of the circuits supplied by each member to route a call, in the form of a fixed commission payable to the company that collected the fees for a call (person-to-person and collect calls). The agreement con-cerned only long-distance calls covering three provinces or more; calls made between two neighbouring companies were not included.

In a symbolic gesture, the TCTS adopted Bell's administrative and ac-counting methods, the companies that had left Bell's fold twenty-five or thirty-five years before having admitted their inability to find another useful management mode. It was the best response to the accusations of overcapi-talization and too-large dividends to shareholders that were so often levelled at Bell: all companies, private or public, that tried to operate their systems differently either failed or had to reintegrate, at one point or another, with the telephone orthdoxy determined once and for all by AT&T and introduced into Canada by Bell.

The TCTS's board of directors was composed of one representative from each of the member companies, with the exception of Bell, which, as a com-pany operating a network straddling two provinces, had the right to two rep-resentatives. All decisions were made on a consensus basis, which meant that a small company had as much power, in theory, as did Bell. In 1931 a management committee was set up to supervise programs devised by the board of directors. Then, operational committees were added to the initial structure. But the basic principles laid out in 1931 – association without re-gard to industrial status, decisions made by consensus – still hold today.

The Canadian transcontinental line was completed in August 1931, and on the following 25 January, the Governor General of Canada, the Count of Bess-borough, presided over the official inauguration of service. It was a highly symbolic occasion, since it took place just a few weeks after adoption of the Statute of Westminster, which provided Canada with its international legal sta-tus, in December 1931. After the railway and the telegraph system, it was the turn of the telephone system to become the emblematic technology of nation building. It is worth repeating that for Bell, which operated the TCTS by remote control, it was the dawn of Canadian national consciousness, and the unoffi-cial, soon to be official, consecration of the status of the company's status as a public utility. Never again would Bell be a company "like the others."

The TCTS consisted of 6,800 kilometres of copper wire strung on 185,000 poles, using Bell's technical quality criteria. Wherever possible, existing lines were used and improved, but more than half of the network was entirely new. For example, the Rockies were crossed 240 kilometres south of CP's rail line, where the old line ran, allowing it to serve isolated towns that other-

In the middle of the depression, and challenged by hostile terrain and extreme weather, it took a monumental effort by Canadian telephone companies to build a transcontinental network. Courtesy Bell Canada.

wise could not have had telephone service. On the other hand, as mentioned above, CP did contribute to the project with its Sudbury–Winnipeg line. The 165-gauge wire used was thicker than the telephone industry was using at the time. The poles were cedar in the west, and cresote-covered pine in the east. The insulators were placed farther apart than the standards in effect in order to avoid crosstalk. Two thirds of the project used carrier currents. All carrier-current circuits were armed with line filters specially adapted for retransmission of radio broadcasts. Nothing was too good for the TCTS.

The trans-Canada line had twenty-two repeaters. It was supposed to have twelve circuits, but, because of the depression, it had only seven (see Table 9). A few feeder lines were added to the main trunk to serve the major cities without going through more than two operators: Vancouver-Calgary, Calgary-Regina, Regina-Winnipeg, Winnipeg-Toronto, Montreal-Saint John. A call from Vancouver to Halifax thus went through only two operators, one each in Montreal and Winnipeg.[13]

Table 9
Circuits in the TCTS system

Line	Number of circuits	Length (km)
Winnipeg-Vancouver	1	2,790
Winnipeg-Montreal	1	2,524
Winnipeg-Calgary	3	1,480
Montreal-Halifax	2	1,424
Total	7	8,218

In total, the seven founding companies had invested the huge sum, for the time, of five million dollars in this ambitious project. The network was even more impressive if one considered that there were just 1.1 million Canadian subscribers, and they made only forty thousand long-distance calls on the TCTS's lines in 1932, the first complete year of operation, resulting in a paltry business volume of $194,000. Even with only seven circuits, there was an overcapacity on the transcontinental line. On the other hand, the rates in force in 1932 did not encourage individuals or businesses to make great numbers of long-distance calls: the first three minutes of a Montreal-Vancouver call cost $8.25. In 1999 the rate was $1.44.[14]

In the end, the TCTS had a significance that went far beyond its technical prowess. As the inaugural speeches of January 1932 repeated over and over from different angles, it was a factor in Canada's unity. For the purposes of the present history, it should be noted that this was the first time that telephony used an east-west axis. Unlike telegraphy, which had been linked with the Canadian national ideology (and the government funding that went with it) starting back in the 1870s, telephony had always been looked upon with suspicion because of the American origin (AT&T) of the main company, Bell. The TCTS turned this perception around completely: telephony had been Canadianized – and without direct government aid.

NATIONALIZING RADIO

Conceived at the end of the "roaring twenties," the TCTS came into being as the depression hit. The number of calls dropped, and then a large number of users cancelled their telephone subscriptions. The Prairies were hit particularly hard. The monoculture zone suffered a terrible drought, dust storms swept the disaster-struck plains, and when the rains finally returned, they brought a scourge of grasshoppers with them. The nationalized telephone companies were once again on the verge of bankruptcy and they turned against Bell, instigating the first serious crisis in the TCTS.

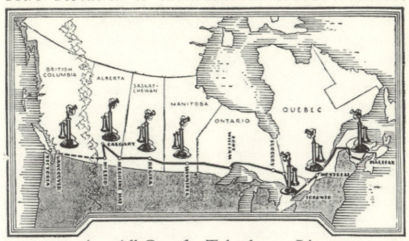

LONG DISTANCE IS INEXPENSIVE – SPEEDY – SIMPLE!

An All-Canada Telephone Line From Atlantic to the Rockies

From August 1st, all-Canada tele-phone circuits will link eight of the Dominion's nine provinces, from the Atlantic seaboard to the Rocky Mountains. By the end of this year, it is expected that the all-Canada line will be established from coast to coast, giving Canadians another solid all-red link.

Long Distance telephone service from coast to coast, of course, is nothing new. Calls between British Columbia points and any of our eastern cities have become almost as casual as local calls. It has been necessary, however, to route such calls partly over United States lines. The all-Canada line will provide a direct

route for handling, entirely within Canadian boundaries, all calls be-tween all points in Canada, east or west.

Canadian telephone engineers have been working on the establishment of this trans-Canada system for the last three years. The line they have now completed from the Atlantic to the Rockies stretches over 3,600 miles. The final British Columbia link now approaching completion will mean an all-Canada line of more than 4,200 miles in length.

Long Distance service is con-stantly improving in speed, reliability and simplicity of use.

F. G. WEBBER
Manager

Advertisement celebrating construction of the trans-Canada telephone network (1931). Courtesy Bell Canada.

Contracts for retransmission of radio broadcasts were the main source of long-distance revenues during the depression. As we have seen, the quality of the TCTS's carrier-current lines had been boosted precisely for this purpose, and the association's directors estimated revenues for the first year of opera-tions at more than $1 million. In fact, the TCTS took in one fifth of this amount.

At the beginning of the 1930s, the telephone companies had about 60 per-cent of the market for retransmission of radio broadcasts. The four western provinces had pooled their telephone networks to retransmit a program called "Quaker Oats" broadcast on the Saskatoon radio station; anything to do with oats was a hit on the Prairies. The success of this program had eased the financial burden of the western companies, and they saw in radio the solution to the catastrophe of the depression.[15]

Nationalization of radio service in 1932 had led to the creation of the Canadian Radio Broadcasting Commission, whose personnel and radio stations were inherited from CN. When the CRBC called for tenders for contracts for network broadcasting, it was predictable that it would favour the telegraph companies from which it had sprung. As well, the call for tenders fell at the worst possible moment for the TCTS: immediately after AT&T and Bell's resounding gaffe that had caused cancellation of the empirewide broadcast of Christmas 1931.

It was immediately obvious that the bids made by the two consortia, CN/CP and the TCTS, were comparable. The telegraph companies made a plea to the government, citing the dire straits in which they found themselves because of the depression, and it seems that they found a sympathetic ear in the Department of Railways. In principle, the CRBC was independent of political power, and its directors wanted a solution combining telegraph and telephone according to technical and financial criteria. Since the telephone companies had a monopoly on the access lines to radio stations, such a solution would guarantee them a share of the contract. Knowing this, the president of the TCTS, who was also the vice-president of Bell, tried to make the best of things by proposing a contract that would give the TCTS 40 percent of the retransmission contract, with the other 60 percent split evenly between the two railways.

The three Prairies telephone companies were not happy, feeling that Bell was not trying to get the most out of the contract, and the old antagonism against the "Eastern parasite" was revived. The Prairie companies wanted the entire contract, and they began their own lobbying effort as a corollary to the TCTS's. Their position was summarized in a telegram sent to the commission in December 1932: "We do not care what division the Bell makes in the East but our system is equipped solely for this purpose. Can give better service than any telegraph lines and desire to insist on what we believe to be our rights in the matter."[16] Finally, in February of 1933, as part of the cost breakdown required by the CRBC, they presented a submission covering the Prairie provinces that undercut the competition by 25 percent.

To understand this persistence, it helps to remember that it was the support of the Prairie provinces for the nationalization of radio that had won over Prime Minister Richard Bedford Bennett and led to creation of the CRBC. These three provinces were, in fact, the owners of their telephone companies, and seeing the prospect of their networks being left out of the contract for radio broadcasting, they felt – justifiably – swindled. This feeling was exacerbated further in Manitoba, where the telephone company was also the owner of the largest radio station.[17]

In April 1933, the CRBC summarily granted the entire contract to the telegraph companies. What had happened? The radio historian Austin Weir, who was also one of the protagonists of the story as a high-level CRBC bureaucrat and a proponent of the solution involving sharing the contract by merit, describes the abrupt decision as follows: Late in February, all negotiations were taken over by Colonel Steel, and I had no further part in them or knowledge of details through consultation or otherwise. Moreover, from that moment neither the western nor the Trans-Canada Telephone interests were given any further serious consideration, nor were they even advised after a contract had been signed with the railways one month later."[18]

It is not known whether this was the result of political pressure, corruption, or simple incompetence on the part of the CRBC commissioners and, in particular, Colonel Steel. A few months later, the CRBC quietly decided to create a second network for commercial programming – supported by advertising – and awarded a contract to a consortium composed of the TCTS and CN/CP, in a 60–40 proportion. As the most popular listening times were already occupied by programming on the first network, the commercial consortium immediately found itself in difficulty.

Things got worse still for the telephone companies. The CRBC refused to tell them when the CN/CP contract for the first network would expire. In May 1935, smelling a rat, the TCTS contacted the commission on this subject. It took two months for the president of the CRBC, Hector Charlesworth, to reply, in a letter dated 24 July 1935, that the contract with the telegraph companies lasted until 1936; it had been renewed just one week before, on 16 July, by Steel.

We will never know exactly the nature of the pressures exerted on the CRBC. Steel, showing remarkable independence for a military man, had omitted to mention the renewal of the contract to his superior in the hierarchy. Later, during hearings by the Parliamentary Commission on the Radio in 1936, he would be summoned to explain why he had excluded the telephone companies. The unflappable colonel had nothing to say.[19]

There was a more serious issue. In the renewal of CN/CP's contract, the CRBC specified that it could use the network to broadcast commercial programs. It goes without saying that this change of course had a catastrophic effect on revenues of the second network, which was abandoned by its advertisers. The telephone companies left the sinking ship in 1936. The Prairie companies, outraged, talked about starting their own telegraph companies, but the other members of the TCTS dissuaded them.

This scandal could have sounded the knell for cooperation between the telephone companies, but the investments already made were too great simply to post them to the profits and losses columns. The TCTS, like Canada itself, was based on a marriage of convenience. Nevertheless, this crisis underlined the collusion that had brought in the three nationalized companies of the West, which were always quick to suspect deep dark conspiracies perpetrated by the East.

12 Bell's Long March to Independence

After the First World War, Charles Sise was gone, but his replacement was no less than his alter ego, Lewis McFarlane; in the shadows of the vice-presidency, Charles Sise, Jr, the founder's oldest son, waited his turn. It may have seemed that nothing had changed, but a decision that seemed to involve labour relations began the movement that, in the end, led to the separation of Bell and AT&T: the sale of shares to employees.

PROPERTY OF BELL TELEPHONE: AT&T LOSES ITS MINORITY SHARE

Bell Telephone had been a subsidiary of AT&T's, but it had never been part of the Bell System. Some old AT&T texts cite Bell Telephone of Canada among the companies associated with the Bell System, but these were simply public-relations documents. Bell's special status meant several things: while the system companies used AT&T patents, Bell owned its own patents, which Charles Sise had gotten back in 1882 (see chapter 5). Bell also had its own long-distance lines, while the system companies used AT&T's. Finally, Canadian patent legislation meant that Bell didn't have the right to purchase telephone equipment from Western Electric; it had to make it itself or purchase it in Canada.

All of these differences, as important as they were, faded before two in-controvertible facts: AT&T owned 38.6 percent of Bell Telephone in 1915, when McFarlane acceded to the presidency, and Western Electric had swept away any thoughts Bell might have had of manufacturing autonomy by purchasing 43.6 percent of Northern Electric. But the full extent of American control was due to a third, rarely mentioned fact: since January 1913, Northern Electric had been responsible for Bell's purchasing. In principle, Bell could freely choose its supplier, but it was Northern Electric's purchasing department that processed the orders and billed them at cost price. Via Northern Electric, Western Electric had a firm grip on Bell. Sise had struggled mightily to gain some margin of manoeuvre, which was very real in the sense that it had created a statutory difference between the parent company and the subsidiary, but in practice, the Bell group's policy in Canada was integrated with the Bell System's. Its independence was purely on paper.

In 1915 Theodore Vail launched a plan for employees to purchase AT&T stock, in a bid to gain their loyalty. Of course, it was also a way of raising capital at a low cost and encouraging the workers to be more productive; as was his habit, Vail was killing a number of birds with one stone. This plan piqued much interest in Canada, and Bell launched its own plan in May 1920. Shares were offered to employees at a preferential rate through automatic deductions from their salary. In spite of the bearish postwar financial market – never had Bell's stock been so low – the plan immediately caught on. After two years, one-third of the employees had subscribed to the stock-purchase plan. During the 1920s, the Employee Stock Plan accounted for three-quarters of the growth in Bell's capital shares. By 1932, in spite of the depression, about 50 percent of Bell employees were shareholders.

This "people's capitalism" was the main factor in decreasing AT&T's share in Bell from 38.3 percent in 1920 to 25.1 percent ten years later. Three other factors also contributed, although to a smaller degree: first, in 1921 the postwar mini-depression pushed Bell to seek other sources of external financing to increase its capital, principally in Canada; then, Bell's purchase of independents and its acquisition of stock in companies in the Maritimes was financed by the issuing of shares; finally, the stock market crash of October 1929 provoked AT&T not only not to exercise its right to acquire shares from stock issued that very month but to sell its rights to purchase. (In fact, yielding to panic, AT&T sold a block of seventeen thousand shares before getting a hold of itself and repurchasing them.)

During the 1930s there was a trend towards stabilization, and Bell's need for capital grew at a slower pace. Following the unfortunate stock issue in October 1929, Bell stopped seeking financing on the stock market, and so AT&T could purchase no more of its subsidiary's shares. Its participation, which amounted to 749,992 shares before the crash, would remain unchanged in absolute numbers until 1975. On the other hand, Bell's capital continued to grow through the Employee Stock Plan. AT&T's participation thus slid slightly in proportion, from 25.1 percent in 1930 to 21.7 percent in 1945. Over the years, in short, everything conspired to dilute AT&T's interest.[1]

What's more, in May 1933 the American government adopted the Federal Securities Act, which obliged all foreign companies that wanted to issue

shares in the United States to disclose certain financial information. Bell refused to submit to this condition, which it judged too severe, and which confirmed its intention no longer to seek funds on financial markets, especially
in the United States.[2]

AT&T's shrinking share in Bell, however, was illusory: Bell had not
gained its independence at this point; its servitude was simply transferred
from the financial to the technological front, for the dilution of AT&T's participation in Bell was not matched by a dilution of Western Electric's participation in Northern Electric, which remained constant at 44 percent from
1914 to 1957. This was the main factor ensuring American domination over
the Canadian telephone industry.[3]

Technology transfers between the United States and Canada had been
fluctuating constantly since the 1892 agreement had become obsolete. Between 1914 and 1919 a series of agreements was signed stipulating that
Western Electric would now maintain ownership of its patents and that
Northern would obtain exclusive Canadian rights in exchange for 1 percent
of royalties on sales of telephone equipment.[4] In 1919 another agreement on
patents was signed between Western, Northern, and Bell Telephone, in
which Bell acquired non-exclusive rights to operation of Western Electric's
patents. The Canadian companies had lost the telephone patents obtained by
Sise in 1882; the Americans' technological hold on them was complete.

In May 1923, the service agreement between Bell and AT&T formalized
relations between the companies along the lines of those already existing between Northern Electric and Western Electric: Bell had to pay for access to
technological, administrative, and financial information from the parent
company. Over the years, royalties were set at 1 percent of Bell's gross revenues and gave Bell access to all patents filed in the United States by AT&T
(in comparison, the telephone companies that were members of the Bell System had to pay 1.5 percent of their revenues in 1929).

The 1923 service agreement did more: it gave Bell access to the expertise
it needed to adapt American patents to the Canadian context, with ample
technical assistance, upon request, with maintenance and repairs. As it also
included data exchanges on operating the telephone system and all the aspects of company management, it consisted of a total technology transfer involving the entire enterprise. Since AT&T was the most advanced telephone
company in the world, Bell had made an excellent deal.

Because of the threat of anti-trust suits and the need to refocus on AT&T's
telephone business, Western Electric had to cut loose its foreign subsidiary,
International Western Electric, which was purchased in September 1925 by
International Telephone and Telegraph, although Western Electric retained its
rights for Canada and Newfoundland. Suddenly, Bell and Northern lost their
exclusivity on the Canadian rights. Moreover, in April 1926 a modification
was added to the service agreement: in case of cancellation, the agreement

would revert to the status of November 1880. The grip was as tight as ever. But who at Bell would have considered ending such an advantageous agreement? Bell's internal newsletter, *The Blue Bell*, clearly expressed the company's position in 1925: "If our Company were now confronted with the necessity of getting these services in some other way it would have to greatly enlarge its own organization and do these things for itself to the best of its ability and to the extent that its resources would permit. There is no question but this would be much more expensive and the results far less satisfactory."[5] Not only was Bell happy with its position as a quasi-subsidiary, but it shuddered at the thought that it might someday have to become independent.

The depression did in fact put this tidy arrangement under reconsideration. The worst year of the depression was 1933. Money was scarce. Bell had to disconnect more than a hundred thousand telephones and sent a twelve-month notice of cancellation of the service agreement. Happily, the crisis subsided; as signs of economic recovery grew, Bell withdrew the notice. Things remained unchanged until 1949, when the service agreement renegotiated with somewhat softer terms.

The service agreement of July 1949 had three parts. The first concerned only relations between Bell and AT&T, which was renewed with an expanded base for royalties: Bell would now pay one percent not only on its own revenues, but also on those of its subsidiaries. The base would grow by a total of 3 percent to become aligned with that of associate companies of the Bell System. All references to the 1880 agreement were eliminated, even for cases in which the service agreement might be abrogated: "While the 1880 agreement was of great value to this Company for many years, its value today is much less and fact open to doubt and this Company requires the much broader rights and assurances that accrue to it under the present agreement."[6] A more prudent understatement could not have been made. But behind the usual convoluted and wooden bureaucratic language of telecommunications was a new demand; "much broader rights and assurances." While this affirmation did yet convey an awareness that Bell should be independent, it was the first inkling of the movement towards repatriation of Bell's strategic interests to Canada. Thus, it was an important event.

The rest of the agreement dealt with the relations between Northern Electric and Western Electric. The second part set royalties for information exchanges at 2.5 percent of Northern's business volume. This was a big hike from the 1919 agreement, and it established the level of payments at 1 percent for both information exchanges and royalties on patents.

The third part of the agreement dealt specifically with patents. Sales to Bell of equipment under Western's licence were exempted from royalties. This affected most of Northern's sales. On the other hand, sales to third parties such as other Canadian telephone companies would be submitted to royalties payments (0.5 to 10 percent). Rights continued to be accorded on a non-exclusive basis.

Finally, the scope of the parts of the agreement concerning Western and Northern was limited to communications equipment and excluded categories such as radio transmitters, sound-recording equipment, and imagery equipment – thus, radar. As we shall see below, the anti-trust division of the US Department of Justice began – also in 1949 – to examine the nature of the relationship between AT&T and Western Electric. Those companies were willing to sacrifice aspects of their relationship with Canada to protect the core of the Bell System in the United States – whence the voluntary restrictions imposed in the agreement.

The 1949 service agreement marked the beginning of a technological undertow. Any desire by the American group to control its northern neighbour was replaced by domestic concerns. The links between the Canadian and American groups continued to forge ahead thanks to the strength of tradition and, above all, Canadian interest.

THE 1956 CONSENT DECREE: CUTTING THE UMBILICAL CORD BETWEEN WESTERN ELECTRIC AND NORTHERN ELECTRIC

But in 1956 "the roof fell in on Northern," in the words of Vernon Oswald Marquez, a Northern Electric president.[7] An American initiative, the Consent Decree of January 1956, signed between AT&T and the United States Justice department, turned the entire Canadian telecommunications industry on its ear. Judicial proceedings began in 1949 investigating AT&T for anti-trust violations, and the Justice department intended to separate AT&T and Western Electric.

An out-of-court settlement, the Consent Decree, avoided the worst since it maintained the Bell System's vertical integration. However, it required Western Electric to limit its activities to manufacturing equipment for the Bell System's companies and the US government and AT&T and Western Electric to accord the same rights on patents to all who requested them. The rights for patents filed before 1956 would be free, and the rest would be subject to reasonable royalties.

The special relationship between Western Electric and Northern Electric disappeared in a puff of smoke, and Northern was reduced to the same status as other Canadian companies. Fortunately, the Consent Decree was not immediately enforceable; the 1949 service agreement would end in 1959. Only the two parts of the agreement concerning manufacturing relations between Northern Electric and Western Electric were cancelled on the spot, and so Bell was little affected by the shock waves.

To give the reader an idea of the disruption caused by the Consent Decree, Northern Electric was using almost 2,400 of Western's patents at the time.

About 1,800 of them had been filed before 1949 and were, from one day to the next, exempted from royalties. The rights on the patents filed after that date, on the other hand, were re-evaluated upward. A new patent agreement was negotiated to provide discounts on royalties for these patents. It had a five-year term, rather than the ten-year term of previous agreements, and was followed by other, increasingly limited contracts in 1964 and 1969. The last link was cut in 1972.[8]

The modest scale of the agreements signed after 1956 attested to the change in relations between Northern Electric and Western Electric. The river of information that had flowed from south to north since Bell's manufacturing division had been created in 1882 was slowed to a trickle. Marquez summarized the situation best:

We are the child of wealthy parents ... This was the boot that kicked us out the door. I've often wondered whether Northern Electric or any other Canadian company would have the courage to take this kind of risk voluntarily and I have some doubts: a Canadian subsidiary of a US company lives a very comfortable life. The technological risk is being taken by the parent. This is the tragedy of Canada: we've made a virtue of imitation. You don't realize how easy it is when you're paying for technology after you use it, instead of waiting five years or more to to get your money back. But I'm confident history will say that this is the best thing that ever happened to us.[9]

On the corporate and financial fronts, the repercussions from the Consent Decree were felt more rapidly. Western Electric asked Bell to repurchase its investment in Northern Electric. In 1957 Bell proceeded with a first purchase of Western stock for almost twenty million dollars, bringing its share in Northern to 89.97 percent, while Western's dropped to 10.02 percent. In 1962 Bell paid another eight million dollars for the rest of Western's stock, which brought its share to 99.99 percent. The remainder belonged to members of Bell's and Northern Electric's boards of directors. Bell bought this balance in 1964, making Northern Electric a wholly owned subsidiary. Thus ended the fight begun by Charles Sise near the end of his life against Western Electric's grip on Northern Electric, which was now repatriated into Bell's hands.

WHO RAN BELL?

The ties between Bell and AT&T had not been severed by the 1956 Consent Decree; the service agreement between the two companies lasted until 1975, although, as we have seen, the financial link had begun to weaken in 1930, when Bell withdrew from the financial markets. Lewis McFarlane had yielded the presidency of Bell to Charles Sise, Jr, five years before, in 1925. Not a single statement, official text, or letter – nothing either man said or wrote – betrayed the slightest desire to be emancipated from the grip of the American company.

When McFarlane became president in 1915, he was sixty-four years old and represented continuity. As manager of the Toronto office of Dominion Telegraph, he was the one who, in August 1876, had allowed Graham Bell to use the company's line to make the first long-distance call, and he had set up the telephone division of Dominion Telegraph two years later. At the beginning of the 1880s he had brought telephone service to the Maritimes, with varying degrees of success. Although he had spent his entire career at Bell under Sise's reign, there was one difference between them: McFarlane had not begun his career as a delegate from the parent company to Canada. On top of that, he had benefited from the battles Sise had fought to wrest power from the Bell System.

McFarlane was a colourful character, with a human warmth that Sise, Sr, lacked. In terms of policy, his presidency marked the break with the unbridled capitalism of the previous era. All of the major social-policy programs were instigated during his reign, as was Bell's cooperative attitude towards regulatory authorities. He injected into the Canadian company the beginnings of a sense of public service spirit as Theodore Vail had defined it in the United States, but this policy would be hindered by his old-fashioned, if craftsmanlike, management style.

Sise, Jr, meticulously continued McFarlane's policy direction and also introduced "scientific management" into the company. Under him, Bell took on the character of a public utility that it retains today. He was an *apparatchik* as grey as the walls of the building at 1050 Beaver Hall Hill that he had had built for Bell's head office in Montreal. In fact, his desire for anonymity was so great that he refused to let his name be engraved on the commemorative cornerstone that was laid in May 1928. Sise, Jr, ran Bell from 1925 to 1945, infusing the management of the company with his sober, studiedly conservative style – one as invisible as the empire he directed. His successor, Frederick Johnson, a similarly colourless British-born accountant, ran the company until 1953.

Like all senior managers at Bell, Johnson was taken by surprise by the postwar economic boom, which created unprecedented demand for telephone service. Because of this lack of planning, the number of requests unfulfilled by Bell reached ninety-four thousand at the end of 1947, even though most of the lines installed at the time were party lines. Northern had reconverted its wartime production in record time, and Bell was on a hiring blitz: 90 percent of its 2,500 newly demobilized employees were rehired, including war wounded. Some ten thousand were hired in two years, almost doubling personnel to a total of twenty-three thousand. This new blood had to be integrated into the company.

Johnson, whose career had been in the accounting department, was not a man of vision, but he had solid financial expertise. Bell's assets doubled during his presidency, proving that Canadian business circles had definitively adopted the telephone. Under his rule, Bell began to issue shares – although not in the United States, because of its long-standing opposition to the 1933 Federal Securities Act. Thus, AT&T could not purchase Bell stock, and its share in the company began to tumble, dropping below 10 percent in 1950, then below 5 percent in 1956.

It was also under Johnson's presidency that service agreements between Bell and other Canadian telephone companies, modelled after the agreements between Bell and AT&T, began to be used. In 1949 MT&T, NB Tel, MTS, SGT, and AGT each signed an agreement giving them access to all of Bell's network-operation practices and its consultancy services for technology, management, and training (six hundred courses and seminars). Absolutely all aspects of running a telephone company were covered; an average service agreement was estimated to cover 250,000 pages of information. Avalon Telephone signed a similar agreement in 1957. Of the member companies of the TCTS, only BC Tel did not sign an agreement with Bell, because of its corporate links with GTE. Overall, Johnson's policy of service agreements was at least as important as the TCTS in unifying the Canadian telephone industry around common standards.[10]

Johnson's successor was Thomas Wardrope Eadie, architect of the unification of Canadian telecommunications. An engineer who was fascinated by radio waves, he quickly grasped the political significance of the equation "telecommunications network equals nation" and obtained from the minister of Trade and Commerce, Clarence Decatur Howe, the contract for the Mid-Canada line. Above all, Eadie made the decisions that turned the telephone into a mass medium: he used microwaves to rebuild the trans-Canada telephone network and replaced step-by-step switches with crossbar switches.

Eadie's strongest suit was politics. He had occasion to build on Johnson's policy of service agreements, as he assumed the presidency of the TCTS at the same time as he did Bell's, thus becoming the only person to hold both positions at once. Building a microwave network from sea to sea was a good thing, but not a goal in itself; rather, it was an opportunity to fill the empty shell of the TCTS. Eadie would put his talents to the task of persuading telephone companies to cooperate, harmonize, and even integrate some of their functions. The incentive he used was money: he advanced funds to Prairie companies that had to assume costs proportionally greater than Bell's, earning himself the nickname "the microwave missionary."

Eadie headed Bell from 1953 to 1963, and it was with him that Vail's ideology reached its apogee. He continued in the footsteps of his predecessors, but he was the first president of Bell since Charles Fleetford Sise to have a corporate vision. Unlike the unbridled capitalism that Sise subscribed to, Eadie's vision fit with the notion of public service. It was no coincidence that the telephone finally attained universality under his rule. With a phone in every home, the objective defined by Vail at the beginning of the century was finally reached.

During the Eadie years, Bell's oft-repeated goal was to equal the service quality of the best member companies of the Bell System, a goal that was reached thanks to purely technological advances. Because it was protected from market forces by its monopoly position, the company's expansion was driven by technology. One sign of this priority was the creation of the long-distance department, with which Bell began to look beyond purely geographical divisions to technological ones.

Eadie's blind spot was the necessity to repatriate research and development to Canada. In the 1950s, the Consent Decree was sending an explicit signal to Canadian management: technology transfers from AT&T and Bell Laboratories were off limits in the short term. But the technocratic Eadie did not see this as an opportunity to launch an autonomous Canadian policy. A small, brilliant team had been conducting research on semiconductors since 1952 at Northern's labs in Montreal and Belleville, Ontario. But Bell was not supporting its subsidiary's efforts – on the contrary: Eadie wanted to avoid any efforts that might duplicate the technologies included in the transfers in the AT&T-Bell service agreement, so that Bell didn't pay twice for the same thing. This was good management, but major policy directions could not be based solely on accounting rules.

Eadie belonged, for better or for worse, to the world of the AT&T branch plant, as revealed in an interview he gave on relations between Bell and AT&T just after he retired: "As far as I'm concerned I am disappointed that their interest has become so small ... but they would never buy stock at above par ... I think that's too bad, it's important that Canada get closely interrelated to the Bell System. I should say that the senior officers of the AT&T share that view and feel that service in Canada is of tremendous interest to the members of the Bell System."[11] It must be said in Eadie's defence that AT&T was the best school in the world for management and research and development, and that it was difficult not to admire the company. As well, Eadie had to face a demand for construction unprecedented in Bell's history. With the building of the trans-Canada microwave network and the Mid-Canada Line venture, his presidency represented a great technological leap forward.

NORTHERN ELECTRIC: RESEARCH AND DEVELOPMENT TAKES OFF

The harbinger of change came from elsewhere. Even before the Consent Decree was signed, Northern Electric's president, Colonel R. Dickson Harkness, was uneasy. During the long presidency of Paul F. Sise (1919–48), the company had been under the direct guardianship of Western Electric. Most of Northern's managers were American, and those who weren't had been trained in the Western Electric mould. If Bell was a subsidiary loosely managed by a distant parent company, Northern was more like the branch plant of a centralized company. As soon as the first threats that the Bell System might be dismantled appeared on the horizon in 1949, Harkness called for a study on the future of Northern Electric.

As luck would have it, the authors of the study were two extraordinary men: Chalmers Jack Mackenzie and Mervin J. Kelly. Mackenzie was president of the National Research Council and a great patron of Canadian research during the wartime and postwar periods; Kelly had just retired from the presidency of the Bell Labs in the United States, having launched the research program on electronic switching – resulting in the distant forebear of the transistor – in the 1930s.

Mackenzie and Kelly toured Northern's installations and took a look at the embryonic research and development under way by A. Brewer Hunt's team in Belleville and Cyril A. Peachey's team in Montreal. Mackenzie signed the report, which contained the seed of the Bell group's autonomy policy. It enthusiastically recommended the creation of an independent research unit in Montreal: "A complex technical industry cannot compete forever on technological blood transfusions or without all of its vital organs … Northern can very quickly bring into being a research and development organization that will be able to attend not only to all the Company's own needs but also become a preferred contractor for Government and other bodies who wish to place development contracts."[12]

The report was portentous in more than one sense: it presaged both the convergence of communications and electronics and the diversification and growing competition in telecommunications. The careful allusion to government contracts should have touched a sensitive chord among Bell top management. But Eadie once again revealed his blind spot by refusing to create a telecommunications-research laboratory. In his own words, "It was just beyond our financial possibilities but also not feasible to recruit the scientists to man such an organization."[13]

But one man at Bell read the report and seized the opportunity on the fly: Alex Lester, Eadie's right-hand man. The day after the Consent Decree was promulgated, he realized that Mackenzie and Kelly's report was pointing in the only possible direction Bell Canada could take. Lester would be to Bell what Hunt and Peachey were to Northern: an indefatigable promoter of the Canadianization of research and development. He didn't hesitate, as we shall see, to outbid Northern to accelerate development of new products.

After 1956, there were no more financial or technological obstacles to the entire Canadian group going its own way under the best possible conditions. Everything was in place for the final act, which was effected in 1975, with the blessing of an obliging parent company and, especially, with the inestimable acquisition of technological excellence developed by AT&T, Western Electric, and their joint subsidiary, the renowned Bell

Laboratories. The private sector thus provided a successful example of "clean" decolonization.

THE MILITARY-TELEPHONE COMPLEX

During the period of postwar effervescence, two men were proponents of co-operation with the government: Thomas Eadie and Alex Lester. Their national vision of telecommunications went hand in hand with proximity to political power. Bell was integrated into the Trans-Canada Telephone System and became an instrument of the government with regard to telecommunications policy.

As the cold war took hold, working for the government often meant working for the military. During the Second World War, the major military telecommunications projects had been conducted by telephone companies outside of the associative structure. Many mobilized Bell engineers and technicians mobilized found themselves in the Royal Canadian Signal Corps, leading to a certain cultural rapprochement, which was concretized after the war in a little-known and unusual institution: the National Defence College in Kingston.

In 1947 Canada wanted to built a close relationship between the army and civil administrators. At the National Defence College, a program of courses was put together for future army and public-utility managers, who were freed from their usual responsibilities for nine months on full salary to attend (the period has since risen to eleven months).

Only the most intelligent were admitted, a total of twenty to thirty people per session. What they learned was the exercise of power and, especially, self-knowledge. After 1949, once the format was established, one or two managers from private industry were admitted per session. The first company to send a representative was Bell Canada. He was a bright young member of the élite – Alex G. Lester. Lester was subsequently involved in every aspect of cooperation between Bell and the military. Afterward, only a handful of telephone men took the college course, including two Bell presidents, Raymond Cyr and Jean Monty.[14]

The goal of the program was to train a political-military-industrial élite. People from all over the world and of all political stripes were invited to make speeches involving highly confidential subject matter. Note taking was forbidden. Then the students were split into small groups and asked to solve real-life problems, such as formulating the Canadian federal budget or preparing for a third world war. Lester recalled that his task was to direct transportation in Europe should a third world war be declared.[15]

Role playing was also involved: one student, for instance, was asked to "become" a communist and learn the subtleties of Marxist dialectics. The high point of the session was a trip abroad: Lester went to visit the directors

Bell Canada's chief engineer, Alex Lester, played a key role in creating the North American anti-ballistic defence systems and, later, in the Canadian telecommunications R&D effort. Courtesy Bell Canada.

of the coal and steel industries in the European Community, which was being formed at the time; then he went on to the Krupp plant in Essen and the Farben plant in Frankfurt, among others. The course Raymond Cyr took at the beginning of the 1970s was oriented more towards the Third World and racial problems in the United States. At the time that it closed in the 1990s, the National Defence College was focusing its studies on the transition to capitalism of the old communist bloc and the rise to prominence of Islamic movements.[16]

Lester's session at the National Defence College involved what might have become a military-telephonic complex. It is clear that telecommunications companies saw in the efforts of the federal government to create a military-industrial complex an opportunity that they tried to exploit to their own ends.

In August 1949 the USSR exploded its first atomic bomb; the Korean War broke out in 1950. In the West, nerves were on edge. Canada's Department of Defence Production decided to create a purchasing department for its electronic equipment; Bell supplied the human resources. In August 1953

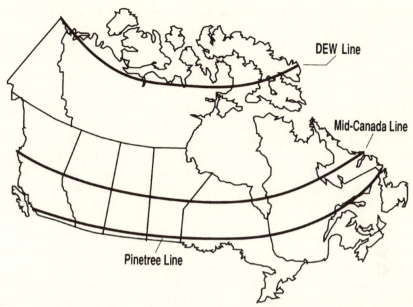

During the Cold War, three lines of electronic defence were set up across Canada.

the USSR exploded a hydrogen bomb, and panic spread among the members of the Atlantic Alliance. In the same year, the United States broke ground on a North American air-defence system. The heart of the system was in Colorado Springs, in the bunkers of what, in 1957, became the North American Air Defense Agreement (NORAD). It was linked to four radar stations intended to detect any enemy bombers.

The subcontinental defence lines were arranged in roughly concentric arcs. This electronic Maginot line corresponded to the stages of a hypothetical Soviet air attack over the North Pole. First was the Ballistic Missile Early Warning System Line, composed of three massive stations bristling with radar antennas each the size of a football field, situated in Alaska, Greenland, and Great Britain. The three other lines were on Canadian soil. North of the Arctic Circle, on the seventieth parallel, the Distant Early Warning Line started in Alaska and crossed the Yukon and the Northwest Territories. The Mid-Canada Line ran along the fifty-fifth parallel. The DEW Line was supposed to give two hours' warning to the American air force; the Mid-Canada Line, one hour. Finally, the Pinetree Line ran along the Canadian side of the Canada–United States border, except at the ends, where it veered north to follow the Atlantic and Pacific coasts.

The TCTS played a secondary role in construction of the BMEWS stations and the DEW Line, but it was much more directly involved in construction of the Pinetree Line, which crossed its territory, and it was the master contractor for the Mid-Canada Line.

The first system to be completed was the line of forty-eight Pinetree radar stations, a joint venture between Canada and the United States. (In fact, the work on Pinetree began in 1950, about three years before the US began planning the global defence system.) The Canadian Department of Defence assigned the telephone companies the task of linking up the radar sites by means of a microwave network called ADCOM.

With the work in midstream, Bell sent Alex Lester to the Department of Defence Production. Lester was a self-made engineer. He had quit high school in Montreal to hire on at Bell and had learned on the job, taking night courses to acquire the theoretical underpinnings he lacked. His mentor was Bell's engineer-in-chief at the time, none other than Robert Macaulay, the builder of the TCTS.[16]

When Lester arrived at the Department of Defence Production, he was horrified by the waste that was involved with building the ADCOM network and tried to get upper-level bureaucrats to stop it. In vain. Before his warning rose through the hierarchy, the minister, C.D. Howe, had already authorized the expenditure of $22 million on a project initially budgeted at $3 million. ADCOM was finished by the army; when, in 1965, strategic developments rendered it obsolete, it was sold to Bell for less than four million dollars.

In the other provinces, the Royal Canadian Air Force was content to rent lines from telephone and telegraph companies. Wherever possible, the Pinetree Line used the existing network, which nevertheless had to be improved or augmented. Bell acted as a subcontractor for the TCTS member companies from August to October 1953, sending some one hundred employees to sites sprinkled across the country. For decades, the Pinetree network was operated by a committee of engineers from CN, CP, and Bell, to the entire satisfaction of the military. It was dismantled only in 1989, when the USSR threat disappeared.[17]

The Atlantic section of the Pinetree Line was the most technically challenging. Dubbed Pole Vault, it was to extend almost 2,500 kilometres. But conventional microwaves had to be repropagated every forty kilometres, and how would the microwave towers be maintained in the vast stretches of the Far North? They would have to have fuel, and the maintenance teams would have to travel by helicopter or boat, creating inevitable delays if something broke down.

Howe turned to Bell, which had just set up its Special Contracts Division to meet the needs of the Department of Defence, for the solution. Lester, the division's director, proposed to link the stations with a completely new system that sent signals to the troposphere, where they were bounced back to earth 240 kilometres away. When they returned to earth, however, the signals would have lost most of their strength, which posed some practical problems.

The tropospheric system in the Far North: the "cannon" in the centre of the antenna emitted a microwave signal that was sent into the troposphere. Courtesy Bell Canada.

The Pole Vault system thus required major research and development, which was undertaken by Northern Electric's research group in Belleville. A young researcher, Wally C. Benger, was the author of a number of fundamental patents on very-low-noise parametric amplification, with which the weak tropospheric signals could be detected. (Benger then played a central role in setting up Northern Electric's research laboratory in Ottawa and in development of the SP-1 electronic switch.) Thanks to Pole Vault, microwave stations could be built precisely on the Pinetree Line radar sites.[18]

The Pole Vault contract was signed in January 1954; in February 1955 the tropospheric network began operating. The southern section comprised thirty-six telephone circuits; the northern section, eighteen. It was the first network of its type in the world, and it came into being in the hostile terrain of the Far North, in a feat of almost superhuman technological prowess. The 120 members of the Bell team had worked sixty hours a week with virtually no absenteeism. According to Lester, it was a "bold and joyous adventure."[19]

The American air force, which was directly financing the project, was so happy with Pole Vault that it asked Bell to extend the Pinetree Line to Greenland, Iceland, and the Azores. Bell agreed to work in Greenland but felt that it wasn't well enough equipped for more exotic locales. It constructed three bases in Greenland, the first time the company had worked outside Canada.

In contrast to the Pinetree Line, the DEW Line was a completely American project built on Canadian soil. The TCTS's contribution was limited to sending eight telecommunications experts to supervise construction of a detection station on the shores of the Arctic Ocean. The radar and communications systems were installed in 1954 and 1955. Unlike the other two defence lines, the DEW Line has never been dismantled. From 1989 to 1994 it was upgraded and is now known as the North Warning System.[20]

The Mid-Canada Line was based not on ordinary radar, like the Pinetree and DEW Lines, but on a detection system based on the Doppler effect. This new technology was developed by the Defence Research Board and the Royal Canadian Air Force, with the help of McGill University, after which it was named the McGill Fence. It was based on the change of frequency that occurs in a sound or electrical wave due to movement of the source of the wave in relation to the observer. Most people would recognize the Doppler effect as, for instance, the change in pitch of a siren as an ambulance approaches and then passes by.

The Mid-Canada Line soon went awry. The minister, Howe, had pushed aside the government agencies in charge of the project in his determination to dismantle the war economy and encourage private enterprise. Federal bureaucrats fought back. Work began in October 1956 and a month later a real bureaucratic "putsch" took place. A Crown corporation called Defence Construction Limited was created and given all responsibility for supplies. The result was that Bell managed the project, but with no control over the quality of the work, let alone the costs. In fact, the contract for executing the work wasn't signed until August 1957 – after the work was completed.

The government's decision to have open calls for tender led to some strange situations: for example, the contract for diesel generators went to a British manufacturer, which bid the lowest – but the price did not include transport. A fleet of Superconstellations, one of the largest planes available, had to be chartered to pick up the generators in Great Britain, bring them to southern Canada for storage, and then take them north to the Doppler stations in the spring. It was impossible to stem the invasion of the bureaucracy. All the decisions of the Mid-Canada Line management committee were adopted by a vote of thirteen to one: on one side, representatives from government, Defence Construction Ltd, and the RCAF; on the other side, Bell's representative, Lester.

The initial costs had been estimated by the army's engineers at $100 million, but this figure was baseless since no one knew the technical specifications for the Mid-Canada Line at the time. Lester revised the figure upward to $161 million, then to $169 million. The final figure was $228 million, in which transport alone accounted for $42 million. Until the end, the RCAF continued to expand the scope of construction.

And that was not all. There was the deadline of 1 January 1957 to which the military was so attached. To fulfil the TCTS's commitment against all odds, in the spring of 1956 Lester went so far as to charter a fleet of 150 ships of all shapes and sizes (including Liberty Ships dispatched from England), which, full of workers and equipment, sat in formation waiting for the ice to break up so that they could rush into Hudson Bay. But it was too little, too late. Two people drowned in the operation, bringing to five the number who died in the building of the Mid-Canada Line (two died in a helicopter accident and one committed suicide). The delivery date was not respected. Only the western part of the line was completed in time, with the other parts straggling in six months later. This work, which was to symbolize Canada's contribution to continental defence, became instead a symbol of waste and administrative negligence.

The Mid-Canada Line was above all the fruit of the labour of teams five thousand strong. Taking account of employee rollover during the two years that the work took, a total of about ten thousand people were involved. With activities in full swing, the Special Contracts Division comprised twelve hundred employees, with the rest of the Mid-Canada team made up of subcontractors.

To have an idea of the effort this represented, it should be recalled that the Pole Vault project was not yet completed and construction on the trans-Canadian microwave network was beginning. This meant that there were three huge work sites under way at the same time. Every Bell engineer and technician worked over time on one or the other of these projects in 1955 and 1956. The blitzkrieg approach had become the norm for Canadian telecommunications, prefiguring Bell's huge contracts in Saudi Arabia in the 1970s and 1980s.

In spite of all its structural handicaps, the Mid-Canada Line was completed, with a total of ninety Doppler stations and eight main bases spread over 4,500 kilometres. Its effectiveness was never put to the test, though it was on alert during the 1961 Cuban missile crisis. In 1964, when intercontinental missile replaced nuclear bombers, the Doppler towers were demolished so that they would not pose a hazard to aviation, and the buildings were padlocked and abandoned.

The Mid-Canada Line represented the apogee of cooperation between the telephone industry and the military. The subsequent breakdown in relations was due principally to cultural incompatibility: to execute a military contract, certain rules had to be followed and political games played. A degree of flexibility was required that Bell's and Northern Electric's managers didn't seem to have. At any rate, there were no more contracts of the scope of Pole Vault and Mid-Canada, for the Canadian government took a pacifist turn in the 1960s.

13 The Other Telephone Companies

For the three Maritime provinces and Newfoundland, still a British colony, the First World War was a period of rapid expansion, as this region was the springboard for the Canadian war effort.

For Maritime Telegraph and Telephone, the First World War was not a distant abstraction. On 6 December 1917, death struck at the heart of Halifax when a Norwegian freighter rammed into a French warship loaded with dynamite in the middle of the port: "At 9:06, the *Mont-Blanc* blew up. The harbour churned. Piers, ships, buildings vanished. The blast uprooted trees, swept away bridges, tossed railway tracks and freight cars. In seconds, a square mile of Halifax was obliterated. Then the air forced out by the blast rushed back to fill the vacuum, sucking lethal debris with it. Hundreds of fires flashed up from scattered coals and live wires. A tidal wave surged from the harbour. A mushroom cloud of smoke and gases towered three miles in the sky."[1] It was devastating: more than two thousand perished and ten thousand were injured. Entire sections of Halifax were razed. For MT&T, it was the greatest catastrophe in its history. Two operators were killed at home, and many were injured. As for equipment, 582 telephones were cut off, the seaside telephone line was destroyed, four exchanges were damaged, and, of course, everyone was trying to make a call at the same time. MT&T had to put ads in the newspapers the next day begging the population to refrain from making any but emergency calls.

Thanks to a particularly effective mobilization of MT&T's staff, the undestroyed part of the system was back in operation the very afternoon of the disaster. Employees, many of whom had lost their homes, were asked to stay in half-demolished exchanges around the clock. Other phone companies lent a hand: New Brunswick Telephone, Bell, Northern Electric, and even New England Telephone sent volunteers to help with repairs. In the Lorne exchange, the damage was so great that the switchboard was unusable. MT&T took this opportunity to install an automatic exchange, becoming the first Canadian company, except for those on the Prairies and a handful of independents, to do so. (NB Tel didn't have its first automatic switch until 1929.)

Even before the *Mont-Blanc* blew up, MT&T was in dire financial straits. Money was tight, interest rates high, and the debt load heavy. Since it was out of the question to raise rates in wartime, MT&T management considered selling its majority share in the unprofitable Prince Edward Island company. The island network was not very well developed and farmers had begun to create their own telephone companies. There were about forty rural companies in PEI at the end of the First World War, limiting expansion of the public system.

Howard P. Robinson, the very dynamic general manager of NB Tel, in association with a group of businessmen from Saint John and Halifax, had just created a federally chartered company, Eastern Telephone and Telegraph, with the goal of merging the telephone companies in the three Maritime provinces, and perhaps even adding Newfoundland. At the end of 1917, Robinson offered to purchase MT&T's entire interest in the Prince Edward Island telephone system. Thrilled with its good luck, MT&T's board of directors hastened to close the deal. But they came to regret the decision: the sale raised a storm of protest among the shareholders, who objected to the "foreign" take-over so strongly that Robinson had to cancel the sale in March 1918.

But now, Eastern, an empty shell, represented a potential threat for MT&T because of its federal charter, which allowed it to operate a telephone system in any territory east of Manitoba. MT&T acquired Eastern in October 1920 for ten thousand dollars and soon afterward began to operate the Prince Edward Island system under Eastern's charter.[2]

In 1929 AT&T was developing its plans for a transatlantic cable, which it considered landing in Newfoundland. But the cable would then have to cross Canadian territory, and since under American law the company could not engage in international activity, it offered to purchase Eastern from MT&T, without the Prince Edward Island system, for $50,000. MT&T refused to sell, offering in its turn to route communications to the American border through its own network. The situation was at an impasse when Charles Sise, Jr, got wind of the affair. He called the president of MT&T to tell him that his attitude might cause AT&T to choose a landing point in the United States rather than Canada. MT&T quickly reversed its position and sold Eastern's charter to AT&T for $75,000. The Prince Edward Island telephone network was incorporated as Island Telephone in April 1929, a name it has kept to this day.[3] AT&T worked on construction of the line via New Brunswick and Nova Scotia until the cable project was interrupted first by the depression and then by the Second World War.

Social benefits came to the Maritimes phone companies starting in the mid-1910s. It was Sise, Jr, Bell's representative on MT&T's board of directors, who had that company adopt its first retirement plan in July 1917 and its first stock-savings plan ten years later. NB Tel adopted its stock-savings plan in 1916 and its retirement plan in 1926. These initiatives generally came from Bell representatives on these companies' boards of directors. Bell had never completely abandoned the Maritimes, and its influence there, as in the West, always brought progress with regard to labour relations and technological improvements to service quality, as well as homogeneity to Canadian telecommunications.

The economic crisis in 1929 presented Bell with its first opportunity to return in force to the Maritimes. Over the years, Bell had let its share in NB Tel dip. In 1931 acquisition rumours coming from the United States provoked Bell to reverse this trend, and its share suddenly rose from 15.8 percent to 54.9 percent. It kept this majority shareholder position until 1947, then let it slide again.

Bell did not, however, act in this regard towards MT&T, even though the company was suffering from a lack of liquidity. Bell did make investments in 1929 and 1931, bringing its participation to 9.8 percent. However, this financial support proved insufficient.[4]

THE SECOND WORLD WAR
BRINGS NEWFOUNDLAND CLOSER TO CANADA

Throughout this period, the Newfoundland network, under British jurisdiction, remained marginal to developments in Canada. After the First World War, there had been two companies in Newfoundland: Anglo-American Telegraph, which operated the telegraph system and, secondarily, the telephone system in St John's; and United Towns Electric, which operated an electrical utility and, secondarily, a telephone system on the rest of the island. In 1919 Anglo-American was purchased by Western Union, for which telephone service was not simply a secondary activity but an impediment.

J.J. Murphy, president of United Towns Electric, purchased the unwanted telephone facilities. In 1919 he grouped together all the exchanges in his possession under the name Avalon Telephone. This was a small enterprise whose operating territory covered, as its name indicates, only the Avalon Peninsula, not the rest of the colony. The first long-distance line on the island was built only in 1921. Newfoundland was linked to Canada in 1939 by radiotelephone. A few months after the official inauguration of this connection by the governors general of Canada and Newfoundland, the Second World War broke out and Great Britain interrupted the service; all radiotelephone circuits between Canada, Newfoundland, and Great Britain were requisitioned for the war effort.

A rudimentary service was re-established in July 1940; it was submitted to military censorship, and the connection was cut at the first questionable word. The war nevertheless had a positive consequence for Newfoundland: in exchange for fifty old destroyers, Great Britain authorized the United States to build military bases there. When the American army saw the primitive state of Newfoundland's telephone network, it raised a hue and cry: how could they conduct a war without communications worthy of the name?

The American bases had to be linked up immediately, and they had to be connected to St John's – a total of eight hundred kilometres of line over rocky, storm-swept soil that was frozen throughout the interminable Newfoundland winter. In short, the same obstacles existed now that had slowed construction of the land-based part of the transatlantic telegraph cable in the nineteenth century. Only Bell was equal to to the task. In April 1942 it sent 330 men on site and recruited another 700 in Newfoundland:

The Newfoundland project dawned as one of considerable glamour but it soon became evident that it would be a challenging task ... There were no ships to transport pooles, so we had to find our own on the island ... Equipment on the railway was in short supply, as were most things at that time and there was no good highway across the island ... The train was always overcrowded ... In many places the train was buffeted by high winds and at strategic spots anchors were buried in the right-of-way so that chains could be attached to the tops of coaches to keep the train upright until the winds abated. When heavy snow accompanied the wind the train was sometimes buried solidly.[5]

The network was ready in one year – record time – and the engineer responsible for the operation was none other than Thomas Eadie. This achievement no doubt was a factor in to his being named to the presidency of Bell ten years later.[6]

A few years before the war, in 1937, the Newfoundland Post Office had started to build and operate a telephone system. The colony's public agency already operated telegraph cables along the railway line and had been mandated by the Royal Air Force to build a line between Gander and a neighbouring hydroplane base. As soon as the Canadian army saw the work undertaken by Bell for the US army, it wanted its own line, so it installed a telephone line crossing the entire island on the Newfoundland Post Office poles.[7]

Even more than the First World War, the Second World War transformed the Maritimes and Newfoundland on the economic and military fronts. The result was an unprecedented expansion of the telephone system, while everywhere else in Canada construction was at a standstill. MT&T and NB Tel proved incapable of satisfying the military demand on their own. Everywhere, Tom Eadie sent construction teams to the rescue.

MICROWAVE PIONEERS

It wasn't until after the war that microwaves were used in telecommunications, and the Maritimes achieved a world first by linking Prince Edward Island to the continent. Until then, PEI had been connected by three underwater cables belonging to the federal government (two to New Brunswick, one to Nova Scotia). The spectacular growth of telephone traffic had rendered these connections obsolete; moreover, ice and strong currents in the Northumberland Strait often broke the cables.

MT&T took the initiative in adopting the brand-new microwave technology. The system chosen had been developed and manufactured in Federal Electric's laboratories in New Jersey, by a Frenchman, E.E. Lavin, who used technology then called pulse modulation. Such a commercial application of microwaves was so unusual that Federal Electric asked for help from its parent company, ITT, when it came to installing the system.

In November 1948 a microwave line between New Glasgow, Nova Scotia, and Charlottetown, Prince Edward Island, was inaugurated with great pomp. This was the beginning of reliable communications between the small island province and the mainland; in fact, only fifteen of the twenty-three high-capacity circuits of the system were needed at first (thirteen for telephone and two for radio). The federal government soon abandoned its now useless underwater cables, which left only one link – an unthinkable situation in telecommunications, where redundancy was always built in. In 1952 a second microwave link was built, this time to New Brunswick.

This double success opened the door for development of microwave technology across Canada. In May 1953 a second microwave system came on line in the Maritimes: a Halifax–Saint John link replaced both the radio system over the Bay of Fundy implemented during the war and the land-based line. The new system had eighteen circuits and it also straddled the Bay of Fundy. But it was the advent of television that gave microwave the decisive edge. In March 1954 the first television station in the Maritimes opened in Saint John. NB Tel and MT&T enthusiastically threw themselves into the microwave venture, and in December 1956 an integrated Maritimes-wide system began operating.

In December 1953, AT&T announced in Canada, the United States, and Great Britain that it was resuming its transatlantic cable project. To cross New Brunswick and Nova Scotia, AT&T would use Eastern's federal charter, which it still held, but not the old telephone line from the 1930s, which would have had to be doubled by a microwave system. Then there was the Atlantic Ocean. It took a total of three years before the gigantic AT&T realized a dream as old as telephony: linking North America and Europe by telephone cable. The line was inaugurated on 25 September 1956.[8]

MANITOBA: THE END OF A DREAM

The three telephone companies in the Prairies were special cases in the history of the industry in North America, since they belonged to provincial governments. Without any tradition in management of publicly owned companies, the governments constantly had to improvise and correct the inevitable errors they had made at the beginning. Between 1915 and 1956, however, administrative order was restored and many of Bell's accounting practices and techniques were adopted.

In May 1915 a new Manitoba government was voted in. Conservative premier Rodmond Palen Roblin, the author of Bell's nationalization, had been forced to resign after a real estate scandal, and the first move of the new Liberal government was to order an assessment of the network. Unlike a similar process taking place at the same time in Nova Scotia, the assessment of Manitoba Government Telephones was intended not to determine a basis for regulation, but to criticize the old government. It was knocked off in just one month.

Among the least surprising revelations was the finding that Bell had been overpaid by $802,336 when the system was nationalized, confiming the criticisms levelled by Francis Dagger at the time. It is impossible to know how much credence this figure should be given. Since Bell had been forced to sell, equitable compensation could be based on the present value of future profits. In any case, MGT would continue to treat the contested sum as "intangible capital" in its books.[9]

The change in government coincided with the First World War and a series of poor harvests. In 1913–14, moreover, MGT had begun to amass reserves to compensate for depreciation of its equipment, which diminished its investment capacity. The intensive connection program came to an end and the dream of universality vanished as quickly as it had come; economizing was the order of the day. The war years were lean, but the company managed to balance its books. In 1915 there was even a small profit from operation of the telephone service. Manitoba was thus the first Prairie government to get its telephone company back on its feet.[10]

On a technological level, MGT threw itself into automating its exchanges, using the town of Brandon as a test site in 1917. In 1919 Winnipeg was the first city in Manitoba to start automating. In this, MGT was following the movement begun in Saskatchewan and Alberta, but it preceded Bell by a number of years.

The postwar period was particularly difficult for Manitoba. First, the Winnipeg general strike paralyzed the telephone system for six weeks. This was a solidarity strike, since the salary demands had been settled before the work stoppage took place. Then, two very severe storms destroyed part of the system in early 1921. Finally, the inflationary spiral throughout Canada forced the company to apply for a 28 percent rate increase. The increase was approved by the Manitoba Public Utilities Commission, but it penalized the people.

The mini-recession of 1921 hit the Manitoba monoculture economy hard, and farmers cancelled their telephone subscriptions in large numbers. Three years in a row, the number of subscribers fell, then it stayed stagnant for two more years before beginning to rise again in 1926. The decrease in subscription levels was due exclusively to the farmers, since the number of telephones in Winnipeg continued to rise slowly (see Table 10). The provincial government released major budgetary credits to help MGT get through the difficult years. It should be noted that the prolonged drop in subscriptions was unique to Manitoba (although New Brunswick experienced a slight drop in 1921 and Saskatchewan in 1922). Everywhere else, the postwar mini-recession resulted only in slower growth, and the recovery was much more rapid.

Still, 1921 was a turning point for the telephone industry in Manitoba. In February of that year, the Manitoba government asked John E. Lowry to take over the reins at MGT. Lowry remained at the head of the telephone company until 1945 and left his mark not only on Manitoba but on all of Canada – as we have seen, he was one of the instigators of the TCTS.

Lowry, an Irishman, had started his career as an engineer in Great Britain's largest telephone company, the National Telephone Commission, and had come to Canada in 1908 to supervise automation of Edmonton's municipal system. After a Strowger switchboard was installed in 1915, he was named general manager of the city's telephone department. Lowry was both a man of action and a natural leader. One of his colleagues later said about him, "[He] made all his workers feel important, treating each as an integral part of the human machine that kept the telephone department working efficiently."[11]

After the war and the mini-depression of 1921, expansion took hold again and Lowry led the Manitoba telephone industry into the industrial age. Under his direction, the Manitoba Telephone System (the name was changed in 1921) quickly began a program to convert all Winnipeg exchanges to step-by-step switches; in 1926 it became the first city in Canada to have a completely automated system. This modernization was not coincidental: Winnipeg was the main source of revenue for the MTS because of its high population concentration.

The gap to be made up came primarily from rural service and, to a certain extent, long-distance service, which was just beginning to expand at the time. The company's annual report shows interfinancing at work in Mani-

Table 10
Number of Telephones in Winnipeg and in Manitoba, 1920–24

	Number of telephones	
Year	Winnipeg	Manitoba
1920	36,654	69,040
1921	37,606	68,749
1922	38,113	67,113
1923	38,413	66,765
1924	39,417	66,958

Source: Britnell, "Public Ownership," appendix, Table 5.

toba: the average rate charged for a rural telephone was thirty dollars, while the average cost of that telephone was $52.70.[12]

Lowry had a taste for technological adventure, so he threw the MTS into the brand-new radio sector in 1923, with the blessing of the federal government. As we have seen, this incursion into radio lasted until after the Second World War, when the federal government changed its mind and contested the right of provinces to be in the radio-broadcasting sector. At this point, the MTS had to sell its two stations, CKY in Winnipeg and CKX in Brandon.

In 1927 the MTS was connected up to AT&T's long-distance network, which gave it access to the United States, Mexico, and Cuba. The following year, thanks to radiotelephone, it became possible to call Europe. The MTS then decided to join other Canadian telephone companies to form the TCTS.

The 1929 economic crisis was particularly severe in the Prairies. In 1930, while the number of telephones in Canada as a whole was still rising slightly, it had already begun to fall in Manitoba. Between 1929 and 1934, Manitoba lost almost a quarter (22.8 percent) of its telephone customer base. The recovery began slowly in 1934 in the rest of the country, while in Manitoba the drop continued. Nevertheless, the reserve accumulated in the amortization fund allowed the company to remain financially healthy throughout the crisis: the corrections of the 1920s had borne fruit.

It wasn't until 1939 that Manitoba once again had the same number of telephones as it had had in 1929; the MTS was making every possible effort to accelerate the recovery. In addition to its participation in construction of the transcontinental network, the MTS began to build northern lines to serve isolated mining towns beyond The Pas. Erecting poles in the rock of the Canadian Shield was one of the crowning achievements of the Canadian telephone venture.

The crisis also provided the opportunity for a major change in the MTS's organization. A 1933 law abolished the position of minister of Telephones and Telegraphs and made the MTS a Crown corporation. The incorporation did not take place until 1940, but Lowry and the government immediately acted as if the corporation already existed – and it therefore did. The name "Manitoba Telephone System" was not endorsed, so the text of the law mentioned only the "telephone commission." But it didn't matter; Lowry confirmed the name MTS. Why did the government act stealthily to change the MTS's legal status? We can only conjecture that Lowry took clandestine action in order to reinforce his independence without making waves. If this was his intention, he succeeded very well.[13]

After the war, there were still only two cities in Manitoba with automated exchanges: Winnipeg and Brandon. The MTS took advantage of the necessary reconstruction of the system to automate small rural exchanges starting in 1949. The network made a new leap northward by means of radiotelephone, thus covering the entire province, and the first microwave line was inaugurated in 1956 between Winnipeg and Ontario. This was the beginning of the trans-Canada microwave system, and it took long-distance telephony into the era of mass communications and television. Not everyone had a telephone yet, but everyone wanted one. Universal service was within sight.[14]

TELEPHONE COOPERATIVES IN SASKATCHEWAN

Saskatchewan was the showcase of Canadian telephone populism. Development of all Canadian telephone networks, private and public, was slowed by the First World War – except for Saskatchewan's. And this was simply because it was the subscribers themselves who ran the system – and not at all because of Saskatchewan Telephone department policy. Indeed, the bewildering annual reports published by the deputy minister of Telephones contained bits of fiction: "there are certain features in this excessive rural extension work which invite consideration. The cost of these systems has greatly advanced within the past two or three years. This advance may be attributed to several causes. Labour is both scarce and high priced. So is material. But with the bountiful returns from their work the farmers of the province do not seem to consider cost. They appear to want the service at any price."[15]

Telephone cooperatives installed telephones on farms and ran lines to the nearest exchanges, installed in towns by the Department of Telephones. But the cooperatives were moving so quickly that the government could not keep up, and each year it set the deadline for applying to register cooperatives earlier. Although this slowed development of cooperatives, the rural system still very often had to wait several months on the outskirts of a village for an exchange to be set up. Since the farmers paid better than did the government, many telephone technicians forsook the department for the cooperatives, which made progress even more uneven.

This frenetic pace continued, aside from a brief lull during the mini-recession following World War I, until the mid-1920s. How did the system, the only one of its kind in Canada, work? The 1920 annual report provides a detailed description of the relationships between the agricultural co-operatives and the Telephone department:

Handling their own service has acquainted rural subscribers with some of the things necessary to establish and maintain service. Amongst other things learned by them is that permanent location of central office is a matter of great importance. Rural companies are therefore supplying a central building to provide not only accommodation for the equipment but as well living accommodation for the operating staff ... In all such cases the department installs and maintains the necessary central equipment. The companies are given to understand that this must not be taken to mean that the department assumes the sole responsibility and expense of operating both the rural system and the government system. That is a matter to be undertaken jointly. The practice followed is to appoint the rural telephone company our agent and to pay to the rural telephone company the usual commissions paid to Commission officers. The rural telephone company hires the operators necessary to give whatever service may be locally wanted. It may be that it is a service beyond what is provided for in departmental regulations. Any case if the commissions earned and paid by the department do not serve to meet the cost of operating then it must be provided for and met locally by the companies. This places in the hands of the users of the service sufficient control to establish a service deemed necessary by them if they are willing to meet the expense. It also distributes responsibility between the department and the companies in looking after the service.[16]

The result of this joint management was that Saskatchewan had a higher penetration rate than did Manitoba and Alberta, even though its per capita personal income was below theirs (see Table 11).

The success of the telephone industry in Saskatchewan was due to its decentralized structure, which allowed farmers to maintain control of development. This was the best part of Francis Dagger's legacy, which had whipped up a sense of urgency about the need to install telephones among an industrious population. On the other hand, the government bureaucracy was manifestly overwhelmed by the scale of the phenomenon it had instigated, and the administration lagged far behind or did not follow at all. This was the worst part of Dagger's legacy. The department did not always put aside funds for equipment depreciation, and in general accounting procedures were anarchic.

Saskatchewan had had Liberal governments ever since it became a province in 1905, but the 1929 election brought a Conservative coalition to power. The new Telephone minister, J.F. Bryant, ordered an audit of the

Table 11
Comparison of Telephone Penetration to Per Capita Income, Saskatchewan,
Alberta, Manitoba, 1926

Province	Rate *(telephones per 100 residents)*	*Average annual per capita income ($)*
Saskatchewan	12.54	437
Alberta	11.68	488
Manitoba	11.15	465

Source: Statisics Canada, catalogues 56–201 and 13–201.

government system's accounts by an independent accounting firm. Applying an amortization rate of 4.83 percent, the audit revealed a deficit of more than $900,000. Many other irregularities also surfaced, going back to the initial nationalization. For example, the poles that the department had given to the cooperatives between 1909 and 1913 had been capitalized for $350,000, then simply erased from the books. A brief debate began in the province's legislative assembly during which Bryant deemed the Liberal policy "a comedy of errors." The debate revealed nothing new except for the telephone department's accounting mess, after which everything resumed as before and confusion took over once again.[17]

On top of this, the economic crisis hit Saskatchewan harder than it did any other province: per capita income plummeted about 52 percent between 1930 and 1933. At the time, there were 1,100 telephone cooperatives, accounting for about 63 percent of all telephones. According to the official statistics, people cancelled their subscriptions *en masse*, especially in the countryside – between 1929 and 1934, there was a drop of more than one-third (36.4 percent) of subscribers. In many cases, however, this meant only that the co-operatives were unhooking their lines from the public network since they were unable to pay the interconnection fee, while the farmers generally maintained limited service that allowed them to communicate with each other.

Recovery from the crisis was very slow. In 1933 no bonds were issued, which meant that construction of the rural network came to a standstill. Saskatchewan then went directly from the crisis to the penuries of war, with the result that the network had to be completely rebuilt in 1947. Since the existing rural network at the time of the crisis hadn't been modernized since it was built, it was actually about a half-century old.

In 1944 the Cooperative Commonwealth Federation, led by Thomas C. Douglas, took power in Saskatchewan with a whole series of new ideas, among them the creation of Crown corporations to replace direct government administration. The CCF rightly felt that a Crown corporation offered a more flexible structure than a government department and that management through a board of directors composed of experts was preferable to one of bureaucrats. A series of laws adopted in 1945 and 1947 determined the form that the new Crown corporations would take and topped them off with a Government Finance Office, which had the power to lend money to Crown corporations, provide them with technical expertise, and regulate them (establish accounting standards, amortization rates, and so on).

The first Crown corporation created by the CCF, in June 1947, was Saskatchewan Government Telephones. The telephone department was retained to regulate the rural telephone cooperatives, which the CCF considered nationalizing for some time. In 1946 an indirect offer was even made to the Saskatchewan Rural Telephone Companies Association, which spurned the overture during its annual general meeting.

However, the government began to intervene more actively in management of the cooperatives, mainly to encourage them to merge so that each town would be served by a single legal entity. In addition, the cooperative groupings were offered an opportunity to hire a full-time technician. This policy caused an unexpected difficulty in towns where two languages were spoken. In the Canadian prairies, immigrant communities did not necessarily subscribe to the "melting pot" ideology current in the United States. A German cooperative might hesitate a long time before merging with a Ukrainian cooperative.[18]

The cooperative policy that had succeeded so well at the beginning of the telephone industry had become counterproductive. Saskatchewan's rural network had attained an almost universal penetration rate ahead of all other provinces, but it was obsolete (see Table 12). Moreover, farmers' needs were changing. At the beginning, they had simply needed to communicate with the closest market. By the 1950s long-distance service had evolved considerably, and it became urgent to align the standards in Saskatchewan with those in the rest of Canada. The alternative model dreamed up by turn-of-the-century populists was now way out of date.

How could the telephone industry be brought into the industrial age? It had been excluded when the CCF government went against the advice of the farmers and nationalized the cooperatives, to which the farmers had a sentimental attachment. On the other hand, the decentralized structure of the rural network precluded any technological change. By the 1950s Saskatchewan's rural network was stalled.

In the government system, the creation of SGT in 1947 had created a conflict. The employees of the old telephone department had acquired a great deal of autonomy over the years. Through the TCTS, they had drawn closer

Table 12
Farms Served by Telephones in the Prairie Provinces, 1956

	Manitoba	Saskatchewan	Alberta
Number of farmers	44,064	82,230	70,058
Number of rural telephones	28,187	52,025	27,925
% of farmers with telephones	61.7	63.3	40.0

Source: Government of Saskatchewan (the number of farmers comprises only those residing on the farm), cited in Spafford, Telephone Service, 66.

to Bell, adopting its working methods and technical standards. Far from providing increased freedom of action, the creation of a Crown corporation had placed the telephone employees under the control of the Government Finance Office. Suddenly, conflicts multiplied between two different sets of logistics: the sociopolitical plan of the GFO and the econo-technological one inspired by the TCTS.

In 1953 SGT was removed from the GFO's jurisdiction. For SGT it was an important victory, since it would finally become a completely independent telephone company. Of course, it was still managed by a board of directors, most of whom were top-level bureaucrats, and headed by a minister. Nevertheless, the increasingly technical nature of decisions rapidly built a protective screen between the political mismanagement and company management.

But who would make the technical decisions? After all, SGT was a skeleton administration charged with operating a few long-distance lines and the systems in the few large towns in Saskatchewan. In 1947, when it became a Crown corporation, it had a grand total of three engineers.[19] The rise of the CCF to power in Saskatchewan ultimately resulted in a rationalization of SGT's activities. The company's accounting was officially separated from the government's (even though, in practice, this distinction had existed since 1912) and, notably, an allocation was set aside for amortization. The public telephone rates in Saskatchewan, which had always been too low to cover real costs, were hiked in 1953.

In the 1950s SGT had developed a general policy aimed at obtaining a rate of return equal to the interest paid on money borrowed. This rallying to the financial orthodoxy of private industry took place under the social democratic CCF regime, which is but one of the paradoxes in the history of telecommunications in Saskatchewan. SGT came out of the experience with a structure able to meet the challenge of growth that continued through the 1950s and 1960s: the problem of the rural network was still to be resolved, but the solution was at hand.[20]

ALBERTA STRUGGLES
TO MODERNIZE MANAGEMENT

In Alberta, the end of World War I was marked by a resumption of agitated political activity. The province's Liberal premier, Arthur L. Sifton, who had nationalized telephone service, had just been called to serve in the federal government as part of the war coalition. The new Liberal government, led by Charles Stewart, was criticized more and more strongly by the Conservative opposition, which had put management of the telephone system on the agenda. The Conservative party was demanding that a Crown corporation similar to Manitoba's be set up in order to force the creation of an amortization fund and put a stop to patronage. This demand for rigour was received favourably by Alberta Government Telephones managers. Finally, in 1918, Stewart's government gave in and ordered a general assessment of the Alberta system by Wray, a Chicago engineering consultancy firm. The assessment began in November and kept two senior consultants busy for more than a year. This was not overly long when one considers the scope of the task:

Wray and Morse had never seen a telephone head office exactly like AGT's. They arrived in winter when there was a rink in front of the Terrace Building and young fellows like Reg Skitch would skate so furiously at noon that their productivity would sag after lunch. The girls were forever arranging skating parties and on many a Friday night the general office was decorated to make a ballroom. There was a telegraph circuit to Calgary to keep company business off the revenue-producing telephone lines, and when Miss Gilliland, the key operator, went to lunch, old Jack Elliott, the construction superintentent, would come in and take her place. When a call came in for a troubleshooter Horner's livery would send an open sleigh for George Baxter and the sleigh would be piled with buffalo skins and hot rocks to keep him warm.[21]

Soon after the consultants arrived, AGT was forced to apply to the new Public Utilities Commission for a rate hike, which was quickly granted. In April 1919 the basic rate rose by 25 percent in Calgary and by 25 to 50 percent in the countryside. When it was published, the Wray report, an impressive 1,500 pages long, stated that rates were still too low to cover costs; the basic rural rate would have to be raised by another 157 percent ... This was obviously impossible to do, so the Wray report recommended establishment of a land tax on telephone service. The Stewart government did not have the courage to implement this recommendation, and it also rejected the report's central recommendation, which concerned creation of an independent Crown corporation.

On the other hand, the accounting recommendations in the Wray report were welcomed by AGT administrators. The 1919 annual report included an amortization allocation for the first time. The following year, there was another improvement: the accounts were divided into seventy coded categories – gross revenue, expenses, system value, and so on – based on AT&T's. One might say that with the Wray report, AGT entered the era of scientific management – however imperfectly. The frankly improvisatory nature of the pioneering era gave way to a modern enterprise that was less colourful but more reliable when it came to managing a public utility. AGT's official historian vividly summarized the shock aroused by the Wray report: "It was painful having to admit that that old rascal of Charles Fleetford Sise had known what he was talking about when he said a telephone system must have certain revenues. In fact it was so painful no one ever did."[22]

The main problem remained to be dealt with: the chronic shortage of funds in the Alberta telephone industry. In spite of the low amortization rate used (4 percent instead of the 6.5 percent recommended by the Wray report), AGT was running a deficit. It gave up its brand-new amortization fund; in spite of this, by 1921 it found itself incapable of paying current maintenance costs for the network. To top it off, the provincial government was in the habit of paying off its political allies by handing out orders for various supplies just before elections. The Liberal government must have had lots of support in the forestry industry, since AGT found itself with an impressive stock of wooden poles ... The result was a systematic overcapitalization of the network.

When the United Farmers of Alberta, the local agricultural union that had not long before been converted into a political party, came to power in July 1921, it put an end to this detestable practice. The new minister responsible for the telephone industry began by selling all the poles and copper wire that had accumulated in AGT's warehouses. The Saskatchewan telephone department and CP Telegraphs purchased the excess stock at bargain-basement prices. Moreover, bids for system construction were now to be tendered to independent contractors and no longer to AGT. These sound economic measures obviously did not address the low rates, especially the rural ones, but how could the UFA double – or even triple – the rates for farmers? AGT therefore continued to accumulate deficits and was unable to respect its amortization rate, even though it clearly recognized its usefulness.

Throughout these difficult years, construction of the rural network was moving ahead. In October 1925 the Alberta system was connected to Montana and, via the AT&T network, to the rest of the continent, including central Canada and British Columbia. This expansion policy, however, had its price. In October 1926 the Public Utilities Commission ratified a 20 percent rate hike for all telephone services, after which AGT's finances immediately improved. In 1927 an amortization reserve equal to 3.5 percent of the system's value was entered in the company accounts. Would AGT manage to avoid the slippery slope? The moment the rate increase was granted, another problem appeared on the horizon.

Table 13
Edmonton Telephone and AGT Compared, 1928

	Value of the network ($)	Reserve funds ($)	Amortization fund ($)
Edmonton	950,000	275,000	1,000,000
AGT	18,000,000	300,000	850,000

Source: Cashman, *Singing Wires*, 334.

The City of Edmonton revived its old demand from 1914 that AGT pay part of its long-distance revenues to the municipal telephone company. The provincial government's response was the same: there was no question of it; on the contrary, it was Edmonton that should be paying AGT for its use of the long-distance lines and contributing to the financing of the rural network. On top of that, the government made clear its intention to purchase Edmonton's telephone system. A meeting that took place in 1928 was cut short. In fact, a comparison of the financial results of the two networks showed that the City had a crushing advantage (see Table 13).

The municipal authorities felt that the excellent results yielded by their system came from their careful management, while the Telephone department felt that the difference was due to the City's lack of social responsibility. This confrontation provoked an unprecedented crisis. The government tabled a bill providing AGT with a right of way in all of the province's municipalities, intending to offer Edmonton the same treatment that it had already served up to Bell and force the City to sell. If Edmonton absolutely refused, a plan to impose interconnection fees equivalent to 20 percent of the municipal network's gross revenues was in the wings.

An indignant city of Edmonton conducted a survey of the roughly ten thousand independent telephone companies in the United States and found that none of them were paying interconnection fees. The government did not back down in the face of this argument and prepared to impose arbitration by the Public Utilities Commission. Things were at this point when the 1929 crisis hit the Prairies. The battle of Edmonton didn't take place then, but it was only postponed – for forty years. In the meantime, the employees of AGT and the Edmonton phone company got into the habit of choosing a winner by their own means, through golf and curling tournaments.[23]

In the 1920s the "hot" technology was radio. This infatuation gave rise in Alberta to a brief experiment with rebroadcasting American programs. In 1929 a knowledgeable amateur named Bill Grant built a reception antenna at Okotoks, near the American border, and linked it to a sound studio in

Calgary. This distant ancestor of cable distribution was an immediate success, until the federal government got wind of the illegal transmission of foreign programs and put an end to it.[24]

When the depression hit Alberta, AGT lost its financial equilibrium, painfully attained with the 1926 rate hike, and the company was back to deficits. It had to give up its amortization fund, without resolving its deficit problem, for people were cancelling their subscriptions *en masse* and many bills were going unpaid, especially among rural customers.

Rather than do battle with its subscribers, AGT sold the rural network to the farmers. Since the equipment was old, it went dirt cheap. This solution born of desperation enabled AGT to save on operating costs, but that was about all: it had to continue to pay interest on the capital invested with an amortization fund that had been whittled away to almost nothing. Once more, the short term had been favoured over the long term. Alberta lost 31.5 percent of its stock of telephones from 1929 to 1934.

The crisis precipitated the rise to power in Alberta of the well-known radio evangelist William Aberhardt, nicknamed Bible Bill. In September 1935 he formed the first Social Credit government in Canada. AGT, like other public administrations, began to pay its employees with unpopular "prosperity certificates." When rumours began to circulate of the imminent sale of AGT to AT&T, employee morale dropped one more notch. Premier Aberhardt himself had to deny the sale; AT&T's terms were too low, even for a radio evangelist.

War broke out while the company was still demoralized and disorganized as a result of the depression. When the United States began building the Northwest Communications System, AGT was charged with constructing the telephone line. From Edmonton to Fairbanks, Alaska, the line was 3,200 kilometres long, the longest naked-wire open circuit ever erected. Since the company lacked the necessary resources, it had to ask a team from Bell to come in. Imagine the surprise of AGT employees when they found out that a Bell foreman made almost as much money as the director of their company ... The work was executed on a nearly impossible schedule with the help of the Canadian army's Signal Corps. During the winter of 1942–43, the frozen ground had to be dynamited to make holes for the poles. But the line was ready in time.[25]

The postwar telephone boom took AGT, as it did all the other telephone companies, by surprise. AGT was without doubt the only company to admit it publicly: "We had foreseen that after the victory of May 1945 in Europe, there would be a fall of about 10 percent in our long-distance traffic. The real figure has been a rise of about 12 percent."[26]

When they began to reconstruct the network, general manager Bill Bruce realized that there wasn't a single university graduate working at AGT – including himself. He made three of his employees take a degree in engineering, and an engineering department was created at AGT in 1945–46. But Bruce became discouraged at how much had to be accomplished; receiving no support from the provincial Social Credit government, he resigned in 1948.

It was to Alf Higgins that the task of "normalizing" AGT now fell. He undertook a complete review of hiring criteria and pay conditions in order to align AGT with Bell. In September 1953 AGT signed a service agreement with Bell involving a massive transfer of expertise in return for 0.5 percent of AGT's gross revenues. What a long road it had been since the arrogant nationalization of 1908. But 1953 was precisely the year in which Thomas Eadie was beginning construction of a pan-Canadian network that overcame regional disparities. Alberta naturally found a place in this grand design.[27]

By the time Higgins's reign ended in 1957, it could be said that AGT had finally achieved the long *aggiornamento* begun in 1926 with the Wray report. Alberta had managed to overcome the handicap of a hastily undertaken nationalization and rapid development of rural telephone service.

A SPECIAL CASE:
BC TEL UNDER AMERICAN CONTROL

At the beginning of the 1920s, British Columbia Telephone was in crisis. The two founders of the company, the ingenious Dr Lefevre and his dirty trickster William Farrell, were dead. Lefevre's widow, Lily Lefevre, owned one-third of the company, which made her the main shareholder, but she wanted to unload her stock.

Over these same years, the other Canadian companies were interconnecting their networks. BC Tel, squeezed between the Pacific Ocean, the Rockies, and several independent companies in the interior of the province, was not part of this movement. The independents were fiercely attached to their autonomy since they were linked to each other by the federal government's Dominion Telephone System, which offered long-distance service at a loss. To communicate with Alberta, BC Tel had to cross DTS territory and share its long-distance revenues with the independents. On the other hand, BC Tel had long been interconnected with the AT&T network and could communicate easily with eastern Canada via the United States. How was this problem to be solved?

The solution unwittingly came from Lily Lefevre, who approached Bell management hoping to sell her large block of stock. For some obscure reason, Paul Sise, the president of Northern Electric, who was acting as Bell's secretary, offered only $125 per share, while they were trading at above $140. Of course, Mrs Lefevre refused and turned to the United States. AT&T's offer was brushed off for the same reason as Bell's, and then financier Theodore Gary made his move. Gary had purchased Automatic Telephone and Electric, the prestigious Chicago company founded by Almon Strowger, as well as a number of independents in the United States and had built the second largest system in the American telephone industry, after AT&T's.

Blocked from expansion in the United States by AT&T, Gary turned to international markets and acquired several British companies. When Mrs Lefevre's representatives knocked at his door, the timing was perfect. Gary created a holding company in December 1926 with English and American interests and proceeded with an exchange of shares with BC Tel. He offered $185 per share well above market value, financed with an issue of BC Tel preferred shares, which cost him nothing. This sleight of hand was not to everyone's taste and among those who objected was BC Tel president George H. Halse's, who resigned two years later.

The transaction was cunning, but it smelled fishy. One of Theodore Gary's partners was none other than Gordon Farrell, William Farrell's son, and the 1926 deal resembled a bad remake of the one in 1898, when Farrell, Sr, partnered with Lefevre, had appealed to British capital. The difference was that this time, there was no long-term strategy. Gordon Farrell had sold BC Tel to the Americans with no hope of recouping it. After Halse left, he himself took over the presidency, becoming Gary's Canadian straw man.

The deal smelled even fishier because the stock issue that had financed the purchase had massively diluted BC Tel's capital. Like Bell, BC Tel was regulated as a function of its earnings per share. The Railway Commission was justifiably opposed to this transaction, which created no wealth for the telephone company. A long conflict followed between the company and the regulatory body; it was not resolved until 1966, when the Canadian Transport Commission simply subtracted the total of the assets.

Once it entered Gary's orbit, BC Tel stopped its policy of purchasing from Northern Electric in favour of Automatic Electric. Curiously, however, BC Tel did not automate; on the contrary, British Columbia was the most backward province with regard to automation, with the exception of Prince Edward Island. The first step-by-step switch was installed only in 1928, and it was installed not in Vancouver or Victoria but in the tiny village of Hammond, in the Fraser Valley. During the economic crisis, the conversion programme was interrupted, while everywhere else the telephone companies maintained or accelerated their program in order to cut labour costs. Vancouver received its first step-by-step switch in December 1940. Why did the Gary group not impose the Automatic Electric technology on BC Tel more quickly? The company's inertia is even more paradoxical because automation of Vancouver was one of the reasons advanced by the Gary group for purchasing BC Tel: "In an interview yesterday with a representative of the *Financial Times* in connection with the above deal, Sir Alexander Roger said Vancouver is the only large system in the world not in the process of conversion or planning to convert to automatics."[28]

There was no doubt, given the low rate of automation of exchanges and the high rate of telephone penetration in British Columbia, that it was an exemplary illustration of the North American concept of technology – cheap and short term versus expensive and long term. Throughout this period, British Columbia had more telephones per capita than any other province in Canada (Ontario took over the lead only in 1940). Deliberately or not, technological progress was traded for universal expansion of an older technology, and long-term investments were spurned for an immediate profit.

This did not mean that the company was stagnating. With the support of the federal government, which agreed to sell its long-distance sections in the interior, BC Tel purchased the independent companies that were impeding its expansion to the interior of the province. It then concluded agreements with CP Telegraphs, and in November 1928 Canada's Pacific coast was finally linked by telephone to Alberta and the rest of Canada.[29]

Economic crisis interrupted this expansion. In 1935, after several resounding failures, the Gary group was reorganized and BC Tel was handed over to a new "Canadian" subsidiary created for the occasion, Anglo-Canadian Telephone, with its head office in Montreal. Behind this façade, however, the company remained American, and BC Tel continued to be managed by Gordon Farrell as if it were a family firm (the Farrell family had stock in the Gary group).

In October 1955 there was yet another reorganization: the Gary group merged with General Telephone and Anglo-Canadian became a subsidiary of what was renamed General Telephone and Electronics Corporation in 1959. The new holding company was a powerful multinational, controlling 10 percent of the telephone industry in the United States and with interests all over the world.

In spite of this, Gordon Farrell's presidency of BC Tel maintained the fiction of a family firm deeply rooted in Canada. Indeed, Farrell had been president since 1928 and his father had preceded him. The Farrell family was to BC Tel what the Sise family had been to Bell, a situation that continued until 1958, when Gordon retired.[30]

Notwithstanding these changes of ownership, BC Tel participated in the creation of the Trans-Canada Telephone System and the construction of one of the system's most difficult sections: the crossing of the Rockies. In 1929 BC Tel created a subsidiary called North-West Telephone, which set up the largest radiotelephone system in North America to link the isolated towns of Powell River (1930), Ocean Falls (1931), and Prince Rupert and Prince George (1932) to its network. This policy of covering the entire province culminated, after the Second World War, in the purchase of the last independent companies (Mission, Kootenay, Chilliwack) by BC Tel and of Dominion Telephone System's facilities by North-West.[31]

ONTARIO AND QUEBEC:
CONSOLIDATING THE INDEPENDENTS

The number of independent telephone companies in Canada had grown constantly since the turn of the century, and there were 2,400 in 1929. The economic crisis then caused large companies to retreat, obliging farmers to pick up the slack if they didn't want to lose their telephone service altogether. There were 3,400 independents by the beginning of the Second World War (1,200 of them agricultural cooperatives in Saskatchewan). Starting in 1941, however, the trend reversed and the number of independents in Canada began to drop.

Quebec and Ontario were special cases. In the other provinces, the independents were regulated by the same authority as was the dominant company; in Saskatchewan and Alberta, they were even the object of stated policy by the dominant company. In Quebec and Ontario, the telephone network was dominated by Bell, which was under federal authority, while the independent companies were under provincial authority. There are thus two telephone histories in these provinces: that of Bell and that of the independents.

Jules-André Brillant:
Dominating the Telephone Industry in Quebec

In the early twentieth century, the independent telephone industry in Quebec was dominated by the Compagnie de Téléphone Nationale. In 1915 it was in the middle of a struggle with Bell for control of the province when freezing rain destroyed all of the telephone systems in its territory, and it was never able to recover. It left Quebec City, Lévis, Charny, and Saint-Romuald to Bell. The systems in Kamouraska and Témiscouata counties were sold to a third party. The company founded by Dr Demers stagnated for some time and was then sold, in June 1927, to a businessman with the intestinal fortitude to revive it: Jules-André Brillant.

Brillant was born in Saint-Octave-de-Métis, where he had witnessed the arrival of Dr Demers's telephones at the age of nine. He had made his fortune very young in electrification of the lower St Lawrence, creating the Compagnie du Pouvoir from scratch and not hesitating to turn to American interests to raise the greater and greater sums he needed to diversify his activities. It was for this purpose that he purchased Nationale, which was renamed Corporation de Téléphone et de Pouvoir de Québec.[32] The head office remained in Rimouski, but the administrative offices were set up in Quebec City.

This complex name expressed well the idea that Brillant had for his little kingdom: everything that was electrical would be in his realm. He multiplied his purchases of telephone companies, including those in Kamouraska and

Témiscouata along the way. The economic crisis did not interrupt his plans; if people cancelled their subscriptions, he had no qualms about simply pulling out the wires and poles in order to save money.

In fact the depression provided an opportunity for a major consolidation. In January 1935 all independents belonging to the Corporation de Téléphone et de Pouvoir were integrated into two subsidiaries: north of the St Lawrence, Compagnie de Téléphone Portneuf et Champlain; south of the St Lawrence, the Compagnie de Téléphone Nationale was resuscitated. Two years later, the administrative offices were moved from Quebec City back to Rimouski. At the same time, Brillant extended his interests to radio, teaming up with Canadian Marconi to serve the North Shore by radiotelephone. He founded Radio du Bas-Saint-Laurent, with the call letters CJBR AM.

Nothing stopped Brillant, not even the Second World War, which turned out to be good business when the federal government made massive investments to construct telephone and telegraph networks along the shores of the St Lawrence in order to connect up all the coastal defence bases. More than twenty Canadian and Allied ships had been sunk in the St Lawrence by German submarines. This forgotten Second World War arena had a direct impact on telecommunications of the lower St Lawrence. First, the Corporation de Téléphone et de Pouvoir participated in construction of the military telephone line from Rimouski to Saint-Joachim-de-Tourelle, under the supervision of the Radio Corporation of America; the Canadian army itself constructed the part from Saint-Joachim-de-Tourelle to Gaspé. Then, with peace re-established, Brillant purchased the government network. In the Gaspé Peninsula, some villages had telephone cooperatives, but these dispensed only local service. The long-distance network gave them a window on the rest of the world.

On the North Shore, the army had constructed a telegraph line from Tadoussac to Red Bay, on the shore of the Strait of Belle-Isle. Brillant had never dealt in telegraphy, but this didn't matter; he purchased the "ligne du Nord" and, the next year, concluded an agreement with CN to operate his telegraph network on the Gaspé shore of the St Lawrence. He thus found himself with a telegraph network on both shores of the river.

Through purchases and construction of new lines, Brillant built a major telephone company. He had only one problem: whom to pass his company on to when he died. In 1952 he tried to bring his son, Jacques, into management, but Jacques was incompetent, and it was Jules-André who won a victory against Bell in 1955 by taking control of Bonaventure and Gaspé Telephone, bringing all of eastern Quebec under his umbrella.

Brillant's last "coup," however, was not his own doing but that of his friend, Quebec premier Maurice Duplessis. The story goes that Brillant went to ask Duplessis for an expropriation right to erect his poles. Duplessis agreed, on one condition: "Change that horrible name, Corporation de

Téléphone et de Pouvoir. It's not even French. Your company is known by everyone as Québec-Téléphone. Why don't you use it? It's simple and it rolls off the tongue."[33] In February 1955 Brillant changed "that horrible name" and Québec-Téléphone was born.[34]

The other "big" independent was the Compagnie du téléphone Saguenay-Québec, which had belonged to the Dubuc family since 1907. The founder, Antoine Dubuc, had built a small industrial fiefdom in the Saguenay-Lac-Saint-Jean region on pulp and paper and public utilities (water, electricity, and telephony). After his death in 1947, the company accumulated debts and was forced to sell out to Bell in March 1955. Bell then liquidated the head office, which had been in Chicoutimi, and simply integrated the company into its network. This cavalier move offended many people in the Quebec government, who didn't appreciate how Bell swallowed up a small company under provincial authority and got rid of a regional decision-making centre.

The Independents in Ontario

No independent company in Ontario played a role comparable to that of Québec-Téléphone. The number of independents reached a peak in 1921, with 689 systems listed, including 5 municipal utilities, 95 municipal companies belonging to subscribers, 284 private companies, and the rest associations of subscribers.

As a municipal utility, telephone service was treated like natural gas or running water. On the other hand, in a municipal company, the subscribers were owners who had requested telephone service from the municipal authority by putting up their land as security. The system was then given in trust to the town. This ingenious system allowed municipalities to amass the funds needed to establish and maintain a telephone system, but the subscribers were the real owners without having to spend a cent. If the network was sold, the subscribers had to be compensated. This was the form of ownership favoured by Francis Dagger, who had been manager of the Telephone Division of the Ontario Railway and Municipal Board from 1910 to 1931.

After the Second World War, the Ontario government felt that the telephone situation in the rural areas served by independents was serious enough to require its intervention. It asked Ontario Hydro, which had been a world pioneer in rural electrification, to conduct a study of the independents. After consulting with Bell, Ontario Hydro concluded that there was a need to create a new regulatory agency devoted exclusively to the telephone system which would also be charged with encouraging development of the utility in remote regions. A law on rural telephony adopted in July 1954 created the Ontario Telephone Authority.[35]

One of the recommendations in the Ontario Hydro report was to group the independents together so that they might rationalize their activities. As it had in Saskatchewan, this recommendation ran into the indifference and amateurism of the managers of the small companies. Every time a group was formed, one company unfailingly broke off in the middle of negotiations and sold out to Bell. The group then lost all reason for being and the other members ended up selling out in their turn. Outside of Bell's sphere of influence, rural telephone service developed little.

Most subscriber complaints came from the Bancroft and Barry's Bay region, west of Ottawa. After the discovery of uranium and the excitement this caused, the dilapidated state of the rural network, which was managed by thirteen small companies, became unacceptable. The provincial government decided to intervene directly; early in 1955 it created the Ontario Development Corporation, with the mandate to purchase, expropriate, and, if necessary, create and operate telephone companies.

The new Crown corporation soon created a subsidiary called Madawaska Company, which purchased the thirteen Bancroft-Barry's Bay independents and modernized the local network. Uranium fever must have been running high for the Ontario government, traditionally indifferent to the telephone industry and an opponent of government intervention, to have made such a move. Five years later, the government resold the brand-new system to Bell for 90 percent of its value. This nuclear-power-scented venture took the taste for the telephone industry definitively out of Ontario's mouth. Since then, the province has never had a telecommunications policy.[36]

THE FAR NORTH: A MODEST START

In 1923 the only line of communication in the North was a single telegraph line between Hazelton, British Columbia, and Dawson City, on the border of the Yukon and Alaska. Otherwise, information circulated by canoe.

Serving the Far North was to be the job of the Canadian army's Signal Corps, and the technology used was the radiotelephone. In October 1923 a first radio circuit was opened between Dawson City and Mayo. The initial team was composed of eight men under the command of the future commissioner of the CRBC, Major William A. Steel (later a lieutenant-colonel). For this first link they rented a small wood cabin in Mayo and borrowed a second, equally modest cabin from the Royal Canadian Mounted Police in Dawson City. This little group built the radio system for the Northwest Territories and the Yukon from scratch.

Once more, Canadian telecommunications wrote a pioneering page, in a punishingly severe terrain. The Canadian army archives contain some clues to this unknown epic of the solitude of life in isolated cabins hundreds of kilometres from any form of civilization. For example, there was the station on Herschel Island, in the Beaufort Sea, which operated from 1923 to 1937. The ship charged with dropping off supplies to the operators that first winter sank *en route*, so the Signal Corps men had to hunt caribou for food and melt ice for drinking water. None of them had gone to the Arctic for fame and glory, and the pay was no fortune. They were motivated by a sense of duty and, no doubt, a fascination with the Far North.

Starting in 1927, a new function was assigned to the Signal Corps: measuring atmospheric pressure, wind speed, and temperature. Twice a day, these readings were sent to the federal meteorological bureau in Toronto.

Within a few years, the Signal Corps's radiotelephone system extended from the Alaska border to James Bay and from Edmonton to the shores of the Arctic Ocean. As the system expanded and received better-quality transmitters, mining companies, traders, and the RCMP began to acquire their own radio equipment and became substations. In this way, the Far North was slowly linked to the rest of the world.

The army operated the Northwest Territories and Yukon Radio System until September 1958, when it handed the network over to the Department of Transport. After that, CN Telegraphs that took charge of communications in the Yukon and the western part of the Northwest Territories (Bell operated the system in the eastern part).[37]

14 Social Benefits and Labour Peace in the Telephone Industry

The dedication of telephone company employees is the theme of stories that are told and retold in each company. The following exemple is typical:

On July 11, 1911, fire raged into the mining community of South Porcupine in northern Ontario. Telephone operator Marie Gibbons stayed at her switchboard, calling the warning, until flames licked at her doorstep. A neighbour burst in shouting, "What the hell are you doing here?" As she seized her shoes and the company cash box and followed him out the door, the blazing stairway collapsed.[2]

It is with exploits of this kind that the legend of the telephone industry in Canada was built. In 1923 Bell institutionalized heroism, handing out awards to five women and three men who had shown extraordinary courage. The managers of telephone companies thus created a solidarity that surpassed simple team spirit: one might call it company patriotism. This feeling of belonging would play a not negligible role in the elimination of militant unions, which did not understand or know how to oppose such a mystique.

The telephone companies and the unions thus entered a wrestling match over social benefits, but the strategic stakes were cultural. Who would succeed in claiming the loyalty of telephone employees – the companies or the unions?

VAIL'S VISION: HUMANISM OR PATERNALISM?

In the postwar period, all social policy in Canadian telephone companies came directly from the president of AT&T, Theodore Vail, who had defined

Canadian telephone companies were eager to create a legend around the new service. The exploits of the "telephone men" were celebrated in corporate newsletters and advertisements. Courtesy Bell Canada.

his concept of industrial relations before the First World War. On the philosophical level, these relations resulted from the purest of market laws: "Employers buy and employees sell service. Perfect service is only to be found when fidelity and loyalty are reciprocal in employer and employee. It is this relationship that brings satisfaction and success to both."[3] In industrial relations, this cult of equal exchange gave rise to the over-arching social policies of North American telephone companies: "The intent and purpose of the employer in establishing a plan of benefits, is to give tangible expression to the reciprocity which means faithful and loyal service on the part of the employee, with protection from all the ordinary misfortunes to which he is liable; reciprocity which means mutual regard for one another's interest and welfare."[4]

This policy was ahead of all American (and Canadian) social legislation. When Vail put on the table more than was required, he had several ideas in mind. Of course, he wanted to nip any unionism in the bud by offering social benefits superior to anything found in private enterprise at the time. But he also wanted to motivate his employees to provide high-quality service and present a positive image of the company to the public.[5]

Vail's social principles were decried by the unions, which charged him of paternalism. And yet, the unions were seeking the very same social benefits. But there was one difference, and it was a major one: the unions wanted to obtain these benefits through collective agreements negotiated by employee representatives. The battle between the unions and the telephone companies was, in fact, a moral one, involving the issue of who was really looking out for the welfare of telephone employees.

In Canada, the first sign of a thaw in social-benefits policy came from Bell Telephone, when the departure of the intractable Charles F. Sise and the rise to power of Lewis B. McFarlane opened the way for Vail's ideas to infiltrate. In 1917 Bell replaced the unacceptable 1911 pension fund for its "good" employees with a universal Plan for Employees' Pensions, Disability Benefits, and Death Benefits. Men could retire at sixty, women at fifty-five. Meanwhile, the work day for system technicians went from ten to nine hours, reducing the work week to fifty-four hours.

Also in 1917, the major telephone companies began to sell the federal government's Victory Bonds through salary withdrawals at source, and they then considered whether they could use this method of raising funds themselves. Since 1915 Vail had offered AT&T employees an Employee Stock Plan based on salary withdrawals at source. In 1916 NB Tel was the first Canadian company to offer a similar scheme. In 1920 Bell began to sell its own stock by the same means – successfully, as we have seen (see chapter 12).

Generally, Bell and the Maritime telephone companies, MT&T and NB Tel, had closely linked social-benefits policies: the pension plans of the three companies were similar, and it was Bell's representatives to the boards of directors of MT&T and NB Tel who usually proposed adoption of social measures. At the other end of the country, BC Tel instituted a new medical insurance plan in 1916, fully financed by the company (the previous one, financed equally by the company and the employees, had proved insufficient). Employees also gained Saturday afternoon off.[6]

It is worth noting that in the telephone industry, social-benefits policies were introduced by the private sector, while the public sector was content to follow its lead. The first Crown corporation to get on board was AGT, which instituted a medical insurance plan in January 1922 and a retirement plan in May 1926. Paradoxically, 20 percent of AGT's employees opposed the pension fund, on the grounds that they would prefer to receive the company's contribution in the form of a salary hike. Among other things, although it covered all men, the plan was optional for women.[7]

THE RISE OF UNIONISM

At first, employees were unhappy with these measures: they didn't want their bosses' charity. Indeed, the end of the First World War coincided with an unprecedented rise in militant unionism. Unionization had made its first foray into the telephone industry at the beginning of the century, with the International Brotherhood of Electrical Workers. As we have seen, the IBEW

The telephone companies took part in the war effort by all means imaginable. Here operators perched on a Bell truck encourage the population to buy victory bonds. The company deducted the cost of purchasing victory bonds directly from its employees' salaries. Courtesy National Archives (PA 2163).

was most active at BC Tel, where it led bitter strikes against management. Elsewhere, its presence remained underground and passive; in Alberta, it did conduct negotiations with management starting in 1913, although it was not accredited.

The war had put unionism on hold, since many of the movement's leaders were conscripted. The feminization of telephone companies reached record levels: 58 percent of Bell employees were women in 1910, and the proportion had risen to 71 percent by 1920. During the first part of the war, this overwhelmingly female labour pool was concerned with survival: inflation, spiralling prices, and food shortages got the better of union militancy.

As the war ended, a wind of change was blowing in the world. The rise to power of the Communists in Russia raised new hopes among the working classes, especially in western Canada, where a population of new immigrants kept an open ear to appeals from Europe. Union ranks were growing and worker organizations were able to improve somewhat working conditions for women, to whom the labour shortage had given new power. The IBEW agreed to end its policy of local auxiliary sections, which had put operators in the same category as wives of electricians, and gave them separate sections, but with half a vote per woman. This "sexual apartheid" was as far as male management of the unions would go, for they feared the prospect of finding themselves drowned in a sea of women workers. It was enough of a concession, however, to give some hope to telephone operators and revive militantism.[8]

The spark came from the United States, where the railways had been nationalized for the duration of the war. To ease the effects of the inflation that had decimated employees' purchasing power, the Secretary of the Treasury, W.G. McAdoo, had substantially increased the salaries of all railway workers in May 1918. The Canadian government followed suit in August for the nationalized railways, including for telegraph operators who were members of the "Big Four" (Order of Railroad Telegraphers). After that, everyone in Canada spoke hopefully of the "McAdoo award."[9]

At the beginning of August 1918, the Toronto telephone operators registered *en masse* with the IBEW and demanded the McAdoo award. Bell accorded a 25 percent increase, roughly equal to what the operators were demanding, but refused to negotiate with the union. The president, Lewis McFarlane, could not understand that the operators might turn for their welfare to a body other than Bell, which had provided such generous conditions. He declared to the press, "There is no necessity for a meeting with the union since there is nothing to discuss."[10]

But there was, in fact, much to discuss. In the febrile postwar environment, the operators were not content just with money; they wanted dignity as well. And dignity would come through recognition of their own institution – the union. The Labour department became aware of the affair in September 1918 and appointed a Conciliation Board, as provided for in the Lemieux Act. The operators' grievances turned essentially on Bell's refusal to recognize the union and on threats being made against union members. In the limelight at the hearings was Kenneth J. Dunstan, who had already been singled out for his tactlessness and hard line during the 1907 strike:

Dunstan: Local 38A does not represent our employees. We cannot have a divided house.

Judge Snider: Your statement shows that you want to discourage the union. Now if you do that I'll be against you from the start.[11]

The language was plain enough, and Bell management yielded right away. In early October 1918, an agreement was made outside of committee between the company and the IBEW. It was the first collective agreement ever signed by Bell. The victory had been won by a local of operators, by the underpaid and scorned women that the IBEW had refused to unionize some years before. The Conciliation Board adjourned without making any recommendations, since the two parties had settled their differences. The following year, the technicians, in their turn, demanded a meeting with the Conciliation Board and obtained a half-day off on Saturdays, which brought the work week in Montreal to fifty hours, and in Toronto to forty-four hours.[12]

The movement was more or less the same for all Canadian telephone companies, with the exception of NB Tel, where a company union had been created in 1918. By simply threatening to strike, the technicians at BC Tel, finally united with the operators, won a closed-shop clause in the summer of 1918. The Manitoba Government Telephones technicians and operators obtained the McAdoo award in the spring of 1919. The union wave was irresistible, and the telephone companies succumbed one by one. In this context, the general strike that broke out in Winnipeg on 15 May 1919 hit like a thunderbolt.

The strike had started with a simple demand from the metal and construction workers for union recognition, but, in a new twist, the other trades walked out in solidarity. Thus, the MGT employees, who had obtained satisfaction on all of their demands, found themselves on strike, though MGT managers, with the help of strikebreakers, managed to maintain minimal essential service throughout the walk-out. The strike order was followed sporadically in other regions, but two exchanges, Transcona and Dauphin, were completely paralyzed. In Winnipeg, the strike committee had the city in its grip for more than a month. Frightened by what it considered to be a usurpation of power, the federal government ordered the Royal Northwest Mounted Police to shoot into the crowd on 21 June, drowning the movement in a wave of blood.

In the first days of the labour conflict, solidarity strikes broke out all over Canada, although most telephone employees, having already obtained satisfaction on most of their demands, did not join in. In British Columbia, the IBEW, dominated by electricians, joined the movement against the wishes of most of the telephone employees. At first, the operators were exempted from striking so that they could maintain a service deemed essential by the organizing committee. However, over time, the situation grew more bitter and the operators walked out in their turn. The general strike in Vancouver lasted a week longer than the one in Winnipeg, illustrating yet again the severity of social conflicts in British Columbia.

Paradoxically, the telephone employees, who had walked out unwillingly, had to continue the struggle two weeks longer than the other trades did because of BC Tel's decision to demote the supervisors who had taken part in the strike. They went back to work on 15 July without having managed to make the company retreat. The BC Tel operators went back to work one day after their technician comrades, and this badge of courage has entered the mythology of feminist struggle. The historian Elaine Bernard remarks that even today the Vancouver operators refer to the events of 1919 as "the strike where the men sold out the women."[13]

The Winnipeg strike was a disaster for Canadian unionism in general and marked the end, or nearly, of the IBEW in the telephone industry. Indeed, after the strike, the telephone technicians asked the IBEW for permission to start separate locals, like the operators' (but with a whole vote per member). The IBEW refused, signing its own death warrant in the Canadian telephone industry.

UNIONISM: MISSING THE BOAT WITH THE TELEPHONE

The IBEW had been founded on the obsolete premise of the "boomers" and treated all electrical workers as a single trade. Its divisions were geographical and not professional, corresponding to a nineteenth-century situation in which one individual might work in an electrical company, a telephone company, or a telephone company and an interprofessional union was needed so that members could change jobs while preserving the benefits acquired through years of paying union dues.

By the 1920s, things had changed. Telephone employees had obtained generally tacit but universal job security. From then on, they constituted a sufficiently strong entity not to see themselves as electricians, and they were proud to be part of the telephone industry. They were less and less willing to sit through meetings where they had to hear about claims that meant nothing to them, and that they didn't even always understand.[14]

In Vancouver, the crisis between the telephone employees and the IBEW took a somewhat different turn. The IBEW local that had led all the battles until the 1919 solidarity strike was excluded from the union because of its radical positions. IBEW management created a new local, which the extremist majority, composed of electricians, refused to join, although the telephone employees leapt at the opportunity to have their own local and joined up *en masse* (revealing the malaise that reigned between telephone employees and IBEW members). Nevertheless, the Vancouver local closed its doors in 1929 for lack of members.

In the 1920s the IBEW practically disappeared from the telephone industry, with one exception. Even in Manitoba, where the union survived the general strike, it was but a shadow of its former self. It lost its members in Manitoba, as it lost them everywhere else; however, it maintained control over 35 percent of the technicians (the operators and other technicians created an independent union, which quickly took over).

The inadequacy of the IBEW's structures for the telephone industry was not the only source of disaffection. The social policy unveiled by Theodore Vail in the United States stole from the unions their principal motivation: social injustice.

The telephone companies had won a total victory: their industrial culture triumphed everywhere. But the events of 1917–20 had taught them that it was not enough to grant social benefits, even before the labour laws were en-acted, to the employees: they had to consult with them – that is, consult with everyone but the unions.

In Canada, the solution was provided by the Royal Commission on Indus-trial Relations, which was formed in April 1919 at the request of the Conser-vative prime minister, Robert Laird Borden. Its final report recommended institution of a wide range of social benefits (an eight-hour work day, unem-ployment, old-age and medical insurance; recognition of unions and of col-lective agreements). Although most of these recommendations remained a dead letter for years, one immediately caught the eye of employers, includ-ing Bell: "That the government take suitable actions to promote the estab-lishment of Joint Plant and Industrial Councils."[15]

The instigator of this form of employer-worker organization was none other than Mackenzie King. In 1915 he had set up a parity committee in the coal mines and steel mills of Colorado Fuel and Iron, and then he had acted as a consultant to the company's owner, John Davison Rockefeller, whence the name of this first committee: the Rockefeller Industrial Representation Plan. King's basic premise was that both parties should be represented equally in order to settle peacefully any problem relating to working conditions.

In Great Britain, Lloyd George's government had recommended the cre-ation of Joint Industrial Councils in all industries in 1917, to prevent work stoppages during the war. In most companies, the workers' side was repre-sented by their union. Once the United States entered the war, its National War Labor Board adopted the British idea, but with a difference: the Joint In-dustrial Councils were created to prevent the formation of unions, not to pro-vide them with a framework. Such committees were set up at AT&T in 1918, during its brief period of nationalization.

In 1919 Bell introduced the Joint Industrial Council system to Canada through what it called Plant Councils, starting with system technicians – no doubt because they were highly skilled employees, for whom training was expensive, and to whom all telephone companies had therefore always paid the greatest attention. They accepted the idea more or less unanimously dur-ing a series of meetings organized in the spring of 1919; even the IBEW members were won over to the representation committees. The traffic em-ployees, most of them operators, spent more time getting organized, in spite of encouragement from management. The difficulty came mostly from the small exchanges, where there were too few employees to allow for stable or-ganizations to take root.

During the 1930s, all of the Plant Councils were grouped into the Em-ployee Representation Plan. A Bell manager from this period gave a perfect illustration of the mission of this corporate structure: "Without delegating in any way its responsibility for managing the business, management by meeting employee representatives in Joint Conferences has been able to enlist, to a greater degree than ever before, the loyal support of all employees in producing a better all-round job."[16]

The Plant Councils met once a month after working hours, and both parties could put questions on the agenda. The employees used this forum to invite members of upper management to speak to them, strongly manifesting a need to see their superiors face to face, question them, and get information from them. For Bell senior managers, who had been trained in the cult of "the divine right of the boss," so close to the heart of the founder, Charles F. Sise, it was a major change. Their naïveté was expressed by M.H. Winter, system general manager, when he described what this meant for the men who were used to managing in the shadows: "One of the main things we have found about Plant Councils is that we have got to give reasons and not excuses when we cannot meet the wishes of the employees; and that even if we have to give a negative decision, we must be honest and tell the truth about the matter."[17]

In the other Canadian telephone companies, there do not appear to have been Plant Councils. Almost all of them, along with Northern Electric, encouraged the creation of company unions, which dominated labour relations until the 1960s. These organs described themselves as associations and not unions, they had no relationships among themselves or with the labour movement, and their activities were concentrated on negotiating salaries and social benefits. In Manitoba, the IBEW survived, as we have seen, in a minority position and became so uncombative that by the end it was acting just like a company union.[18]

Whether it was the Employee Representation Plan or company unions, the result was the same: employees rallied to the corporate culture. In 1922 a mini-recession obliged most companies to reduce salaries. After the years of expansion that had gone before, it was a bitter pill to swallow, yet there was no trace of recrimination in the new institutions set up in the telephone industry. Management and employees were one hundred percent in agreement.

Was it a triumph of paternalism? Of course it was, but the declining IBEW offered absolutely no alternative. In Vancouver, the union had no better idea with which to mobilize the operators than a promise to create a "Jolly Girls Club" for them, featuring a meeting room furnished with a piano and decorated with vases of flowers and coloured curtains; every two weeks the club held a dance there.[19] The telephone companies did the same thing, but they also offered very tangible benefits, such as job security (though not in writing), social benefits, and higher salaries. They were bound to win this game, and in doing so they obtained a half-century of social peace with no strikes, or almost none (there was a short walk-out lasting a few days in Manitoba in 1950).

Simply the mention of a company union makes most unionists see red. Nevertheless, it was in this way that telephone employees in Canada and the

United States had their first experience of being organized. The company unions took precisely the opposite route from the IBEW, adopting the same structure as the telephone companies. On the positive side, the new unions were able to identify closely with the telephone employees. However, their great weakness was in the pyramidal structure of their organization and the resulting compartmentalization.

In the telegraph industry, the situation was similar. During their fight to obtain the McAdoo award, the "Big Four," and therefore the Order of Railroad Telegraphers, had obtained maintenance of a united parity worker-employer organization (the Canadian Railway Board of Adjustment), whose role was to settle salary issues and interpret collective agreements. The employees had six representatives: four for the "Big Four," one for the maintenance workers, and one for the telegraph operators.

This cooperation, which fit perfectly with Mackenzie King's philosophy, culminated at the beginning of the 1920s with the arrival at the head of CN of Sir Henry Thornton, who created committees to increase productivity. For more than twenty years, the representatives of the "Big Four" worked with management at every level – local, regional, and national – greatly improving both productivity and working conditions. It is estimated that 78 percent of the proposals made by the committees came from the union ranks. Everywhere in the telecommunications industry, a North American version of corporatism reigned.[20]

THE CRISIS OF THE 1930S

During the depression, the telephone companies' activities contracted sharply. For the first time, the number of subscribers fell; in Canada as a whole, there were 13.6 percent fewer in 1933 than in 1929. Also for the first time, the number of employees decreased. Bell had eighteen thousand employees in 1929 and just ten thousand in 1934. Most of this drop was obtained through a hiring freeze and various incentives (severance pay, early retirement). To keep its skilled personnel, the company reduced the technicians' and managers' work week by 20 percent, with a corresponding reduction in salary – an additional 10 percent drop in salary was imposed later – thus limiting the number laid off.

Operators accounted for most of the permanent lay-offs. The depression coincided with automation of the exchanges, and it was the conjunction of the two events that provoked this bloodletting. In 1926 Bell had begun to warn operators when they were hired that their job was not permanent.[21] As a result, the proportion of men in the company, which had reached historic lows in 1920 and risen slightly since, reached a peak toward the end of the 1930s.

Automation of exchanges led to a decrease in demand for operators, and the crisis transformed this decrease into deep cuts. The company insisted that

Table 14
Bell Personnel, 1920–40

Year	Men		Women	
	Number	Percentage	Number	Percentage
1920	3,426	29	8,706	71
1925	4,546	33	9,313	67
1930	5,599	38	8,910	62
1935	4,481	47	4,980	53
1940	5,037	47	5,641	53
1945	6,083	38	9,766	62

Source: Bell annual reports.

the most of the severance pay be given to married women so that single women could keep their jobs, but nevertheless the crisis was a catastrophe for the economically most vulnerable part of the labour force. It wasn't until the end of the Second World War that the number of women at Bell recovered to its 1930 level.

Nor was the situation rosy for the technicians, but the figures don't tell the whole story. In practice, much of the skilled labour was rehired immediately in TCTS mobile construction teams. These weren't permanent jobs, and they required workers to be separated from their families. Nevertheless, the magnitude of the TCTS venture enabled these telephone employees to get through the first years of the depression without too much hardship.

Each telephone company survived the crisis in the same way, although the modalities varied depending on the circumstances. At BC Tel, the company union, true to the West Coast tradition of confrontation, refused reductions in the work week, which obliged the company to lay off more technicians than it wanted to.

From a general point of view, telephone companies tried their best to keep their qualified workforce. They employed more technicians than were actually needed and placed them mainly on network construction, allowing them to function as a team, however skeletal. When the worst of the depression was past, in 1934 or 1935, the companies gave priority to rehiring former employees.

But the telephone industry in Canada did not again reach its 1929 level until around 1939. In the Prairies, where the depression hit harder than elsewhere, the recovery took until the end of the Second World War. From the employers' point of view, the corporatist structures had functioned to the

"I want a JOB"

You know me as a telephone. Actually, I'm a man-of-all-work, and I want a job.

What can I do? Well—

I can do your errands at the stores for groceries, household supplies, and all kinds of little items.

I can guard your home against emergencies.

I can help protect the children.

I can be a good night watchman, for I never sleep.

I'm good company, even when I'm quiet, for you know that I'm there when you want me. You won't be lonesome with me around.

I'll bring more visitors to you.

I'll work for a few cents a day, all day and every day. I'm more than worth the little you pay for my services.

Give me a trial, and see if I don't make good.

Come in or call our Business Office today to ask about service. You can call us without charge from any Public Telephone.

F. G. WEBBER

Manager

The 1929 economic crisis hit the telephone companies very hard. For the first time since invention of the telephone, the number of telephones in service dropped. A Bell advertisement used the image of the crisis to promote telephone service (1933). Courtesy Bell Canada.

extent that they had enabled the parties to avoid all social conflict. It is worth noting, for instance, that the telephone companies got through worst crisis the Canadian economy had ever seen without a single day of strike. From a labour point of view, things are more difficult to evaluate: would a powerful union have better enabled more workers to preserve their jobs and salaries?

During the interwar period, salaries in the telephone industry rose by 8 percent; this was modest but clearly higher than the Canadian average, which dropped by 7 percent over the same period.[22] In 1944 the average work week in the manufacturing sector was 48.2 hours, while in the telephone industry it varied between 44 and 60 hours for male employees (it is impossible to be more precise because of variations in traffic and in conditions from company to company).[23]

Before the Second World War, pension plans were almost unheard of in the private sector, while the telephone companies offered full coverage. In terms of paid holidays and health insurance, the telephone companies also offered benefits superior to the industry average and well above the standards of the social legislation. Within the telephone industry, it was the nationalized companies on the Prairies that, after a difficult beginning, offered the best working conditions because of the pact made between telephone employees and bureaucrats.[24]

Finally, in the 1940s telephone employees began to benefit from the fruit of their labour. At the beginning of the century, an operator had to work for five weeks to earn the equivalent of a year's telephone subscription; by 1921 this work time had dropped by about half; and by 1949 it had dropped to nine days of work. This meant that operators began to be able to afford telephone service after the Second World War, while the technicians could afford to do so in the late 1930s. One could say that at this time the telephone stopped being an alienating industry.[25]

These figures are unsettling, for they indicate that the social balance sheet of corporatism in the telephone industry in the interwar period is positive. Employees were better served by collaboration between the classes than by confrontation. This should encourage labour historians to think twice before they condemn company unions and corporative institutions like Employee Representation Plans. The telephone industry shows that, over this period at any rate, social peace was profitable.

It had taken the wake-up call of 1917–20, when an "international" union expressed in a confused but vehement fashion the malaise of telephone employees, to attain this surprising result. At that time, the beginnings of what would become a constant in the industry were sketched out: the telephone companies rarely act on their own, but react to events. Their lack of initiative is partly compensated for by a capacity to understand their social and economic environment.

MODERNIZATION OF SOCIAL LEGISLATION

The Second World War saw a radical overhaul of social legislation, manifestly inspired by the Wagner Act, which came into force in the United States in April 1937.[26] It forbade employers to maintain company unions and set limits on the right of association, and its enforcement was entrusted to a new agency, the Labor Relations Board. This was the New Deal brought to labour relations.

The result, surprisingly, was that the company unions did not disband but were progressively transformed into a true national union. The National Federation of Telephone Workers was officially created in 1939 from AT&T's old company unions. This still fairly loose federation called a national strike – which it lost – in 1947, after which, in 1949, it reinforced its central organization, adopted its current name, Communications Workers of America, and affiliated itself with the Congress of Industrial Organizations.

In Canada, an Order in Council issued by the federal government in February 1944 stipulated that no employer could dominate a workers' union or an association of employees.[27] This was the end of both the company unions and the Employee Representation Plans. The new legislation established the right of workers to form unions and institute any procedure of definition of negotiating units, accreditation, obligatory collective negotiation, and right to strike. The agency charged with enforcing the new law was the Wartime Labour Relations Board. Canadian labour relations thus, in their turn, entered the modern era, and the telephone companies lost the initiative in the matter of social policy.

In contrast to the industry in the United States, Canadian telecommunications were decentralized, and provincial labour legislation played an essential role, especially in the Prairie provinces. Saskatchewan's Trade Union Act (1944), the Alberta Act (1947), and the Manitoba Labour Act (1948), though quite different in inspiration, had effects similar to the federal law. Everywhere, unionism arose from the ashes.

Bell was a case apart since it had no real company union. The Employee Representation Plans were transformed first into company unions and then into independent unions – although the distinction between a company union and an independent union was tenuous. In 1943 the Plant Employees' Association was created in Ontario in response to a provincial law. With the advent of federal legislation, the embryonic association fell under the authority of the WLRB, which forbade it from receiving travel money for its members from the company. To defend itself, the association was forced to redefine itself as an independent union. There were thus four "independent" unions at Bell: the Plant Employees' Association, the Traffic Employees' Association, the Accounting Employees' Association, and the Commercial Employees' Association.

In spite of accreditation by the WLRB, there was no break between the old Employee Representation Plans and the new associations, which initially behaved exactly like company unions, holding separate negotiations with the company. The evolution towards a more independent form of unionism took place slowly. In May 1949 three associations merged to create the Canadian Telephone Employees' Association (only the traffic employees did not join). It seems that, at the beginning at least, the founders of the CTEA wanted a common front with other unions, as their participation in conferences seeking to create a national union testifies. According to them, it was the dogmatism of the labour unions that was responsible for the failure of this timid initiative.[28]

The advent of independent unionism coincided with a reduction in work time, although it was Bell that first instituted this social benefit. The five-day work week came into effect in May 1946, when the first collective agreement was signed between the company and the plant employees. The number of hours worked per week dropped to forty, a reduction of four from the previous regime, without loss of salary.

THE STRUGGLE FOR A UNION UNIQUE TO THE CANADIAN TELEPHONE INDUSTRY

In British Columbia, unlike elsewhere in Canada, the situation evolved along American lines. In the spring of 1944, the company union that had represented technicians since the 1920s merged with a new association of operators to form the Federation of Telephone Workers of British Columbia.

One of the first moves made by the FTW was to contact other unions and associations in the telephone industry throughout Canada and convene a national congress in Vancouver, which took place in November 1946. All of the unions attended except those in the Maritimes, which were too far away and, no doubt, too poorly organized. Bell was represented by the Plant Employees' Association; Alberta Government Telephone by the IBEW, which had been revived in that company; the Saskatchewan Telephone department by the United Telephone Workers of Canada, which was affiliated with the Canadian Labour Congress; and Manitoba by the Independent Brotherhood of Telephone Workers.

The FTW wanted to create a national telephone union in Canada on the model of the National Federation of Telephone Workers in the United States. The 1946 discussions, however, seemed to create a polarization between hard-liners and moderates. The latter wanted to set up a decentralized grouping of unions, while the former were adamant that the unions affiliate with a central labour union.

And even the hard-liners were divided. The Alberta IBEW was affiliated with the Trades and Labour Congress. Saskatchewan's United Telephone

"The BELL is a good place to work"

Year after year hundreds of young men and women find congenial jobs at the Bell where an ever-expanding business offers unlimited opportunity for advancement. The good fellowship of telephone people, their pride in their jobs, and their ability to work together are some of the reasons why people say "The Bell is a good place to work."

THE BELL TELEPHONE COMPANY OF CANADA

Bell advertisement vaunting the working conditions in the company (1949).

Workers had become the best-organized union in the industry since a social democratic government had been elected in that province in 1944, but it was affiliated with the competing central, the Canadian Labour Congress. Under these conditions, the hard-liners could not win. The conference ended with a compromise fashioned by the FTW: first a grouping of all unions, then an affiliation with one of the two Canadian centrals.

Unfortunately, this solution pleased no one. The moderates, led by the Bell employees, could not endorse such a motion, which alluded to accreditation, and their rank and file rejected it by a wide margin.[29] One year after the conference, they gathered their objections in a resolution, which they submitted to the other unions. It was impossible to do anything without the Bell employees. As for the Alberta and Saskatchewan hard-liners, they were uninterested in a federation that they felt was too soft and that, if it had yielded to their demands, would have had to choose between their respective centrals and thus exclude one or the other of them.

The FTW kept up its contacts with Bell's Plant Employees' Association. At one point, the FTW's directors were approached by the powerful American CWA, but this bold flirtation was rejected by the rank and file, which did not want to be swallowed up by a foreign central. They voted three times, and each time they rejected affiliation. The CWA had more success in Saskatchewan, where it gained affiliation with the United Telephone Workers.

In 1957, a full eleven years after the first conference, a new national conference was organized in Winnipeg on the issue of a single union. Bell employees were represented by the CTEA and the Traffic Employees' Association. But the wonderful structural and cultural fluidity that had sustained the enthusiasm of the Vancouver meeting no longer existed. The workers' unions had become fossilized within dogmatic and intransigent bureaucracies. On paper, all of the conditions needed for success were there: the TLC and the CLC had merged, so there was no longer an institutional obstacle to a national group. But the CWA had manipulated the organization of the conference so well that the IBEW delegates had been excluded.

The independent unions, led by the CTEA and the FTW, then decided to organize a truly national conference the following year in Toronto. This third conference brought together representatives of all the workers from all the companies, and it failed for precisely this reason: the CWA and the IBEW could not come to an agreement. In fact, their only interest in a single union was to raid the ranks of the independent companies. The dream of national unity among Canadian telephone unions had ended.

15 The Canadian Regulatory Model

After the First World War, no one in Canada doubted that the telephone industry was a natural monopoly. The Board of Railway Commissioners' historic 1911 decision in the Ingersoll case (see chapter 5) had locked up the market for good. This exceptional situation made the creation of a neutral and specific mechanism for setting "just and reasonable" rates even more urgent, so that telephone companies could raise the capital they needed to operate, and so that subscribers would be protected against the arbitrariness of a monopoly. To that end a new mechanism based on control of the rate of return was being tested.

In the United States, regulation of rates charged according to rate of return had been established for the first time in 1898 during the Smith v. Ames lawsuit. The courts ordered the regulatory body concerned to calculate the rate of return starting from an assessment of the replacement cost of assets less depreciation.

We know the difficulties that the Board of Railway Commissioners faced when it wanted to apply this method to Bell. It was Nova Scotia's Board of Commissioners of Public Utilities that opened the way to "scientific" regulation in 1914 when it ordered a thorough assessment of the assets of Maritime Telegraph and Telephone. According to the Nova Scotia commissioners, "Under the circumstances the only safe ground from which rate-making can start is the existing value of the property employed with the addition of such intangible capital as may be properly considered."[1]

Assessing the replacement cost of MT&T's property proved to be a considerable project, which lasted from September 1914 to January 1917. However, the exercise proved fruitful. Soon afterward, the company made an application to the board for a rate increase, and all the commissioners had to do was calculate the replacement cost of its system less the depreciation and apply an 8 percent rate of return (set by the government for all regulated companies). In its decision rendered in June 1918, the board accepted the rates proposed by MT&T and specified that all surpluses were to be shared between the company and the provincial government in a proportion of one to three. This method transformed regulation completely: empiricism, with its approximations, yielded to objectivity with its claim to being scientific.

But the postwar period was difficult for the telephone companies. The increased cost of raw materials during the war had caused swelling investments in the network, galloping inflation had pushed up operating costs, and militant unionism had provoked brutal salary hikes. In 1919 MT&T had to go before the Board of Commissioners of Public Utilities once again, claiming that the 1918 rates had been calculated on the basis of the 1914 assessment. A second rate change was authorized in July 1919. It was not really an increase, since the main change concerned the introduction of local measured service – paying for each call instead of paying a flat rate – to the business customers. This was an unpopular measure. Manitoba had adopted LMS in 1911, arousing great controversy. In the United States, New York had adopted it in 1894, but this mode of payment remained the exception in North America, although in Europe most of the PTTs used it.[2]

MT&T did not go before the Nova Scotia Board of Commissioners of Public Utilities again until May 1952. Another world war and a new round of inflation finally made it necessary to impose a generalized and massive (26.2 percent) rate hike. This long lull illustrates the advantages of regulation by calculating a rate of return on the basis of capital.

RATE INCREASES AT BELL

The Board of Railway Commissioners had not been as strongly proactive as had its Nova Scotian counterpart, and so it was caught off guard when Bell did something absolutely unexpected for the time: in September 1918 it made an application to raise rates. Up to then, third parties had dragged the company unwillingly before the regulatory agency. This time, the company itself was taking the initiative: it wanted to raise rates by an average of 20 percent. This request was fiercely opposed both by traditional opponents such as the municipalities and the media, to the unions whose salary demands were the main reason for the hike. Bell refused to proceed with an assessment of its assets because of the urgency of its situation. Lewis B. McFarlane, the president at the time, simply wanted an increase without a study preceding it.

In its decision, made public in April 1919, the Board of Railway Commissioners accorded an increase averaging 10 percent. It also reduced somewhat the amortization rate requested by Bell, setting it at 5.7 percent (instead of 6.5 percent). In other words, the Board had accepted the urgency argument advanced by Bell but had divided the financial burden between the subscribers and the company. This reasoning was explained by one of the commissioners: "That provision should be made, not in the interests of the shareholders, except incidentally, as above, but of the public, whose interests are to suffer by the weakening of the company by the adverse conditions mentioned, endangering the breaking down or impairment of the service in which the public is interested. The interests of the public are not distantly related in this respect to those of the shareholders."[3]

It should be noted that the commissioners had almost entirely blocked local rates from rising. On the other hand, long-distance rate increases were accepted with no discussion. Since local rates had to be kept as low as possible, notwithstanding the costs of local service, long-distance rates alone had to support Bell's financing needs. The commissioners thus created a rate imbalance, but Bell was not in a position to protest, for it had refused to unbundle its expenses in order to allow for a fair method of regulation.

Municipal administrations did not understand the board's social reasoning, and they appealed the decision to the federal government. Meanwhile, Bell's directors anxiously noted that the company's financial situation was continuing to deteriorate. They issued shares in 1919, but the measure was insufficient. With this issue, the enterprise reached its ceiling of authorized capital. To issue more stock, it had to ask Parliament to raise the limit in June 1920. The following month, Bell again went before the Railway Commission to apply for another rate increase. The appeal by the municipalities regarding the previous increase had not yet been heard. All of this created chaos in Bell's hitherto well ordered house. How could the company reassure anxious investors about the ravages of inflation if not by raising rates?

The new request contained an element that fanced the flames of popular opposition. As in Nova Scotia, Bell was asking to replace the flat rate with LMS in the business sector; but people were afraid that LMS would subsequently be extended to residential subscriber.

In April 1921, the Railway Commission accepted a new hike: ten percent for local rates and twenty percent for long distance. But it rejected unanimously the LMS proposal and lowered the amortization rate to four percent.

The board's April 1921 decision confirmed the broad principles established in April 1919 and kept the hike in basic rates as low as possible while letting long-distance rates rise more. It should be noted that these two successive increases in charges for basic service were the first in Montreal and Toronto since 1891 (except for a marginal adjustment in 1907). The annual cost of local service was now $38.50 in Montreal and $33 in Toronto for residential service, compared to $55 and $50 in the 1890s. In terms of the work time needed to pay for subscription to basic service, its cost had dropped by half. The postwar rate hikes thus did not impede the continual decline in the rates of telephone service expressed in actual value.[4]

After this second rate hike, Bell proceeded with a second stock issue in the summer of 1921, but reaction from investors was lukewarm. In general, the increase accorded by the Railway Commission was insufficient for the company to attain the authorized rate of return of 8 per cent. Bell, in a bind, applied to the Board of Railway Commissioners for a third rate increase.

This time, pressure groups and media fury could not be ignored in the board's hearing-room. Even though the hearing went on as usual – that is, as amateurishly as before (Bell's assets still had not been assessed) – the outcome was entirely different.

In February 1922 the commissioners turned down the rate hike because they felt that Bell had not undertaken all the measures it could to save money before asking for an increase in revenues. They also objected to the anachronistic character of the company's rate structure: "There seems to have been no effort [...] to adjust the rates in any scientific way to the value of the telephone service to the subscriber, having regard to the population of the telephone area, the number of stations, or the cost of service therein. The proposed rate increases, over the present rates in those places, serve to accentuate the inequitable and obsolescent features of the existing rate."[5]

The decision hit Bell like a cold shower, ending its repeated applications for rate hikes. However, the company's financial situation wasn't improving; the all-out expansion of the postwar years had imposed a pace of investment that was hard to match. Everyone wanted a cheap telephone, so Bell was forced to multiply its party lines. As well, it was more and more difficult to keep expenses down because of its social policy. The social peace sought and obtained by Bell had come at a price, and in 1925 the company registered a deficit that was small but alarming to management.

THE BEGINNINGS OF "SCIENTIFIC" REGULATION

The Board of Railway Commissioners had severely scolded Bell about the anarchy that reigned in its accounting methods and rate structure. Obviously, the company had to clean up its act. One of McFarlane's last moves as president was to order a complete inventory. The task was a major one, consisting of identifying and measuring each item in the network, from the furniture to the wires and poles. An inventory engineer was appointed, and he soon called in AT&T's experts. Five work units were created, one each for the company's major divisions: network, traffic, marketing, accounting, and engineering. Since an error was always possible, a control sample of 5 percent of the network was checked a second time. The entire process took from January to October 1924.[6]

The company was finally ready for the big confrontation. When the request was made to the Board of Railway Commissioners in January 1926 the directors of the companies met the press – a new move – to explain why. The innovation proposed in the request was to institute rate groups. Previously, each town had its own rate. Now, all towns would fall into one of seven rate groups according to the number of subscribers.

On the first day of hearings, one of the company's witnesses gave a long explanation of the general principles of rates for basic telephone service, making the point that the rate was based not on its cost of production, but on its value. And this value was a function of the number of subscribers that each one could call without paying long-distance charges. Therefore, the greater the number of subscribers, the higher the basic rate. Value was also a function of the nature of the service: business rates would be higher than residential rates. Bell had learned the lesson of 1922.

This new policy of openness bore fruit. Most of the English-language newspapers deemed the rate-increase request reasonable, while most of the French-language newspapers remained neutral; only *La Presse* continued to speak on behalf of the municipalities, but without the conviction with which it had met previous requests. In fact, opposition to Bell came mainly from the government of Ontario, Toronto city hall (with some from Montreal), and the Union of Municipalities. This last bastion of turn-of-the-century populism demanded no more nor less than a return to the 1919 rates, and the lawyer for the government of Ontario even threatened to send Bell's board of directors to prison. The company was asking for an overall increase of about 9 percent, with basic service bearing the brunt of the rise in rates. In fact, long-distance rates had begun to drop, and Bell intended to continue down this road.[7]

Debates once again focused on Bell's depreciation rate and its links with Northern and AT&T. A detailed analysis on automation of the exchanges was also brought forward: how much savings would automation bring? Bell was forced to make the first full public disclosure in its history. Its accounts were gone through with a fine-toothed comb. In his always colourful language, the Ontario representative snarled about Bell's executives, "They should come here absolutely naked. Even if all their cards were on the table that would not satisfy me. I should still want to see what was trumps, kings or knaves, and always in a case of this kind I have found it to be knaves."[8]

Bell decided to take its opponents literally. The last-ditch populist demagoguery was buried under an avalanche of facts and figures. Bell's inventory was, of course, added to the pile of documents. The parties engaged engineers and experts and asked for a number of adjournments to give them time to digest the mass of materials. The hearings produced a total of seventeen thousand pages of testimony and one hundred and eighty exhibits, some of which were hundreds of pages long. The Board of Railway Commissioners finally followed the example of Nova Scotia's Public Services Commission

and jumped feet first into the scientific era. Between adjournments and post-
ponements, the hearings lasted more than a year, and the sessions began to
lose steam. The newspapers, which had at first reported most of the sessions
verbatim, gradually let the issue drop, as it had become too technical.[9]

In its decision rendered in February 1927 the Board of Railway Commissioners approved most of the increase ($2 million instead of $2.7 million)
and depreciation rate (5.34 percent instead of 5.41 percent) requested, and
legitimized the connection between Bell and AT&T. So that it could track the
impact of the new rates, the commission required Bell to submit monthly operations reports. The stream of information that would flow between Bell
and the Commission from then on was perhaps the best guarantee of democracy in the management of telephone service. Finally, the commission
adopted the theory of the value of the telephone advanced by AT&T and
taken up by Bell. The regulatory body and the telephone company thus
agreed on a rates philosophy.[10]

The major advance of the 1920s was the establishment of objective rules
of the game between the regulatory body and the regulated company. The
Board of Railway Commissioners forced Bell to put its accounting methods
and rates in order, and Bell was scrupulous in doing so. The result was a regulatory lull that lasted until 1949, except for an isolated incident in 1938
when a member of Parliament demanded that Bell reduce its rates by 25 percent. The commission turned down the request, which it felt was based on er-
roneous facts.

It should be noted that in 1938 the National Transportation Act broadened
somewhat the powers of the Board of Railway Commissioners, which was
renamed the Canadian Transport Commission. This change in powers did
not concern the telephone companies.

BRITISH COLUMBIA FALLS UNDER
FEDERAL JURISDICTION

British Columbia was a special case because of the long rearguard battle led
at the beginning of the century by British Columbia Telephone, in league
with British Columbia Electric Railways, to keep the provincial government
from creating a public-utilities commission. When BCER was won over to the
idea of regulation in 1915, BC Tel saw the writing on the wall. The company's president, William Farrell, feared provincial regulation, which in his
view would be manipulated by populist agitators. In April 1916 he federally
incorporated a company called Western Canada Telephone. This company
leased BC Tel's installations in September 1918 and adopted in its turn the
name BC Tel in November 1919.[12]

The result of this manoeuvre was to place BC Tel under the jurisdiction of
the Board of Railway Commissioners. In theory, Farrell had arranged things

so that BC Tel could return to provincial jurisdiction (the British Columbia government had created its own public utilities commission that year) simply by cancelling the lease for the telephone system. In practice, federal regulation proved to be irreversible; in January 1923 the two companies that then bore the same name merged.

In March 1921 BC Tel applied to the Board of Railway Commissioners for its first rate increase. Until then, BC Tel's rates had been ruled by a legislated rate ceiling. The company had always set its rates below this maximum and thus could modify them freely. But all of this changed under federal regulation. In the midst of the postwar inflationary spiral, BC Tel prepared its argument carefully. Bell's 1919 application had been gone over with a fine-toothed comb, and BC Tel's rates were based on Bell's but were somewhat more moderate: instead of a 20 percent hike in local rates, BC Tel wanted 12 percent; it did not request rate increases for long-distance calls; and its proposed amortization rate was 6.22 percent. BC Tel also had an advantage over Bell: between 1916 and 1918, before leasing its network, it had conducted a complete inventory of its assets. Curiously, however, the board paid no attention to this inventory, preferring to rule as a function of "reasonable" rates. Regulation by rate of return was not yet a consideration, so in June 1921 the board accorded BC Tel an increase of 10 percent, equal to that of Bell's, and an amortization rate of 6.04 percent.[13]

BC Tel did not go before the regulatory agency again until June 1949. Like the rest of Canada, British Columbia had a long regulatory lull, followed by a cascade of requests during the postwar reconstruction period.

LOCAL RATE CHANGES

Telephone rates obeyed very specific economic laws. Bell had always stated that the basic rate had to rise with the number of subscribers; since its 1927 request, it had officially adopted the principle of usage value. The greater the number of users a subscriber could call without paying long-distance calls, the greater the value. But how to put a figure to the usage value? Traditionally, it was the maximum the subscriber was ready to pay before cancelling the service.

What is less well known is that this evaluation of basic service roughly corresponded to the cost price, which also varied with the number of subscribers. When the telephone industry started up, there was a very simple reason for this: the number of calls grew in proportion to the number subscribers. An operator could answer only a certain number of calls, so the number of operators required was proportional to the growth in the penetration rate of telephone service.

For instance, if one operator could answer five hundred calls a day in a district comprising five hundred subscribers making an average of four calls per day, there would be a need for four operators. If the district was ten times larger, with five thousand subscribers, the average number of calls per subscriber rose to eight per day, bringing the total to forty thousand calls and requiring eighty operators. Thus, in a small district, the operator/subscriber ratio was 1:125; in a large district, it was about 1:60.[14]

Bell, along with the American and European telephone companies, always advanced value usage as a founding principle for setting rates simply because it was nearly impossible to define the cost price for basic service. The local system routed both local and long-distance calls; if the companies had wanted to base the price for basic service on the cost price, they would have ended up with as many different rates as there were telephones. Or else they would have had to introduce a form of LMS. In short, it would have introduced a complexity unknown (with few exceptions) throughout the North American rate structure.[15]

After having established this fundamental principle, telephone companies adopted cross-subsidization between the different categories of service: "Cost should not be the sole basis of rates. If in a system embracing several cities and towns of various sizes the rates in each of them should be based solely on the cost of service in that particular place, the rates in somes places might be prohibitive, and through having no telephone in such places other places would suffer, possibly to such an extent that the devlopment in them would be seriously interfered with."[16]

To simplify matters, the telephone companies, as we have seen, instituted rate groups; then, for efficiency's sake, they made it so that high-income groups subsidized low-income groups. On the whole, urban subscribers subsidized rural subscribers. Companies also made a distinction, within these rate groups, between residential and business service – the latter is always more expensive since it gives the subscriber greater value and also yields, income tax deductions – such that revenues from business service subsidized the money-losing residential-service sectors.

Cross-subsidization was a thundering success, and universal service seemed to be within grasp in North America in the 1920s. As we have seen, the work time necessary to pay for basic service had dropped constantly since the start of the telephone industry. In the 1920s Bell instigated a campaign promoting subscription, aimed at ever-widening sectors of the population. A great number of people subscribed to party lines, which became the tool for penetration into working-class areas on the eve of the 1929 economic crisis.[17]

IMBALANCE IN LONG-DISTANCE SERVICE

Long-distance rates evolved quite differently. Before 1919, they were uniformly proportional to the distance covered. After 1919, a higher rate was adopted for distances below twenty-four miles, after which the rate remained uniform for each additional mile. In 1926 the first distance stage was raised to forty-eight miles and two other regressive rates were established at 80 and 150 miles. What this meant was that the longer the distance called, the lower the rate in terms of distance units.

It was in long-distance calling, as opposed to local service, that the effects of technology were felt most keenly and the savings were the highest. Unlike local service, long-distance service allowed for economies of scale: it is obvious that the unitary cost diminishes in a fully utilized network. Long-distance traffic had constantly increased since the quality of transmission had been improved with the establishment of the TCTS in 1931. It was thus possible to lower the rates on almost the entire schedule of distances, resulting in a generalized decrease in long-distance rates in absolute figures (see Table 15). The only exception, a major one, was very short distances, which involved the largest volume of calls. But this slight deviation nevertheless corresponded to a decrease in relative cost.[18]

Long distance as a whole had become the most profitable sector of the telecommunications services market. This was not the case at the beginning of telephony, as we saw earlier, and it took all the insistence of Theodore Vail and Charles Sise, Sr, to build long-distance lines. The situation had reversed between 1910 and 1920 with technological progress. The arrival of electronics in the form of vacuum lamp repeaters in long-distance lines was a key factor.

Long-distance rates in Canada were aligned with those in the United States until July 1941, when Canada stopped following American rate decreases in favour of a more moderate policy. There are no documents to explain this sudden unlinking, but it seems to have been due to the general price freeze during the war. Since they could not hope to increase rates for local service, the Canadian companies slowed the drop in long-distance rates, although they were aware of the danger that the imbalance could cause, as this statement by a TCTS director in 1946 reveals: "Nevertheless, regardless of the reasons which we may be able to produce to support a higher level of toll rates here than in the United States, too wide a spread would seem to be dangerous."[19] What an understatement! The unlinking was the basis for an imbalance between the two countries that in the 1980s and 1990s would provoke a revolt of the business world against Bell and against Canadian rates in general and was one of the main arguments in favour of competition in the long-distance sector.

Curiously, the 1941 decision was made by Bell and BC Tel, not by any regulatory agency. The companies filed the long-distance rates with the agency, which accepted them without interrogation, examination, or hearings, and they were then applied across the TCTS. The creation of TCTS in 1931 had opened a lengthy regulatory vacuum. As we have seen, starting with the 1919 decision, the regulatory agency never paid much attention to long-

Table 15
Long-Distance Rates, 1920–50*

Year	Montreal– Ottawa ($)	Montreal– Toronto ($)	Montreal– New York ($)	Montreal– Vancouver ($)
1920	0.70	2.05	2.30	14.60
1930	0.65	1.90	1.65	8.00
1940	0.65	1.75	1.45	6.75
1950	0.80	1.75	1.25	4.70

Sources: Canada, Statistics Canada, *Historical Statistics of Canada*, for Canadian data;
BCA, rate files, for US data.
*Daytime rates, station to station, for a three-minute call.

distance rates. This indifference was understandable when long-distance service was a negligible part of the business; however, it is estimated that
between 1928 and 1958 the share of long-distance calls in Bell's total revenues rose from 26 percent to almost 39 percent.[20]

Long-distance service became an increasingly important part of the industry thanks, in particular, to the creation of the TCTS. But, with no assets or
staff and an informal structure, the TCTS was invisible to inquisitive regulatory eyes until 1977, when it did battle with the Canadian Radio-Television
and Telecommunications Commission over Telesat. In the interval, it could
be said that long-distance rates were set by Bell in consultation with the
other members of the TCTS, and the regulatory agency had nothing to do
with it. These rates and the very important issue of sharing the resulting revenues between companies constituted no less than the building blocks of
Bell's leadership in Canadian telecommunications over this period.[21]

UNIVERSALITY BECOMES A REALITY

The objective of all telephone companies and regulatory agencies was to realize Alexander Graham Bell's dream of having the telephone system linking houses "like natural gas or water – and [for] friends ... to converse with
each other without leaving home."[22] In the 1920s this goal looked close at
hand. But then came the 1929 economic crisis and a step backward between
1930 and 1933, during the height of the depression. The end of the crisis was
slow in coming, and it wasn't until 1940 that the total number of installed
telephones exceeded the 1930 level; on the basis of penetration rate – the
only true measure – it wasn't until 1943. After the Second World War, a

Table 16
Telephone Penetration Rate per 100 Residents, 1915–56

	1915ᵃ	1920ᵇ	1930	1933	1940	1945	1950	1956
Newfoundland	–	–	–	–	–	–	6.1	8.9
Prince Edward Island	2.5	5.4	6.7	6.0	5.9	8.3	11.9	14.0
Nova Scotia	4.6	6.5	8.4	8.3	9.3	11.2	16.4	30.8
New Brunswick	5 .1	6.9	8.2	7.1	7.9	9.7	14.0	18.8
Quebec	4.7	5.7	11.1	8.7	10.0	11.9	18.5	25.6
Ontario	9.8	12.5	19.0	15.4	17.5	20.6	28.0	35.2
Manitoba	10.4	11.5	11.6	8.6	10.6	13.6	18.9	26.1
Saskatchewan	6.5	12.0	10.7	7.8	9.2	12.3	15.5	21.7
Alberta	9.4	9.3	11.9	7.6	9.3	11.1	15.8	24.6
British Columbia	11.2	14.5	21.6	16.4	18.5	18.9	24.4	33.3
Canada	7.6	9.8	14.1	11.2	12.8	15.3	21.1	28.2

Source: Statistics Canada, telephone statistics, 56–201 (except as noted).

Notes: a. *Telephone Stations, Dominion of Canada, 31 December 1915*, L.B. McFarlane
(BCA 34 5645).

b. Author's estimate from Bell Canada and Statistics Canada data.

giant leap forward propelled the penetration rate toward universality. In 1956, if the goal was not yet attained, all indications were that it was about to be. The Great Depression had postponed the achievement of Alexander Graham Bell and Theodore Vail by about twenty years.

As Table 16 shows, the highest penetration rates were in Ontario and British Columbia, where the telephone systems were managed by private companies under federal regulation. At the other end of the scale were Newfoundland and Prince Edward Island, whose service was provided by private systems under provincial jurisdiction. The nationalized networks on the Prairies were in the middle position. More troubling was the contrast between the excellent performance of Ontario and the lower penetration rate in Quebec (with a relative improvement after the depression). These two provinces had a common system, belonging to a private company under federal regulation, and widely disparate results. There was thus no correlation between penetration rate and company ownership or regulatory style.

One could analyze the differences in the rate of telephone penetration from the perspective of the economic prosperity of the societies involved: changes in per capita personal income paralleled those in subscriptions.

HIGHWAYS OF THE VOICE

In 1940, Bell's advertising invented the concept of the voice highway … presaging the information superhighway. Courtesy Bell Canada.

Every year, the provinces with the highest per capita income and the highest penetration rate (except for 1915, in Ontario's case) were Ontario and British Columbia. Almost every year, Prince Edward Island was on the low end of the scale, and this province had the fewest telephones. When Newfoundland began to appear in the statistics, it took last place from Prince Edward Island. Similarly, the relative poverty of Quebec and wealth of Ontario provide one explanation for the disparity in the penetration rate in these two provinces.

However, these results could also be looked at from the point of view of technology – that is, the presence of party lines in rural areas and, marginally, in the cities (see Table 17). British Columbia had the highest proportion of party lines in Canada; thus, the high penetration rate in this province is the result of a cheap technology, in the truest North American tradition. Ontario had a party-line rate consistently higher than that in Quebec, for the same reason. However, there was a much weaker correspondence between the penetration rate and the proportion of party lines. Some companies relied on party lines, while others did not. Prince Edward Island, New Brunswick, Nova Scotia, and Saskatchewan multiplied their party lines in a context of hardship. British Columbia and Ontario did so in a context of prosperity, and these were the provinces with the highest rates of telephone subscription.

On the other hand, the same gap between Quebec and Ontario had persisted over time. The Quebec system was of better quality than the Ontario one, and this must have contributed to slower development in the poorer province. But how to explain that Bell installed a relatively expensive technology in a poor province? Nothing in the company archives indicates that there was a deliberate policy in favour of party lines in Ontario. In the absence of any other explanation, one must turn to a cultural reason. Did Quebec's francophone Catholic society prefer no telephone to a party line? Was the desire for confidentiality stronger in Quebec than in Protestant anglophone Ontario? If so, the explanation for the disparity would be sociological as well as economic.

In general, the technological factor did not radically modify the correspondence penetration rates and personal income. It amplified or diminished gaps, but it did not cancel them out. Telephone service spread in a very unequal way across the country, and this inequality corresponded almost perfectly to the relative wealth of the various provinces.

THE GREAT TELEPHONE RUSH (1945–56)

After the Second World War, Bell had to confront two problems: the first, system reconstruction, it was well aware of, while it had underestimated the second – explosion in the demand for hook-ups. The term "reconstruction" might seem surprising, since Canada had not been devastated by bombs.

Table 17
Party Lines, 1939–56 (percent of party lines in all lines)

	1939	1945	1956
Newfoundland	–	–	38.90
Prince Edward Island	54.97	59.04	51.76
Nova Scotia	41.56	47.59	44.75
New Brunswick	45.55	52.06	50.49
Quebec	33.91	40.80	41.84
Ontario	45.96	51.69	43.86
Manitoba	28.49	29.74	44.89
Saskatchewan	54.95	49.38	32.56
Alberta	24.39	22.20	10.41
British Columbia	58.17	62.02	63.88
Yukon	77.78	84.21	89.29
Northwest Territories	–	–	–
Canada	42.89	47.25	42.87

Source: Canada, Statistics Canada, Annual Catalogue 56–203.

However, systematic underinvestment for five years had allowed the network to become disorganized to the point that all company directors spontaneously used the European image of reconstruction.

Furthermore, salaries had risen by 70 percent since 1939, and Bell made applications to the Canadian Transport Commission for successive rate increases in 1949 and 1951. The CTC inaugurated a method for regulating earnings per share, consisting simply of setting profits at a level that enabled the company to sell stock on the financial markets. The return required by investors was then equal to the total amount of dividends paid out, while during the interwar period, regulation had simply endorsed the historic return on dividends and added a growth component.[23]

In November 1950 the commission allowed, in its entirety, the first increase requested; in February 1952 it accorded 80 percent of the second increase. In spite of the regulatory innovation of taking into account earnings per share, the two hearings unfolded like a bad remake of those of the 1920s. The only intervenors were the municipalities, along with a few Ontario chambers of commerce. Although they did not question Bell's reconstruction program, they criticized its breadth, inquired about the urgency

Bell advertising after the Second World War. The number of hookup requests exceeded Bell's capacity to provide connections and universality of service was in sight. Courtesy Bell Canada.

of the backlog of requests for service, demanded a reduction in the dividend of two dollars and of the surplus of forty-two cents that accompanied it – and last but nor least, they asked that the employee pension plan be cut!

For its part, Bell refused to disclose the respective cost prices for local and long-distance service, claiming that these figures had nothing to do with rate determination. Strictly speaking, this argument was correct, but it was an indication of the company's ill will in the face of the requirements of regulatory transparency. It is true that Bell could point to extenuating circumstances: the City of Montreal was demanding this comparative balance sheet for the sole purpose of opposing cross-subsidization of isolated regions by cities, a practice it considered discriminatory.[24] Thus, the intellectual weakness of the intervenors' arguments reinforced the Bell directors' feeling that they were right and everyone else was wrong.

The CTC's 1950 and 1952 decisions were technical and rejected the purely economic argument of the municipalities in favour of long-term development of the system, the logic of which was that the public was demanding and deserved to obtain modern and efficient telephone technologies.[25] Thus, the commission was denying municipalities representing the general public the right to speak in the public interest. Bell was *ipso facto* appointed privileged representative of the subscribers, which was crystal-clear proof of the inadequacy of the public hearings process.

The intervenors did have win on one point, however: long-distance rates. In the 1952 request, Bell asked only for an increase in local rates. The intervenors asked for and obtained that the increase be spread uniformly over local and long-distance service in order to keep local rates as low as possible. Bell had opposed any increase in long-distance rates because of the growing gap in these rates between the United States and Canada. However, the fact that this danger was completely real was not enough for Bell to separate the respective costs of local and long-distance service and proved that long-distance calls were already overbilled.[26]

As important in another way was the CTC's 1950 decision regarding BC Tel, a subsidiary of the Gary group, which was found guilty of purchasing at an inflated price equipment made by another Gary subsidiary. Although the CTC did not have the right to regulate the supplier or cancel the disputed purchase contract, it subtracted the excessive expenditures from the regulatory base. Of course, this measure was enough to make BC Tel see the light; in 1953 it began to pay normal prices once again.[27]

This action underlined the beneficial role that federal regulation played. In spite of its detractors' complaints, the regulator was far from making common cause with the industry. On the contrary, the light-handed and evolutionary regulation effectively played a watchdog role that was absolutely essential in a sector composed of monopolies. The BC Tel precedent proved that the relationship between Bell and Northern Electric had always been well balanced, since the regulator had found nothing untoward in their relationship in many examinations that had taken place since 1907.

Throughout Canada, the 1950s were characterized by telephone companies going before their respective regulatory bodies (SGT had its rates approved directly by the provincial government). Given the urgent need for reconstruction, generous rate hikes were systematically accorded. This was the first wave of increases since the 1920s – in some cases since the 1910s – but it was not the last.

COMPETITION SNEAKS IN THE BACK DOOR

Since the 1930s, the telephone and telegraph companies had rediscovered that they were competitors due to the use of carrier-current technology for long-distance lines. With the advent of microwaves in the 1950s, a new opportunity for conflict arose.

When the Canadian Broadcasting Corporation submitted a proposal for a trans-Canada microwave network, it found the competitors lined up as follows: the telephone companies, allied within the TCTS, against the associated telegraph companies CN and CP. The Department of Transport, which issued operating permits for wave lengths, was hoping that the two consortiums would agree to make a common submission. Bell opposed this, for it wanted the infrastructures to belong exclusively to the TCTS, even though the telegraph companies might later operate it jointly with the telephone companies. The federal government intervened and settled the dispute on 24 September 1953: "The Cabinet agreed that no one person or corporation should receive a monopoly to operate a microwave relay system"[28] Of course, the network would be designed to transmit television signals, but since the technology in question could also route telephone signals, things were unlikely to remain as they were. The government thus opened the door to competition in the telecommunications networks.

In the adjacent sector of terminals, the first challenge to Bell's monopoly took place in 1955. A small Ontario company, United Sterl-A-Fone, had invented and manufactured an apparatus designed to destroy with ultraviolet waves microbes and bacteria that could become lodged in a telephone receiver and transmit disease to its various users. Farfetched or not, Bell took it very seriously, refusing to allow the apparatus to be connected to its wires. United Sterl-A-Fone appealed to Ontario Superior Court, which found that its apparatus contravened Bell's general rules as approved by the CTC. Regulation nine, in particular, provided that no equipment, telephonic or otherwise, could be attached to a Bell apparatus or to the system unless there was a special agreement between the company and the user. United Sterl-A-Fone had not obtained such an agreement from Bell; exit Sterl-A-Fone.[29] This incongruous episode took place at the same time as a similar episode involving a device called the Hush-a-Phone in the United States. Although the conclusion of the two affairs was the same – confirmation of the monopoly – it showed that certain users were uneasy with the existing market structure.

In fact, the 1950s saw a proliferation of agreements between Bell and certain users with special needs. The military connected up its radar and other Doppler stations to the public network with the consent and cooperation of the TCTS. Hospitals began to connect up electrocardiograms to the telephone system: it was unthinkable for the telephone companies to refuse this type of service. But technology multiplied needs and it became more and more difficult for Bell to satisfy them all. This naturally posed the question of the monopoly of one company over a network that was becoming less involved with telephony and increasingly diversified in terms of telecommunications.

16 Electromechanical Technology Hits Its Peak

After the First World War, there was no telephone network, in the true sense, in Canada. There were manual telephone exchanges linked to each other in an unsystematic way, while long-distance lines were still the exception and long-distance communications were expensive and of poor quality. The telephone was perceived as a technology with a local community-based vocation.

However, all the technologies that were to bring telephony out of this restrictive framework had already been invented: the vacuum-lamp repeater could allow the voice to be carried over theoretically unlimited distances, radio could take voice communications across oceans, and switches could automate telephone exchanges. In the interwar period, the universal telephone network was born.

AUTOMATION ACROSS CANADA

Canada backed into the era of automated telephony. The reason, as we have seen, was Bell's refusal to abandon manual exchanges. After the First World War, there was an absurd situation in which most Canadian telephone companies had begun to automate their exchanges except for the largest ones – BC Tel and, of course, Bell. Moreover, the opposition of the only company that had the resources to carry out a complete automation development program had slowed the change-over in other companies.

After the *Mont-Blanc* exploded in the port of Halifax in 1917, Maritime Telephone and Telegraph decided to reconstruct one of the destroyed exchanges with an automatic switch. The Bell and New England Telephone

Bell's first automated exchange, the Grover exchange in Toronto, opened in July 1924. A special group of employees called all subscribers served by the exchange to explain the nature of the change. Courtesy Bell Canada.

experts tried to dissuade MT&T from automating on the pretext that the technology was not perfected. But they did not take into account the determination of A.J. Barnes, an ex-New England Telephone employee who had been working at MT&T since 1911 and had risen to chief engineer. Barnes was convinced that automation was the solution of the future, and he did not hesitate to defy the opinion of Bell and even the omnipotent AT&T. He won over company management, which mobilized the company's meagre resources to conduct the indispensable studies, and in 1919 it purchased Automatic Electric step-by-step equipment. A switchboard began operating in February, 1921, in the Lorne exchange. Thanks to Barnes's farsightedness, MT&T was the first non-independent and non-nationalized company in Canada to adopt automatic switching.

The Prairie companies were far advanced in terms of automation: the Saskatchewan Telephone Department had started the trend in 1912 in Saskatoon and by 1921 it had six automated exchanges.[1] Alberta Government Telephones also had six, and Manitoba Government Telephones had two. In addition, step-by-step switches had been installed by independent companies all over Canada, among the largest of them Edmonton Telephones, which had had an automated exchange since 1908. New Brunswick Telephone seems to have had one automated exchange in Woodstock, which it acquired when it purchased an independent company. But any efforts at automation remained marginal because of the absence of the main actor: Bell.

The historic turnaround was finally made by AT&T in 1919. In that year, the company's annual report announced its decision to automate its new exchanges. Automatic Electric's step-by-step equipment had been vastly improved since its first attempts in 1892. Moreover, AT&T already had some Strowger switchboards in its system due to acquisition of independent companies.

LARGE-SCALE AUTOMATION AT BELL

In Canada, Bell announced its conversion to automation in its 1920 annual report. The company was hesitating, however, between a mix of technologies, such as AT&T's in the United States (step-by-step in rural zones and small towns, panel in large cities), and a homogeneous step-by-step solution. For the first exchange to be automated, the Grover exchange in Toronto, all studies pointed to installing a panel switchboard. Western Electric was exerting considerable pressure in that direction and Northern Electric had already modified its assembly lines to manufacture it when Bell's chief engineer, Robert V. Macaulay, the future builder of the Trans-Canada Telephone System, convinced top management to reconsider and choose the step-by-step system. This was an important decision. It was the first time that Bell said no to its parent company when it came to technology. The all-Canadian solution was dictated by the needs of the market. Only Montreal and Toronto (and Vancouver in the case of BC Tel) were large enough concentrations of population to have justified adoption of the Panel. Furthermore, the choice of the step-by-step switch would standardize the equipment and achieve appreciable economies of scale in terms of supply and maintenance.[2] One amusing detail was that in September 1924, Bell hired MT&T's chief engineer to oversee automation of the exchanges; A.J. Barnes thus had the satisfaction of seeing the audacity that had led him to go against Bell's warning and automate Halifax in 1921 rewarded by the very company he had defied.

Thus, the first switchboard was installed in July 1924 in the Grover exchange in Toronto; the second, in the Lancaster exchange in Montreal, in April 1925. In 1933 Quebec City became the first completely automated city in Bell's territory.

Fearing a negative reaction to what it still perceived as a reduction in quality of service, Bell deployed a massive advertising campaign to inform employees and, especially, subscribers of the change. In the company's internal documents, the "burden" of having to dial a number was constantly mentioned. For subscribers, automation of the exchanges meant the introduction of the first telephones with dials. Until then, the telephone had always evolved towards greater simplicity: elimination of the acid batteries that had to be replaced regularly, elimination of the crank that had to be turned to activate a generator. Suddenly, a new element was being added to the apparatus.

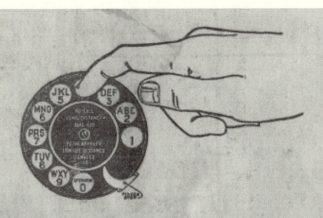

Do you know
how to Dial?

Telephone subscribers in the AMherst area are now being instructed at their own premises in the use of the automatic telephone.

We hope it may be convenient for all AMherst subscribers to receive our instructors when they call. If that is not possible, we invite you to call at 118 Notre Dame W., where we can demonstrate in a few minutes the proper use of the new telephone.

AMherst Automatic Exchange will go into operation at midnight, Saturday. August 1st.

Why not learn to-day how to dial?

F. G. WEBBER,
Manager.

Advertisement explaining to Montreal subscribers how to dial a telephone number. Courtesy Bell Canada.

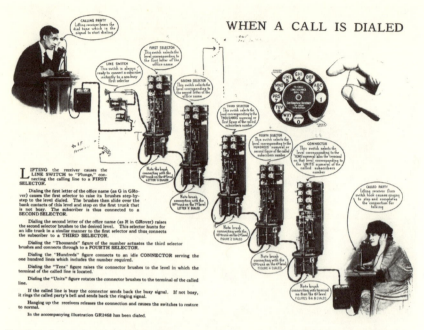

Brochure for employees explaining how to use a dial telephone (1923). For a long time Bell refused to automate its exchanges, to avoid imposing the "burden" of dialing a number on subscribers. Courtesy Bell Canada.

Bell deployed a massive advertising effort to explain to employees and customers how to use the dial telephone. Here, Bell employees tell Montreal subscribers about the change. Courtesy Bell Canada.

For subscribers, automation of exchanges meant the appearance of the first dial telephones. Courtesy Bell Canada.

Aside from this functional modification, the technical make-up of the telephone apparatus remained more or less the same. The receiver had been added in 1928, but the bell was always separate from the instrument (attached to the wall); it was not integrated into the telephone case until 1937, when the black telephone was born. Like the famous Ford Model T, telephones were available in only one colour, and usually in only one model (though there were, in fact, two: the desk model and the wall model). In the telephone industry, as elsewhere, it was the time of mass production, which meant uniform and cheap products.

It wasn't until the 1950s that this assembly-line philosophy was slightly modified. In 1952 Bell launched the model 500, designed by Western Electric and made by Northern Electric, as the ultimate black telephone. Three years later, a thunderbolt hit the telecommunications world: Bell threw

In the 1930s, telephones were available in one colour and two models. Shown, the Model 100 table model. Courtesy Bell Canada.

caution to the wind and offered the model 500 in several colours (though still only one shape). An era had ended and the consumer society was born.

Automation of exchanges also brought a change in telephone numbers. Two letters were added to subscribers' four-digit numbers, such as LA for the Lancaster exchange in Montreal. Thus, LA 2222 designated subscriber 2222 served by the Lancaster exchange. Why letters and not numbers? Because memory tests conducted by AT&T at the beginning of the century had shown that the average person could not remember numbers above a certain length. This conclusion typified AT&T's constant extreme concern with human factors and its propensity to underestimate the potential of the man/machine dialogue.

But the main impact of automation was on the operators. Starting in 1926, Bell began warning newly hired operators that it could not guarantee the permanence of their jobs. As long as growth continued at a steady rate, there was no problem; the number of operators continued to rise slowly. Then the depression hit. In Quebec City, the number of operators dropped from 130 to 60 in 1933. Forty-six operators were laid off, but since they had all been hired after 1926, they were categorized as temporary employees. The others were transferred to other departments, where they were declared "surplus."[3]

On a purely technological level, the arrival of step-by-step switchboards in the large cities posed the entirely new problem of their coexistence with manual systems, raising the problem of making connections between the two different types of exchanges. Having the subscriber in an automatic exchange speak to the operator to connect to a subscriber in a manual exchange was rejected out of hand; that callers be required to use a different method according to the group of subscribers they were calling must be avoided at any price. Instead, a call indicator was developed that posted on the telephone switchboard the number dialled by the subscriber, and the operator then established the connection manually. In the other direction, from a manual telephone to a dial telephone, the method was as it had always been: through an operator.

It should be noted that an operator's work station with a call indicator was very expensive (around $14,000), but the notion of service quality held by the telephone companies was such that they did not shrink from such an expense. This effort was even more praiseworthy given that the number of work stations with call indicators reached its apex in the mid-1930s, when the depression was at its worst.[4]

AUTOMATION TAKES HOLD IN CANADA

Once Bell committed itself, automation of exchanges accelerated greatly across Canada. In 1940 the number of telephones linked to an automated exchange surpassed the number linked to a manual exchange (see Table 18). However, the speed of conversion varied greatly from one company to the next, with BC Tel bringing up the rear among the large companies for a long time. Manitoba and, to a lesser extent, Alberta continued to lead the race to automation, while in Saskatchewan, which had launched the movement on the eve of the First World War, many independent companies did not see the need to make the change or did not have the means to do so and remained faithful to the manual system until they were bought out by a large company.

Table 18 shows a glaring anomaly: the gap between Quebec and Ontario, two provinces served by the same company and subject to the same regulation. Bell consistently favoured automation in Quebec. This difference is so accentuated that it cannot be simply the result of on-the-spot decisions. Yet, as for party lines (see chapter 15), I was unable to find in Bell's archives a single document explaining this decision. The reason may have been technological: the high number of existing party lines in Ontario most definitely slowed automation.

The rapid automation of Quebec could also have been the consequence of Bell's reluctance to Frenchify its operations – in particular, the operators' service. The silence surrounding the decision to automate Quebec on a high priority is strangely reminiscent of the company's systematic silence on language questions in general.

Table 18
Automation of Exchanges (percent of telephones linked to an automatic exchange)

	1929	1934	1939	1945	1956
Newfoundland	–	–	–	–	83.99
Prince Edward Island	0.00	0.00	0.00	0.00	66.52
Nova Scotia	12.01	34.53	44.88	51.43	75.85
New Brunswick	9.83	28.49	46.77	47.91	63.94
Quebec	33.84	67.20	69.83	71.44	85.97
Ontario	20.20	35.06	45.48	53.98	77.59
Manitoba	68.70	71.41	70.19	71.64	81.07
Saskatchewan	22.22	29.06	32.98	39.25	59.20
Alberta	55.37	64.40	62.36	60.40	78.18
British Columbia	0.24	14.83	17.98	34.62	64.54
Canada	25.59	42.55	49.40	56.09	77.31

Source: Canada, Statistics Canada, Annual Catalogue 56–203.

AUTOMATION ON THE INTERNATIONAL SCENE

In industrially advanced countries in Europe, automation was proportionally more rapid than it was in North America. But, it bears repeating, it involved a much smaller stock of telephones. As Table 19 shows, the rate of automation of exchanges was inversely proportional to the penetration rate, clearly underlining the difference between countries that had opted for a high-quality technology for an élite (Germany, Italy, the Netherlands) and those that preferred a cheap technology for the masses (United States, Canada, Sweden).

It should be noted, however, that Canada's automation rate was somewhat higher than that in the United States and Sweden, although its penetration rate was nearly equal to that in the other two countries: thanks to the innovative policy of the Prairie provinces and Bell's resolution once it had opted for automation, Canada seems to have put together the best of two worlds. On the other hand, France and Great Britain had a technology that was both élitist and deficient – a failure on all fronts.

While most European countries were besotted with the advanced but necessarily expensive concept of the automated exchange, North American companies were prudently waiting for the price to drop. But the low rate of automation in North America is misleading: more than half of all dial tele-

Table 19
Automation of Exchanges and Telephone Penetration Rate, 1938

Country	Automation rate of exchanges[a] (%)	Penetration rate of telephones[b] (%)
Germany	88	5
Italy	85	1
Netherlands	75	5
Belgium	66	5
Great Britain	55	6
Canada	47	12
France	46	4
Sweden	41	12
United States	41	15

Notes: a. Automation rates, with the exception of Canada's, are taken from Libois, *Genèse*, 269. For Canada, see Canada, Dominion Statistics Office (now Statistics Canada), *Taux d'automatisation pour le Canada, 1940*.

b. The penetration rates are taken from AT&T, *Telephone Statistics of the World* (annual publication).

phones were in the United States and Canada, due to a massive effort that affected a population already largely served by the telephone. Countries in Europe (with the exception of Sweden) automated their exchanges more easily because they had so few to begin with.

Finally, it should be noted that automation affected only local switching. When subscribers wanted to make a long-distance call in North America, they had to go through the operator. Here, the gap between North America and Europe was much smaller. In the interwar period, it was mostly small countries (Belgium, Switzerland) and well-defined regions (Bavaria, the French Riviera) that were served by very limited-range automated long-distance service. Nevertheless, North America was particularly late in automating long-distance service for continental calls (1950s) and even later for overseas calls (1970s).

Private branch exchange or PBX, developed in parallel to the exchanges. Automation enabled users in a company to call from one extension to another within a building without going through an operator; on the other hand, to call outside, human intervention was still necessary. The first model of this type, the 605A system, seems to have been installed in 1926 in CN's offices in Montreal.[5] But many companies kept their manual switchboards, no doubt feeling that the need to employ an operator to make external calls took away much of the attraction of "automatic" PBX. The technical evolution of the private switchboard was thus slow, and the technology changed so little that it was difficult to adapt it to the crossbar when this new generation of switches came on line.

Automation might seem to be a venture of the distant past for readers inundated with computers and digital technology. However, as of the mid-1990s, some rural subscribers were still being served by a switch installed in the 1920s. This exceptional longevity testifies to the durability of the equipment developed by the Automatic Electric engineers with the support of Northern Electric and Western Electric. Never again will a technological generation aspire to a useful life of seventy years. Moreover, the retirement of the step-by-step systems today is not due to breakdowns or obsolescence; they are being replaced by digital systems for which the maintenance cost is lower and which have other advantages (long-distance maintenance, for telecommunications companies; availability of added-value services, for subscribers). Initially, Bell planned to phase out its step-by-step technology by 1996, but the advent of competition in long-distance service changed this plan. Bell focuses its investments in large cities, where competition is most active. Some other Canadian companies (in the Prairies and NB Tel) have taken a more radical course, but they are less affected by competition. As of this writing, no one in Canada has predicted when the last analogue switch will be unplugged.

SETTING TELEPHONE NUMBERS

The introduction of automatic long-distance calling would not have been possible without the adoption of uniform telephone numbers in Canada and the United States. Up to then, telephone numbers had been assigned in a completely disorganized way and had anywhere from two to seven digits. Most telephone companies assigned numbers exchange by exchange without an overall plan or without taking account of the practices of other companies.

In 1945, a numbering system was established for North America, requiring all numbers to have seven digits and all area codes three digits. This grand total of ten digits would allow each subscriber on the continent to have a different number. Once again, the plan was first applied in Toronto, in 1951; each subscription was assigned two letters and five digits – for example, EM3-6362, for Empire 3-6362. Montreal followed in 1956. (Companies that had not automated their exchanges did not have to adopt this system.) The system was in effect for fifteen years; then letters were replaced by numbers because using letters limited the possible number of combinations to the words with which the letters were associated.

Telephone numbers included letters to make them easier to memorize.

It was established that the region covered by an area code could not cross a national or provincial border, but a country or province could contain several area codes. The 1945 North American system had eighty-six area codes, including eight for Canada (Prince Edward Island was linked to Nova Scotia, and Newfoundland was not yet part of Canada). In 1976, because of a saturation of telephone numbers in some area codes, a new system of numbers and area codes, similar to the present one, was implemented. Canada's number of area codes increased to fifteen; by 2000, three more had been added.

By the early 1990s, the entire system was threatened with saturation, thanks to massive use of cellular phones and fax machines. The North American numbering system was limited to a maximum of only 152 area codes because the first digit of the code could never be a 1 or a 0 (reserved for long-distance calls and operators), while the second number had to be a 1 or a 0. In 1995, the network was upgraded to accept 2 through 9 as a middle digit, exponentially increasing the number of area codes available and easing the bottleneck.

THE CONTINENTAL TOLL-SWITCHING PLAN

Improvements in long-distance communications led to an explosion in the number of telephone calls in the 1920s. To avoid having each long-distance call go through too many exchanges, point-to-point long-distance lines were multiplied, resulting in a situation reminiscent of the first years of telephony, before the invention of the exchange, when each user had to link up to all other users.

The solution for long-distance communications was similar to that for local communications: long-distance switches were installed. But not all local exchanges were linked to a single switching centre; levels were established from the local switch to a switch highly placed in a hierarchy of exchanges through transit switches. This complex architecture, built as a function of geographical imperatives and calling habits, aimed to minimize the number of physical long-distance connections and transit switches between two points. In 1928–29, the North American Toll-Switching Plan was created. The switches were divided into four categories.

1 At the top of the pyramid were ten regional centres that were linked up by high-density transmission lanes. Eight of these centres were in the United States; two were in Canada (Montreal and Regina).
2 More than thirty sectional centres were linked to their respective regional centres.
3 About 125 primary centres were in the heart of the clusters that formed the regular long-distance centres.
4 About 2,400 ordinary long-distance centres connected to "local centres" in the same city or a nearby location served the entire North American continent.[6]

The first three categories were long-distance transit centres. Switches evaluated the availability of lines and chose the best way to route a call.

The advent of long-distance telephone service overturned customs and traditional concepts of space. Courtesy Bell Canada.

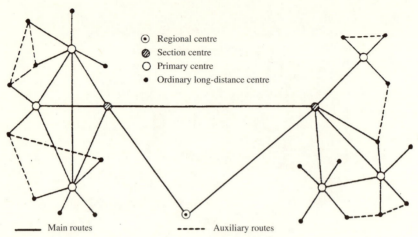

Regional centre
Section centre
Primary centre
Ordinary long-distance centre

——— Main routes   ----- Auxiliary routes

Canada-wide plan for long-distance switching.

Thus, if certain lines were overloaded, it was possible to reroute a call to-wards other lines that were geographically less direct but free. This routing method was inaugurated manually at the end of the 1920s, but it did not be-come truly significant until the 1950s, with the introduction of crossbar switches and automation of long-distance service.

In Canada, the plan was adopted within the framework of the TCTS. This carefully hierarchized toll-switching plan enabled callers to reach any-where on the continent transiting through a minimum of exchanges, which was a valuable savings when one considers that long-distance switching was manual until the 1950s. From this time on, one could truly speak of a telecommunications network; the telephone industry as a whole was acting as an organic entity in which activation of any particular element touched off a perfectly controlled chain reaction to any other point. Each telephone was potentially in contact with all other telephones in North America. This hierarchized system was maintained more or less intact until March 1991, when it was replaced by dynamic call routing.[7]

Automation was due largely to the invention of the crossbar switch. Based on a Swedish-American design, the crossbar system represents the highest technology achieved by electromechanical switching. It is based on the princi-ple of separation of the selection circuit from the connecting circuit. Instead of blindly exploring all lines of the exchange until it finds one free, the crossbar system stores stores all the digits that are dialled. A common control made of relays marks the position of the called number. Once the position is marked, all the other elements of the common control equipment can be freed up to help other communications.

At the core of this system is the relay invented long ago (Samuel Morse actually had incorporated the relay principles in his telegraph patents).

The first users of transatlantic telephones were media outlets. Courtesy Bell Canada.

You can now
telephone England direct
by All-British channels

By co-operation of The Bell Telephone Company of Canada, Canadian Marconi Company and the British Post Office, an All-British telephone service is now established between Canada and the British Isles. Telephone calls between the two countries will now span the Atlantic direct instead of being routed via United States points.

This All-British Service is a fitting complement to the recently inaugurated Trans-Canada Telephone System and, as a great new instrument of Imperial unity, coincides appropriately with the Imperial Economic Conference at Ottawa.

The new All-British route to England consists of The Bell Telephone Company's Long Distance lines through Montreal to the Canadian Marconi transmitting station at Drummondville, Quebec, thence through the air to the British Post Office receiving

station at Baldock, near London, England, and from Baldock by Long Distance telephone lines through London to the destination.

From England, the route is now by Long Distance telephone lines through London to the British Post Office transmitting station at Rugby, thence through the air to the Canadian Marconi receiving station at Yamachiche, Quebec, and from Yamachiche by Bell Telephone Long Distance lines through Montreal to the Canadian destination.

Travelling with the speed of light (186,000 miles per second) the voice flashes around this complete route in the fraction of a second, giving instantaneous exchange of speech. Telephoning from Canada to points in the British Isles is now as simple as the ordinary Long Distance call. Rates remain the same; a call to London, England, costs $30 for three minutes' conversation.

Canadian Marconi Company
The Bell Telephone Company of Canada

A joint advertisement by Bell and Marconi encouraging subscribers to use the new transatlantic telephone service. Courtesy Bell Canada.

When an weak electrical signal enters a relay, it forms a contact that triggers a local source of energy, which allows the signal to be retransmitted at full power. Thanks to the relay, the electromechanical succeeded in performing various logical operations well before the advent of computers. Once the entire telephone number has been dialled and registered in the common control equipment, the physical connection is established by the crossing of two bars, a moving horizontal bar, corresponding to the calling line, and a fixed vertical bar, corresponding to the number called (whence the name "cross-bar"). Most mechanical movements of the step-by-step system are eliminated and the energy required is reduced proportionately.

The crossbar system, a true electromechanical switch, brought switching into the era of massive economies of scale. In 1955 the first crossbar switch was built in Toronto by Northern Electric under the name Number Five System. When Northern Electric launches its first research and development program, it will be to build small crossbar systems under the name SA-1. The first Northern plant built abroad will be a plant in Turkey that builds SA-1 switches. This technology was the keystone of direct distance dialling, which was available in Toronto as of 1958, in Montreal in 1961, and then throughout Canada.

RADIOTELEPHONY CROSSES THE ATLANTIC

Previous experience with radiotelephony had clearly established that this technology could not replace wires and copper cables, so telephone companies settled for introducing radiotelephone links in locations where it was not economical to build physical links. The major application of radiotelephony was no doubt transatlantic connections. North America was linked to Europe commercially in January 1927, but the inaugural rate between New York and London was seventy-five dollars for the first three minutes, an extravagant sum beyond the reach of most ordinary mortals. The first paid call took place between the editors of the *New York Times* and the London *Times*. But most communications on the first day dealt with banking transactions, which totalled a respectable sum of six million dollars. Immediately, the same user groups were formed as had used telegraphy at the beginning: the press and financial institutions.[8]

The first transatlantic circuit used longwaves, but interference made these communications unreliable. At the same time Marconi himself successfully introduced shortwaves into radiotelegraphic communications. As telecommunications historian David Headrick rightly puts it, "his work on shortwave put him in the most select company of all: people who have changed the world twice."[9] When AT&T launched a second radiotelephone systems, it used shortwaves and quality became acceptable. Let us note that this success led to a temporary halt in another project: the one-circuit transatlantic cable studied for several year by AT&T. The shortwave radiotelephone system had four circuits.

In Canada, transatlantic service was established in October 1927 via New York, at a tariff comparable to AT&T's (about seventy-two dollars for the first three minutes). A direct Montreal-London service was inaugurated in July 1932, for the British imperial economic conference taking place in Ottawa that year, using transmission and reception stations built by Canadian Marconi at Drummondville and Yamachiche, Quebec, respectively.

The transmission station in Drummondville, Quebec, through which all radiotelephone traffic between Canada and Europe was routed. Courtesy Bell Canada.

Also in 1932, the Trans-Canada Telephone System began routing telephone calls across the country, and the Drummondville-Yamachiche circuit became Canada's window on the world. However, it was not a very well used window: in 1932 on Christmas Day (a major day for calling traffic), the Bell-Marconi circuit routed ten overseas calls. The charge for calling between Montreal and London soon dropped to thirty dollars for the first three minutes, but the cost was still a disincentive. Moreover, the sound quality of communciations remained mediocre, since shortwaves were prone to interference from the aurora borealis and the call-routing procedure was still very slow because of the many human interventions needed to establish a connection.[10]

In total, on the eve of the Second World War, North America and Europe were linked by five channels. This modest beginning created an international market for telephone service by linking Canada to some sixty countries. But radiotelephony remained a substitute for "true" telephony, hampered as it was by the limited number of channels available and the vagaries of the weather. It wasn't until the transatlantic cable was laid in 1956 that the technological bottleneck was finally resolved and the telephone network became truly worldwide.

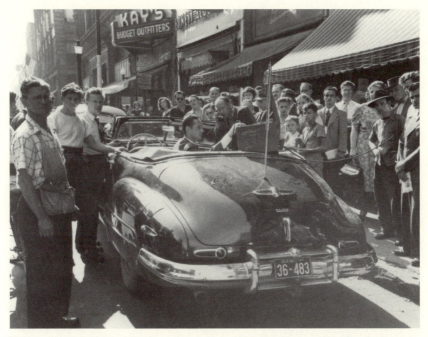

The manager of the Montreal radio station CKVL had one of the first mobile radiotelephones in Canada (1947). Here he calls emergency services to report a road accident. Courtesy Bell Canada.

The second major application of radio was mobile communications. In 1921 the Detroit Police Department equipped its cars with mobile radios, and the concept became very popular among public utilities. But it was mobile radio, not mobile radiotelephony, and it was not linked to the telephone system.

In April 1930, as we have seen, CN inaugurated a true radiotelephone service in its trains, but the experiment did not survive the onset of the depression. It was not until after the Second World War that mobile radiotelephony was marketed commercially.

In June 1947 Bell introduced mobile radiotelephony to Montreal and Toronto. It was a Western Electric system, which operated on a very narrow frequency band of 150 MHz and consisted of a central transmission antenna situated in the middle of the service zone and one or several reception antennas. In Montreal, there were four reception antennas at strategic locations in the city, all of them linked to a mobile-telephone control terminal that selected the strongest reception signal and routed it into the telecommunications network. Interconnection between the radio system and the ground-based system was achieved by an operator.

The quality of such a system depended on the power of the transmitter; the farther from the centre, the weaker the communication. Groups of buildings and hills created "shadow" zones. Moreover, mobile radiotelephony was no more confidential than a party line, since any ham-radio operator could tune in to the public frequencies and listen to conversations, or even join in. In spite of all these inconveniences, there were far too many requests for service for the twenty-four channels available in a given area. The number of subscribers quickly reached its ceiling of two to three thousand per city, and there were waiting lists. For the most part, mobile radiotelephony remained limited to emergency services and a few privileged professional groups. In 1947 the first users of the new service were photographers from the Toronto *Globe and Mail*. It was not surprising that its influence in the public imagination always surpassed the reality. It became a symbol of power, fascinating but out of reach for ordinary citizens.

The last application of radiotelephony was serving distant regions. As we have seen, in the interwar period, the Canadian Army Signal Corps opened up the Yukon and the Northwest Territories with this technology. In the same era, BC Tel inaugurated a series of links with isolated towns in British Columbia and Alaska. The radiotelephone broke the silence of the Far North.

In all three cases, expansion of this transmission vector remained limited. It was, in fact, a technology in waiting.

MICROWAVES: A HERITAGE OF THE WAR

Microwaves were the peacetime offspring of radar, which had been developed in Great Britain just before the Second World War by the Scottish physicist Robert Alexander Watson-Watt. Like radar, microwaves use the principle of reflection of certain radio waves: although medium and long-waves can penetrate walls (after all, one can listen to the radio inside one's home), extremely short waves bounce off the slightest obstacle, even a small object such as an airplane. It was this discovery that helped the Royal Air Force win the Battle of Britain in 1940. In fact, most British radar systems were built in Canada for obvious security reasons. Watson-Watt liked to say that Canada was the "radar arsenal of the western world."[11]

Microwaves use frequencies above a gigaHertz (billion cycles per second). In comparison, shortwave radio programs use the band that goes from 6 to 21 megaHertz (million cycles per second). Microwaves are thus much shorter than shortwaves, and shorter even than those used by FM radio and television. They are fragile and travel only in straight lines. So that they are not blocked by buildings or natural obstacles, transmitters and receptors are placed atop derrick-shaped towers, with an average spacing of 50 kilometres.

The first microwave systems, built in series, were British and American; they followed the advance of the Allied forces during the Second World War on all battlefields. In North America, AT&T tested civil microwave systems between New York and Philadelphia in 1945. But it was Canada that put the

Microwaves, the pacific offspring of radar, transmit in a straight line. There are towers every fifty kilometres to allow transmission of the signal over obstacles (buildings, trees, hills, etc.). Courtesy Bell Canada.

first commercial system into use between Prince Edward Island and Nova Scotia in November 1948 (see chapter 13).

During the same period, the TCTS was showing signs of exhaustion. One part of the most important sections of this network was still composed of open wires, and was thus completely inadequate to respond to the postwar explosion in demand for telephone service. Some sections were hastily equipped with carrier-current systems, while elsewhere CN's or CP's lines, and even sometimes American lines, were leased; the overall plan of the

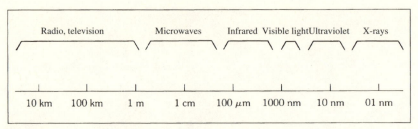

The electromagnetic spectrum and telecommunications.

1930s had been catch as catch can. It was in this chaotic situation that Thomas Eadie, Bell's vice-president of operations and soon to be president of the TCTS, intervened. Eadie was sure that the time for half-measures was over; the trans-Canadian line would have to be rebuilt from stem to stern using shortwave technology.

Indeed, in the 1950s telecommunications were changing in nature. Of course, there was the unstoppable wave of telephony, which was reaching a quality threshold as it approached universality of service. But there was also the advent of television, with the CBC's first broadcasts in 1952. The network that was already carrying text and voice suddenly had to deal with images and become a truly polyvalent system. Transmitting a video signal required a minimum capacity of 2.8 million cycles per second, or 2.8 MHz, while a telephone signal required only 3,400 HZ. Since capacity is measured in terms of bandwidth, television required what was called a broadband network. Only microwaves could handle this quantity of data.

THE TRANS-CANADA MICROWAVE NETWORK

The friendship between Thomas Eadie and Joseph-Alphonse Ouimet, the Canadian Broadcasting Corporation's chief engineer, certainly played a decisive role in the microwave venture. Bell began construction of a 650-kilometre network comprising fourteen microwave stations in the Windsor-Montreal corridor, catching CN and CP up short and keeping these companies from doing with microwaves and television what they had done with carrier currents and radio. Not that the railway companies were idle in this respect: at the same time, they were constructing a microwave line between Toronto and London. But they were incapable of keeping up with the infernal pace of Bell and its microwave missionary, Tom Eadie.

For its first system, Bell turned, naturally, to technology used by AT&T, which at the time was building a transcontinental network from New York to San Francisco (which went into service in September 1951). The TD-2 system, developed by the Bell Labs, could route a maximum of twelve unidirectional radio-frequency (RF) channels. One RF channel could route one video

Construction of a microwave station as part of the trans-Canada microwave network.

signal or six hundred voice signals (two channels were needed to route six hundred telephone conversations, one channel for each direction of the transmission). The Windsor-Toronto-Ottawa-Montreal line was completed in May 1953, just in time to retransmit the images of Queen Elizabeth II's coronation. Most important, the artery was connected to the AT&T system via Buffalo and served as a gateway to broadcasts of American television programs. Were telecommunications companies once again turning to the north-south axis?

In 1953 Eadie became president of Bell and Ouimet became general manager of the Canadian Broadcasting Corporation. The following January, the Crown corporation called for tenders for a contract for cross-Canada retransmission of television programs. (As we have seen, the CBC had stopped calling for public tenders for radio broadcasts in 1932.) This contract alone could justify construction of a transcontinental network. Two scenarios were possible: the TCTS could meet the challenge as a unit, or Bell could turn its back on the other telephone companies and join CN and CP. The two railway companies, whose telegraph revenues were plummeting, lobbied in favour of a consortium with Bell in which each party would hold one third of the assets.

Bell's position was thus very strong. In either case, it was assured of a share in the contract. The other members of the TCTS found themselves facing a choice: should they launch themselves into an uncertain venture (the memory of what had happened with the radio contracts for the first trans-Canada network still smarted) or risk losing control of long-distance communications to CN and CP?

CN and CP's lobbying effort particularly concerned the provincial government in Alberta, which would have preferred to see private enterprise build the microwave network. Its fear of losing control of long-distance communications – image and voice – won the day, and Alberta rallied to the TCTS solution. The social democratic government in Saskatchewan hesitated for a long time before committing itself to Bell, which stank too strongly of the sulphur of private enterprise. It took the premier himself, Thomas C. Douglas, endorsing the project for the province finally to agree to it. The TCTS presented its united submission and won the contract in March 1955. This decision was a catastrophe for the railways, which saw their hope of re-entering the telephone industry, closed to them since 1880, come to nought and confirmed the pre-eminence of the telephone companies in the telecommunications sector.[12]

The president of the TCTS's engineering committee was none other than Alex Lester, Eadie's confidant. Fully aware of the geopolitical stakes, Lester threw all his prestige (and Bell's) in the balance to make sure that the microwave network would be entirely within Canada, even though, in his words, "the pull of traffic was North to South to a very great extent and if you were looking in terms of overall North American economics you probably can prove in having a series of vertical lines from North to South, from the main centres in Canada to a Northern Transcontinental line across the States."[13]

Obviously, it would have been easier to cut south of the Great Lakes through the United States, but, once more, Canada used technology to build a national superstructure defying economic considerations. On 1 July – Dominion Day – 1958, the transcontinental microwave network was inaugurated with great patriotic pomp. Sixty-two hundred kilometres long and with 139 stations, it was the longest microwave line in the world at the time. At

The trans-Canada microwave network was comprised of 139 stations spread over 6,200 kilometres. Built between 1955 and 1958 by Canadian telephone companies, it is the longest
microwave network in the world. Shown, a microwave station in British Columbia. Courtesy
Bell Canada.

first, only two channels were put into service, out of the TD-2 system's potential of twelve. Since the capacity of each RF channel was doubled by the
first Windsor-Montreal network, however, it could transmit 1,200 telephone
signals. Its cost, fifty million dollars, was assumed entirely by the telephone
companies, although it could be considered that the CBC contract played an
indirect para-governmental stimulative role. The following year, a section
was built to link Newfoundland to the main network.

In March 1962 the CN/CP consortium in its turn began construction of a
transcontinental microwave network. The TCTS protested vigorously to the
Department of Transport, which at the time was reponsible for telecommuni-
cations, arguing that construction of a second network was "useless, economically not judicious, and would go against the public interest."[14] But it
was no use; in May 1964 CN/CP inaugurated a second network. Unlike the
TCTS's TD-2 system, which had been developed by Western Electric in the
US, CN/CP's MM-600 system was Canadian, designed and manufactured by
RCA Victor Limited in Montreal. RCA was, of course, the subsidiary of an
American firm, but the worldwide mandate for telecommunications had been
given to the Montreal plant because of the experience it had acquired in radar
during the Second World War. Thus, RCA Montreal ended up selling MM-600
and, later, MM-1200 systems throughout the world, including to the United
States. In 1965 it built a microwave route from Ankara through Tehran to
Karachi for USAID.[15]

The CN/CP system was actually a third network. The golden rule of tele-communications was never to put all eggs in one basket, and so all infra-structures were doubled. Soon after the TCTS's microwave system was put into service in 1958, its engineering committee had begun to construct a support backbone. What was thereafter called the TCTS microwave system was thus composed of two parallel routes, called the north and south routes.[16]

The Canadian telecommunications scene was profoundly changed by this system. The trans-Canada network had seven circuits, augmented, in some places, by new carrier-current circuits, which were considered an appreciable improvement. Suddenly, microwaves provided a capacity of 1,200 circuits per route, not including unexploited potential, and there were three parallel routes. The race to high capacity was under way. Telecommunications left the era of shortages to step straight into a time of abundance.

THE FIRST TRANSATLANTIC TELEPHONE CABLE

Construction of the transatlantic cable had begun in 1929, when AT&T purchased an obscure little Prince Edward Island company called Eastern Telephone and Telegraph from MT&T (see chapter 13). The 1929 project more or less followed the path of Cyrus Field's old transatlantic cable via Newfoundland, which thus had to be linked to the American network via New Brunswick and Nova Scotia. Land-based lines began to be laid in 1930, but this project, already slowed down by the relative success of the radiotelephone system, was interrupted the following year because of the depression. Two line sections were already built, and they were ceded to NB Tel and MT&T, which integrated them into their long-distance network. Instead, the number of radiotelephone circuits was increased.

Work on the transatlantic cable was not stopped. However, the cable planned for in 1929 had a very small capacity (a single telephone circuit) because technology had not advanced to the point where underwater repeaters were possible. By the end of the war, this problem was solved with repeaters strong enough to resist the water pressure of the depths and small enough to be incorporated into the structure of the cable.

By the early 1950s overseas telephone traffic had exploded. The number of radiotelephone calls between the United States and Great Britain had risen from twenty-seven thousand in 1927 to more than a hundred thousand, and it was difficult to increase the capacity of the radiotelephone systems. AT&T, the British Post Office, and the Canadian Overseas Telecommunications Corporation (today called Teleglobe Canada) concluded a cooperation agreement to construct and operate the transatlantic cable. All the work was to be done by the Americans and Britons, while Canada was designated a transit country and future user. This sharing of roles was reflected in the division of ownership of the cable: AT&T had 50 percent, the BPO 41 percent, and the COTC a mere 9 percent.

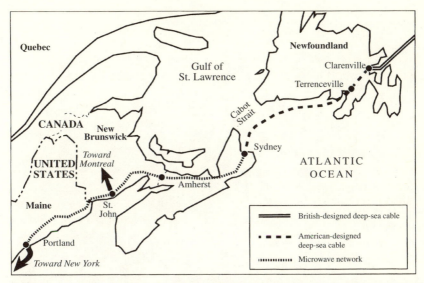

The First transatlantic telephone cable was landed in Canada in 1956. The Canadian part of the system was a microwave network built by a Canadian subsidiary of AT&T (Eastern Telephone & Telegraph).

A rejuvenated Eastern Telephone and Telegraph constructed a microwave line from Sydney, in northern Nova Scotia, to the American border. The BPO laid an underwater cable between Nova Scotia and the southern shore of Newfoundland. A buried cable then had to be laid from Terrenceville east to Clarenville, which was to be the North American terminus of the transatlantic cable. The AT&T teams encountered the same difficulties as had Frederick Gisborne's men a century earlier, but they had, of course, vastly improved equipment, including automated trenchers.

Finally, the transatlantic section was laid in two parts, in 1955 and 1956, by the largest cable ship in the world, the British *Monarch*. There were two parallel 3,600-kilometre coaxial cables, laid thirty kilometres apart. Each unidirectional cable was sheathed in a reinforced structure to resist the pressure of the depths and had fifty-one repeaters so compact that only an expert eye could detect the bulges and three vacuum valves capable of operating uninterrupted for twenty years with a power feed by the cable itself. Indeed, once at the bottom of the ocean, the system had to work on the first try.

On 25 September 1956 the first transatlantic communication by cable finally took place between Ottawa, New York, and London. Considered a masterpiece of AT&T technology, the cable operated for twenty-two years, until it ruptured just ten days before it was to be officially retired from service, in November 1978. The system, called TAT 1, had a capacity of thirty-six telephone circuits, the communication quality was vastly improved, and the

The spectacular arrival of the transatlantic cable in Newfoundland; a crew hoists the cable onto land. Courtesy AT&T Archives.

price was much more affordable: twelve dollars US between Montreal and London for the first three minutes.

Finally, telephony became a mature technology in terms of social penetration and geographical expansion. Radiotelephony never was truly integrated in the telephone networks. Therefore, it falls to TAT I to have actually put North American and European networks into contact. This would lead to a series of new problems that could be solved only through international cooperation. TAT I did respond to precise international standards established by the International Telecommunication Union. International cooperation, which for Canada was limited to adaptation of American standards to local context, comes out of the old continental framework. It achieved universality with an electromechanical analogue technology.

17 The International Scene

What were Canadian telecommunications worth up to 1956? It is impossible to answer this question without looking at what was going on elsewhere in the world. I therefore compared the penetration rates in different countries, then on different continents. The result showed a resemblance (predictable) between Canada and the United States, but also a commonality (unexpected) between Canada, Scandinavia, and Switzerland. Telegraphy occupied a place on its own as a stagnating technology.

If any foreign country influenced the development of the Canadian telephone industry, it was obviously the United States. The rest of the world followed a different path, characterized by government intervention.

Finally, the international dimension of telecommunications also included Canada's involvement in international organizations. This involvement was all the more important since it gave rise to a new entity charged with international communications.

CANADA'S POSITION IN THE WORLD

Canada has always had one of the highest telephone-penetration rates in the world. In Table 19, Canada is compared to the United States and to a sample group of industrialized countries. Canada was in second place, behind the United States, in 1914, and it slid to third place, behind the United States and Sweden, during the 1930s, and to fourth place between 1938 and 1946, since New Zealand had made some progress over these years. After the Second World War, Switzerland was briefly ahead of Canada. In fact, among countries not included in Table 20, it is also worth mentioning Denmark and

Table 20

Telephone Penetration Rate in Selected Countries, 1914–56
(number of telephones per 100 citizens)

	1914	1921	1931	1934	1939	1946[a]	1950	1956
Australia	2.8	4.3	8.1	7.3	9.1	10.9	13.2	17.5
Canada	6.5	9.8	14.0	11.1	12.1	14.4	19.6	26.3
France	0.8	1.2	2.8	3.2	3.8	4.7	5.6	7.2
Germany	2.1	3.0	5.0	4.5	5.2	3.3	4.4	7.6
Great Britain	1.7	2.1	4.3	4.8	6.7	8.2	10.2	13.5
Japan[b]	0.4	0.6	1.4	1.5	1.9	1.3	2.0	3.5
New Zealand	4.6	7.0	10.2	10.0	12.7	15.6	18.2	24.6
Sweden	4.1	6.6	8.7	9.5	12.7	17.7	22.8	30.4
Switzerland	2.5	3.8	7.3	8.8	10.7	14.7	18.2	24.3
United States	9.7	12.4	16.4	13.2	15.4	21.0	27.1	33.7

Source: AT&T, *Telephone and Telegraph Statistics of the World*. Figures were calculated on
1 January of each year, unless otherwise noted. Starting in 1946, the German figures
concerned only the Federal Republic of Germany.

Notes: a. Because of the Second World War, few statistics were published in 1945. The
            figures for Japan were calculated only in 1947 and those for Germany in 1948. The
            figures for some other countries were also distorted, those for Canada were
            calculated in 1944, and those for Great Britain and Sweden in 1945.
         b. Japanese statistics were calculated as at 31 March of the current year.

Norway, which followed the leading group by a bit, and which constituted,
along with the above-mentioned countries, a community of nations whose
telecommunications evolution was similar.

Only North America and Oceania saw telephone-penetration rates fall
sharply during the depression of the 1930s. During this period, the only
country in Europe whose telephone-penetration rate was affected was Germany, but to a much lesser extent. All other European countries saw their
stock of telephones rise constantly.

When we compare evolution of telephone penetration continent by continent (see Tables 21, 22, and 23), North America was in the lead throughout
the period under consideration, with more than half of the telephones in the
world. The overall weakness of Europe masked enormous disparities between Scandinavia and Switzerland, which were more or less at the same
level as North America, and Eastern Europe, which was at the level of under-
developed countries. In 1956 Poland had a penetration rate equal to that in

Table 21
Number of Telephones per Continent, 1914–56 (000)

	1914	1921	1931	1934	1939	1946	1950	1956
North America	10,121	14,302	21,836	18,107	21,617	30,100	43,424	60,421
Central America	–	–	–	–	–	–	524	733
South America	166	287	619	652	895	1,290	1,657	2,568
Europe	4,013	5,290	10,589	11,307	15,305	16,980	20,000	29,090
Asia	306	494	1,250	1,421	1,913	1,500	2,367	4,411
Africa	65	102	247	271	406	430	806	1,411
Oceania	217	376	794	737	954	1,200	1,522	2,366
World, total	14,888	20,851	35,335	32,495	41,090	51,500	70,300	101,000

Source: AT&T, Telephone and Telegraph Statistics of the World. Figures as at
1 January of each year.

Table 22
Distribution of Telephones by Continent, 1956

	(%)
North America	59.8
Central America	0.7
South America	2.5
Europe	28.8
Asia	4.4
Africa	1.4
Oceania	2.3

Note: The total does not add up to 100 due to rounding.

Tunisia and Malaysia but below the average in South America. It should be noted that the rise to power of communist regimes changed nothing, either for the better or for the worse, in Eastern Europe.

The rest of the world was more or less without telephones, with the exception, as one might expect, of colonial administrations, a situation that resulted in utterly unequal access to what was not yet being called information. Control of telecommunications had enabled the West to manage the world with a relatively small number of administrative personnel until the 1950s.

Table 23
Penetration Rate per Continent, 1914–56 (number of telephones per 100 citizens)

	1914	1921	1931	1934	1939	1946ᵃ	1950	1956
North America	7.5	9.8	13.0	10.3	11.8	15.4	26.5	33.1
Central America	–	–	–	–	–	–	1.0	1.3
South America	0.3	0.4	0.7	0.7	1.0	1.3	1.5	2.1
Europe	0.8	1.2	2.0	2.0	2.7	2.9	3.3	5.1
Asia	0.04	0.1	0.1	0.1	0.2	0.1	0.2	0.3
Africa	0.05	0.1	0.2	0.2	0.3	0.2	0.4	0.6
Oceania	0.4	0.6	1.0	0.8	1.0	1.1	11.3	16.0
World, total	0.9	1.2	1.8	1.5	1.9	2.2	3.0	3.7

Source: AT&T, *Telephone and Telegraph Statistics of the World.* The figures are calculated as at 1 January of each year.

When it gained independence in 1947, the imperial administration in India had 1,250 British bureaucrats. No doubt only the presence of a modern, and élitist, infrastructure had enabled so few people to govern so many.[1]

THE AMERICAN MODEL

When Theodore N. Vail retired in June 1919, the telephone industry in the United States had already taken its "definitive" form, which lasted until 1984. The operational structure of every telephone company was based on three divisions: the network division was responsible for construction and maintenance of telephone lines; the traffic division managed the telephone exchanges and supplied service to the subscribers; and the marketing division was responsible for public relations and promotion (during the period under consideration, this involved mainly encouraging people to use long-distance service).[2]

In addition, the resolutely technological orientation of the companies conferred a crucial role on their engineering departments, which set standards and established development plans for the system and traffic departments. The chief engineer was the second most important person in the company, as was demonstrated in the 1930s by Robert Macaulay's part in building the Trans-Canada Telephone System, and after the Second World War by Alex Lester's leadership in research and development at Bell.

This structure worked marvellously, and Vail's successors at the head of AT&T continued to function within the institutional and ideological framework he had set up; the result was the best telephone service in the world. Bringing together the best of the entrepreneurial culture of the private sector with the sense of government civil service, the model was copied by Canada and a number of European administrations.

In 1925 Western Electric's engineering division split off to form a separate company, Bell Laboratories. (In 1934, AT&T's research-and-development division joined the new company.) It was the beginning of a great technological venture in the United States, leading to development of sound movies, television, and transistors, as well as confirming the Big Bang theory.

At the same time, under pressure from the Department of Justice, AT&T gave up it foreign investment in exchange for a tacit acceptance of its monopoly in the US. It sold International Western Electric to a small, little-known company, International Telephone and Telegraph. AT&T kept North America and ITT got the rest of the world – the plants in Great Britain, Belgium, France, Spain, Italy, and the Netherlands, as well as minority holdings in several other countries.[3]

Direct regulation of the telecommunications industry in the United States began in 1934, when the Roosevelt government created a new body, the Federal Communications Commission. Almost immediately, the FCC began studying AT&T's activities; after three years, a preliminary report ferociously attacked the company's policy of systematically purchasing products from its subsidiary, Western Electric, without asking for bids. However, the final report made no precise recommendations.[4] It was not until 1949 that an antitrust suit was brought against AT&T by the Department of Justice.

But this was the beginning of the cold war. Western Electric began to cooperate with the Department of Defense to produce nuclear bombs on a large scale. A new corporation called Sandia was created in 1945; it would produce atom bombs at cost for the Department of Defence. Later on, Western Electric produced missiles and guidance equipment, but for profit. AT&T took advantage of this to seek a friendly arrangement, resulting in the 1956 Consent Decree, which had major consequences for Canada, as we have seen.

THE REST OF THE WORLD: GOVERNMENT ADMINISTRATIONS

In 1956 more than 83 percent of the stock of telephones in Europe belonged to national governments, which provided telecommunications services usually within a PTT structure. And yet there was a small group of diehard countries that had a private-sector telecommunications industry. Denmark, Finland, Portugal, Italy, and Spain were served by systems that were entirely or mostly privately owned, while only small parts of the systems in Holland, Norway, and Sweden were privately owned.

Aside from Denmark and Finland, the entirely or mostly privately owned systems belonged to countries with a low rate of telephone penetration. It is thus impossible to say categorically that private enterprise provided good management and government provided bad management. The public telephone administrations in Sweden and Switzerland were always models of good management.

Great Britain's telecommunications evolution had a few twists and turns: in 1928 it privatized all extra-European telegraph services by transferring them to Cable and Wireless, which thus dominated half of the world network. In 1950, however, Cable & Wireless did not escape the wave of postwar nationalizations. The telephone industry had been entirely under government control since 1912, with the exception of the municipal company in the town of Hull.

THE ITU: BIRTH OF MODERN COOPERATION

After the First World War, when the League of Nations was created, a plan to unify all electrical communications was put forward, including a proposal to merge the International Telegraph Union with the group of countries that had ratified the radiotelegraphic conventions (sometimes called the International Radiotelegraph Union). One aspect of the plan was to make more provisions for the telephone, which was beginning to extend its tentacles beyond international borders. The United States wanted two separate organizations to break up the British system based on underwater cables and radiotelegraphy. The European nations generally wanted a unified organization. The result was a decentralized agency with three consultative committees. Three advisory committees were set up: the International Telegraph Consultative Committee in 1924; the International Telephone Consultative Committee in 1925; and the International Radio Consultative Committee in 1927.[5]

Finally, at the Plenipotentiary Conference in Madrid in 1932, the International Telegraph Union and the International Radiotelegraph Union merged. Telegraphy, telephony, and radio were finally contained within a single organization, which was named the International Telecommunications Union. The word "telecommunications" had been coined at the beginning of the century by Édouard Estaunié, a French writer and engineer. The Madrid conference brought the word into the current language and gave it an official definition: "Any telegraph or telephone communication of signs, signals, writings, images, and sounds of any nature, by wire, radio, or other systems or processes of electric or visual (semaphores) signaling."[6]

For Canada, Madrid was a turning point, since it finally ratified the international convention that was the legal basis for the ITU. The convention took effect in January 1934, marking the beginning of Canada's official – but not very active – participation in the ITU. Canada was not only late in joining but cautious: it limited its membership to the Radio Regulations but did not join the Telephone and Telegraph Regulations that accompanied the convention, on the pretext that its telecommunications systems were operated by private companies and it was not the government's place to undertake commitments on their behalf. In doing this, Canada modelled its policy upon that of the United States, which also joined the ITU in 1934 but opted out of these two sets of regulations. In 1937, as a small second step, Canada countersigned the Telegraph Regulations but not the Telephone Regulations.

The ITU then entered the dark tunnel that led to the Second World War. Because its head office was in Bern – in the capital of a neutral country – it emerged from the war without disappearing; although its activities were reduced to a bare minimum, the Bern office provided correspondence between member nations. It took up its full role again at the Conference of Plenipotentiaries in Atlantic City in 1947, at the instigation of the United States. Thereafter, the US presented itself as the champion of international cooperation with regard to telecommunications. A similar awakening to international realities was perceptible in Canada, which sent a twenty-three-person delegation to Atlantic City.

At the request of the United States, the ITU obtained the prestigious status of a specialized agency of the United Nations. To accelerate decision making between conferences of plenipotentiaries, a board of directors was set up that had complete authority over the operation of permanent organs. The old office, staffed by some thirty employees of Swiss PTTs, was replaced by an independent secretariat composed of a growing number of nationals from all participating countries. To mark the change, the head office was moved from Bern to Geneva, where it has remained ever since. Atlantic City marked the beginning of the modern ITU.[7] Another result of the Atlantic City conference was the creation of the International Frequency Registration Board, which allocated available frequencies to all nations. It held the equivalent of almost a supra-national judicial power.

TELECOMMUNICATIONS AND THE COMMONWEALTH

The ITU was not the only international telecommunications organization. The British Commonwealth provided a framework for international cooperation within which Canada was much more directly involved.

As we have seen, Canada had been linked to Europe by radiotelephone since 1927. However, the communications were provided by AT&T and the British Post Office, and therefore had to transit via New York and London. For reasons that were both practical and sovereignty-related, Canada decided to negotiate directly with Great Britain. The Commonwealth was in a

The telephone companies had always conveyed a pacifist ideology for communications. In 1938 a Bell advertisement stated that a telephone call between Chamberlain, Mussolini, and Hitler had preserved the peace. Courtesy Bell Canada.

position to process telecommunications both by radio and by cable, and the Imperial Cómmunications Advisory Committee, comprising representatives from Great Britain, Canada, Australia, New Zealand, and India, was set up in 1928. Its head office, naturally, was in London.

The Imperial Wireless and Cable Conference of January 1928 had merged Marconi and Eastern Telegraph and its associated companies, which owned most British Cables into a new company, Cable and Wireless. The external telecommunications of Commonwealth member countries were in the hands of a single British company that operated both the radio links and the telegraph cables. Most international communications thus transited through London, without regard to distance. For example, a telegram sent from Australia to California went through India, Great Britain, and Canada, using the Cable and Wireless network as much as possible. The Commonwealth members found this system frustrating, for it gave them no control over their own international communications.

During the Second World War, Australia decided to end its communications dependency on the Commonwealth. In the fall of 1942, it asked for a meeting of the Commonwealth Telegraph Conference in Canberra, which then recommended the creation of a new centralized organization. The idea arose among the Commonwealth directors, who, to respect the equality among member nations required by the Westminster Statute, said that each had to take charge of its own installations. The theoretician of this new strategy was a Canadian, Sir Campbell Stuart. He later recalled: "With the entry of America into the war it was found to be neither practicable nor desirable to maintain the pre-war policy of regarding the British Empire as if protected by an unscaleable wall with its main gateway to the world through London."[9] In 1943 the ICAC was replaced by the Commonwealth Communications Council to reflect the growing importance of the Dominions; however, in order not to impede the Allies' war effort, the Dominions agreed to delay all changes until peace was restored.[10]

The postwar period and the beginnings of decolonization revived interest in the Commonwealth. But the structures in place, which resembled a private club, were inadequate in telecommunications, as in other areas, and reform was needed. In July and August 1945 the Commonwealth Telecommunications Conference agreed on the principle of nationalization of Cable and Wireless. Nothing firm was done yet. We had to wait until May 1948 and the Commonwealth Telegraph Agreement, which was concluded between the member nations so that each could see individually to nationalization of installations providing international telecommunications on their respective territories; designation of an existing ministry or creation of a Crown corporation to operate these installations; and appointment of a representative to the Commonwealth Telecommunications Board, which was created in March 1949 to replace the CCC (Canada sent a civil servant from the Department of Transport).[11]

A CROWN CORPORATION
FOR INTERNATIONAL COMMUNICATIONS

When Bell heard about the 1948 agreement, it tried to intervene by proposing to extend its activities to international communications, but it wasn't to be. The agreement reflected the Labour ideology then prevailing in Great Britain, and Canada followed its recommendations to the letter, creating the Canadian Overseas Telecommunications Company in January 1950.[12]

The COTC's mandate was to acquire the assets of Cable and Wireless, just nationalized by the government in London, and of its subsidiary Canadian Marconi. These assets consisted of the terminal stations of the two transatlantic telegraph cables in Harbour Grace, Newfoundland, and Halifax, and of the transpacific cable in Bamfield, British Columbia.

On the radio front, the COTC acquired the Drummondville transmission station and the Yamachiche reception station, both in Quebec. This made a total of three radiotelephone and thirteen radiotelegraphic circuits for links with Europe, Saint-Pierre and Miquelon, the West Indies, and Australia via the United States.

All of these installations were acquired by the COTC in June 1950. The company's first president died after one year in office; his replacement, Douglas F. Bowie, held the position for more than twenty years. This ex-manager of Cable and Wireless built the company with a public service focus. The Crown corporation also inherited three hundred employees from Cable and Wireless and Marconi.

In spite of this modest start, the company had to grow quickly, matching the fearsome pace of postwar reconstruction. First of all, Canada repatriated radio communications between Montreal and Australia thanks to the opening, in November 1956, of new transmission and reception stations at Cloverdale and Ladner, British Columbia, respectively. Then it abandoned the Bamfield station, where the old transpacific cable inaugurated by Sandford Fleming in 1902 terminated, for more modern facilities at Port Alberni and Vancouver.

The symbol of growth in international communications remained, however, the laying of the first transatlantic telephone cable, the TAT 1 system, in September 1956. The COTC did not operate the cable's terminal station, but it was part of the consortium that managed it. Suddenly the traffic in international communications exploded, and the COTC's revenues quadrupled in its first year of operation, even though the cable was in use only during the second half of the fiscal year.

The COTC was overloaded. As soon as the cable was opened to the public, its usage reached a level that had been forecast for only after a year or two of service. Voice-processing apparatuses had to be set up in the terminal stations to increase capacity. The number of telephone circuits went from six to thirteen, then to twenty. At the same time, the COTC introduced an international telex service, which was launched in December 1956. For the COTC, it was a total success.

Conclusion:
How Telephony
Changed the World

After this long voyage through the history of Canadian telecommunications, we have seen telecommunications change in nature several times. Telegraphy formed its central infrastructure in the early years, while telephony was already a fascinating prospect, though as yet a marginal tool. The telegraph cable under the Atlantic became the symbol for the reorganization of geographic and political space by a technology.

Then the telephone, in its turn, conquered distance and achieved universal service. It also became an infrastructure but without occupying a dominant position. It was an aid or complement to face-to-face conversations or to transportation. The importance of the telephone's impact came more from its omnipresence in daily life than from its structural effect on the macroeconomic level.

In relation to the technologies that came before it (railway, steamships, telegraphy), telephony had a different nature: it entered the home. For centuries, the home was the most self-sufficient unit of economic and social life: in it, people lived, worked, educated their children, were born, and died. Suddenly, in the first half of the nineteenth century, the household began to be wrapped in an increasingly ineluctable spider's web of complementary networks: gas lines, water pipes, and sewers invaded the home. By the end of the century, telephones and electricity were added. Alexander Graham Bell summarized the global nature of the phenomenon in a letter to his father dated the very day he invented the telephone: "And the day is coming when telegraph wires will be laid on to houses just like water or gas – and friends will converse with each other without leaving home."[1]

The explosion of public infrastructure into the formerly cloistered universe of the family home was a true revolution. The house lost its autonomy and was integrated into a greater social entity: the city.

This revolution was paradoxical. The nineteenth century marked the triumph of individual rights over collective rights. Professional corporations, communal properties, and hereditary rights were dismantled in favour of the "rights of man." But at the same time as the ideal of individualism was triumphing, individuals were losing their physical autonomy – the self-sufficient home – which was giving way to the great societal projects of the twentieth century. In this regard, Patrice Flichy, a French telecommunications researcher, established a revealing similarity between Saint-Simonian socialism and the development of the notion of network.[2]

The telephone marked the vanguard of the collectivization of private life. In effect, the notion of switching, which is at the heart of the telephone network, is technological and symbolic re-creation of the urban space, which can be perceived as a place of encounters and exchanges of all types. It is not surprising, then, that in Canada telephone service was often greeted with a mixture of anxiety and hostility. We have seen how citizens took axes to the first telephone poles put up in front of their houses. It took twenty-five years of juridical and political battle for Bell to gain the legal right to set its poles in public places. It took intervention by no less than the Privy Council in London to decide in favour of public service.

During the smallpox epidemic of 1885, the rumour spread that the disease was propagated by the telephone lines. A crowd bearing torches and cudgels went to Bell Canada's head office in Montreal to set it afire. A regiment of the Canadian army had to be called out to save the building from destruction.[3] This attitude was not a simple Luddite reflex. As I noted above, the telephone represented the intrusion of a collective technology into the private stronghold of the house. To this was added the fact that the telephone was initially developed on an élitist model. In contrast to the ideal expressed by Bell in his letter to his father, the businessmen who succeeded him managed to erase completely the "sociable" aspect of the telephone.

Telephone service had to make money, but above all it was not to be used for pleasure. This reductionist conception of the telephone had a very concrete impact on marketing: all efforts were directed towards the business sector, to the detriment of the residential sector.

Up to the First World War, the marching orders were "Educate the public on how to use the telephone." American and Canadian telephone companies shared this preoccupation. People had to be encouraged to use the telephone at the office and the plant. The minority who had a telephone at home were to use it parsimoniously to communicate with the workplace, emergency services, or suppliers. In short, the business phone was not to be used for conversation but for giving orders and expanding the authority of

To Speed Up Our Industries
—Use Long Distance!

"*I* AM reorganizing business methods — speeding up the mechanism of industry, multiplying the number of sales, reducing the cost of selling, and making it possible to accomplish more in the business day.

"I give you the right of way into Everyman's office. In the midst of a conference, listening to your rival's salesman, reading a telegram or special delivery letter, he will stop — and listen to Long Distance.

"I am the quickest Messenger on earth, reliable, never sick or on holiday, never too tired to work all night, always ready to serve you. 13,000 trained telephone employees make it possible for me to render such service.

"From the office I direct the operation of branches — buy and sell everything — make new customers — stimulate and encourage salesmen — forestall competition — verify credits — pacify customers — collect bills. I have been called 'the greatest business-getter in the world.'

"For the factory and warehouse, I speed up and direct incoming and outgoing shipments — take advantage of a favorable turn in the market to secure coveted material or supplies—order new parts to replace broken equipment.

"I am the wonder-servant of the age — your cheapest and most faithful employee, The Long Distance Telephone."

Every Bell Telephone is a Long Distance Station

The early telephone reproduced the social hierarchy inherited from the Victorian era: an instrument of businessmen, it was used to manage companies and give orders – in short, for "serious matters" and certainly not for personal use (1927).

Bell's first advertisement promoting the "pleasure telephone" for long-distance purchasing, personal safety, calling friends, and making "social engagements." (1911 advertisement.) Courtesy Bell Canada.

those in power. The British telecommunications specialist Colin Cherry summarizes the situation thus: "Every use suggested serves for master/ servant relations."[4]

There is nothing surprising about this early usage: telephony reproduced very exactly the hierarchical structure of Canadian society in the Victorian era, and interaction through the telephone was reduced to a minimum. It was simply a substitute for the messengers that crowded city streets and business corridors.

Bell's advertising only promoted the telephone as a management tool.[5] When it evoked chatting on the phone, it was to discourage it – until 1911, when Bell's first "sociable" advertising appeared. Of course, there was a technical reason for this: the telephone exchanges were often overwhelmed by the rapid growth in numbers of subscribers. Nevertheless, a clear position was taken against using the "pleasure telephone."[6]

What is more, in newspapers of the era, chatting on the telephone was most often attributed to women. Industrial society had withdrawn the economic and social functions from the family home by sending the man, and sometimes the children, out to earn a living elsewhere. As for the woman, left alone in a deserted house, deprived of her traditional roles as educator and manager, she must not try to form a new connection with society: she might tie up the telephone exchange!

Because the technology had a force of its own, beyond the intentions of Bell and AT&T, such a policy was soon condemned. In both Canada and the United States, this condemnation took the form of competition from independent companies: cooperatives, municipal services, and private enterprises founded by local notables bypassed the restrictive yoke of urban élitist telephony.

The social class that upset the established order was the petite bourgeoisie involved in commerce – in other words, retailers. Their influence was decisive in Ontario, but it extended throughout Canada. In Montreal, the very name of the main independent company – Merchants' Telephone Company – conveyed this social origin. Their immediate objective, of course, was to encourage people to place orders by telephone. In doing so, they took telephone use beyond the limited context of big-business management. Thus, making grocery orders by telephone moved telephone use into a new sphere – use by the general public.

As we have seen it in Chapter 5, in Montreal there was an additional reason for founding an independent company: language. The fact that Bell's telephone operators answered in English discouraged part of the population from subscribing. From 1892 to 1913, the Compagnie de Téléphone des Marchands offered service in French to a population that was notably underequipped with telephones in comparison to English speakers.

How can this complete turn in favour of the telephone be explained? The vast physical size of Canada is part of the reason. But geography alone does not explain the passion for the telephone, or it would have been part of the mix at the very beginning. An active element is needed, such as can only be found in ideology. In North America, this movement was not socialism but populism – specifically, civic populism. The movement was led by a historian and poet, William Dough Lighthall. Mayor of Westmount, a wealthy suburb of Montreal, he had founded the Union of Canadian Municipalities at the turn of the century.

Civic populism was advocated mainly by the petite bourgeoisie anxious about the rise of large capitalist consortiums, symbolized in their eyes by the railway, electricity, and telephone companies. For populists, the solution to all of the ills of urban life was the strengthening of common liberties and the creation of municipal public services.

This minimalist program drew people from all quarters, particularly from progressive and union circles. In the countryside, it found a strong echo among farmers. In short, civic populism led to a "sacred union" against the excesses of unfettered capitalism. Nevertheless, the movement fiercely advocated private enterprise, small business, and the traditional values of religion and family – in which it differed radically from socialism.[7]

Populism thus led to an acceleration in the supply of telephones to the Canadian population. But the quality of populist telephone service was not comparable to that offered by the large telephone companies such as Bell Canada. Service by independent companies relied on primitive technology. Their networks used single wires with ground return, cables were not buried, and a single line might serve up to eighteen or twenty subscribers. In the countryside, farmers built their networks with fencing wire attached to trees.

It was primitive telephone service, but it could be built rapidly and serve a large number of people. In it could be seen the typically North American equation "democracy = technology accessible to all" so dear to sociologist Eugene S. Ferguson. It also demonstrated very well how, in the United States, technology had a moral dimension: it opened the path to happiness, and everyone had a right to be happy. In many minds, this notion is still alive in North America.[8]

Thus, the conflict between the populists and the telephone companies had somewhat philosophical roots: it was a struggle between a democratic service with a "cheap and dirty" technology and a technology of excellent quality with service reserved for the élite. In telephony, the large companies, such as AT&T and Bell Canada, subscribed to the élitist tradition. The independent companies represented "American-ness."

The ingenuity of North American society was that the populists solved the situation by founding independent companies with local capital and cheap technology. The telephone companies could then buy up the independents and exploit the market that they had created out of whole cloth. Europe, on the other hand, had to be content with many years with high-quality telephone service – and then, not in all countries – and low distribution.

Long Distance

Brings the Farm to Town

Now that he can order it at once from his favorite dealer in village, town or city, nothing is too good for the farmer.

Within the same convenient radius of your store are scores of thrifty housewives who would be pleased to have a merchant solicit their trade by Long Distance.

Even if a sale does not result, you have advertised your business in a very effective way. They will probably call when they come to town, or order from your advertising. It's human nature.

Long Distance can bring suburban prospects within sound of your voice. Turn it into a salesman and keep it busy. Every other means of communication is a substitute for your voice—the voice is YOU.

Every Bell Telephone is a Long Distance Station

THE BELL TELEPHONE
COMPANY OF CANADA

Telephones highlighted the contrasts between urban and rural life: far from stabilizing the rural population, the culture shock unravelled the social and economic fabric of the countryside.

And then it was children's turn to use the telephone: "Yes – children today will never know a world without telephones, will not remember when a telephone call was a miracle, nor when the first connections were made across continents and oceans. To them, 'home' will never be farther away than the nearest phone." (1931)
Courtesy Bell Canada.

Bell Canada hesitated a long time before connecting its high-quality lines to the independents' iron-wire networks. Starting in 1905, regulation required them to do so. The result was lightning-quick penetration of the telephone into small towns and the countryside. North America's lead over most of Europe with regard to telephone infrastructure, acquired in the first fifty years of the twentieth century, was the result of this penetration.

But it is important to note that success was attained at the cost of institutionalization of a two-tier network. Today, many rural areas still have party lines – thus lower-quality service than that in the cities. In 1997 about 145 thousand subscribers were still served by party lines in Canada.

The rapid success of the telephone in the countryside had a paradoxical side. In the vast Canadian expanses, where isolation ruled, rural telephones often raised great hopes. Aside from establishing connections between farms, this technology put rural areas in more direct communication with the city. Farmers could follow stock-market fluctuations from day to day and obtain a better price for their harvests. But what was the impact of the telephone on agricultural life? The newspapers of the time were full of ecstatic articles on its benefits. In 1905 the following appeared in the magazine *Telephony:* "With a telephone in the house, a buggy in the barn, and rural mail box at the gate, the problem of how to keep the boys and girls on the farm is solved."[9] Such statements were a leitmotiv in all writings of the era.

However, the telephone did not prevent the Canadian countryside from depopulating at the same pace as in all other less well equipped countries. At the turn of the century, when telephones appeared in the countryside, 33 percent of the Canadian population was rural; the percentage has declined constantly ever since.

The telephone alone cannot be blamed: magazines, the railway, then the radio and automobile gradually put rural areas in direct contact with the fascinations of urban life. But the telephone played a unique role: its sociability and its sensorial extension of the individual no doubt had much to do with the culture shock between the city and the countryside. Far from stabilizing the population, the telephone contributed to the rural exodus.

This leads us directly to one basic impact of the telephone: the cultural impact. The telephone propagated urban values. The dilemma between centralization and decentralization is a false one: whether it is used to manage city centres crowded with skyscrapers, expand suburbs, or maintain isolated farms, the telephone conveys the same urban values. Far from protecting rural life, the telephone brings the city to the countryside.

The Canadian communications theorist Marshall McLuhan summarizes this urban vision very well: "Visual technology creates a centre-margin pattern of organization whether by literacy or by industry and a price system. But electric technology is instant and omnipresent and creates multiple centres-without-margins."[10] McLuhan is correct to stress the instantaneous character of telephony; he is wrong, however, to consider it as an extension of traditional oral civilizations. As an urban medium par excellence, telephony privileges the ephemeral and contributes to the erasure of memory to the extent that it becomes a substitute for written messages. It rehabilitates speech – not speech loaded with myths that are part of a historical, even metaphysical, continuum but a mechanized speech, devoid of its physical importance, relativized by equality of access.

A telephone ringing opens the acoustic space to a loved one, a plumber, or a wrong number. It gives rise to a magical anonymity on which literature – especially detective novels – have drawn. Far from taking us back to the space-time relationship of early humanity, this mechanized acoustic space takes the dictatorship of linear time, subdivided into smaller and smaller segments, to an extreme. The French economist Marc Guillaume gives a perfect description of the cultural implications of this technological phenomenon: "The effect of switching is also a change in social time. It accelerates discoveries, encounters, and accomplishments. Eliminating errors or hesitations that might be fertile, reducing the time spent walking and searching, it causes short circuits, often productive, but sometimes dangerous."[11]

It is not by chance that telephone operators were stripped of their language and their time. Since the beginnings of telephony, they were taught to respond to subscribers with stock phrases in order to eliminate any hesitation on their part and reduce the processing time of each communication. This Taylorization of language led to the creation of a "model employee," whom Katherine Schmitt, manager of AT&T's operator department, defined at the beginning of the century in stark terms: "The operator must now be made as nearly as possible a paragon of perfection, a kind of human machine."[12]

Thus, the telephone did not bring back the Hellenic world in which heroes mixed with gods in a temporal framework that stretched into eternity. On the contrary, during the period under study here, technology threw into sharp relief the unbalanced nature of industrial society. Human beings cannot be separated from their bodies with impunity.

It is true that Schmitt's "human machine" is not the only issue associated with telephony. It was nevertheless the exacerbated symbol of a degradation of social relations, which were found in less-acute ways in other impacts of the telephone – in particular, as a tool of hierarchical management, in the situation of the woman at home, in the rural exodus, and in the destruction of inner cities to the profit of suburbs. All of these impacts helped the early industrial order to triumph.

The situation was beginning to change, however, at the point where this book ends. Universal telephone service, after universal radio service, made available a common tool to all of society. The dislocational effects of telecommunications ceded to a social, economic, and cultural integration. Involved in the effort of postwar "reconstruction," Canadian society believed in a future based on economic progress and social stability: everyone with electricity, running water, central heating, and a car parked in front of a sub-urban bungalow. Telephone, radio, and, soon, television would unify the ele-ments of the Canadian mosaic into a harmonious whole. It was the 1950s. As we know, history did not end there – and neither did the history of telecom-munications.

Notes

INTRODUCTION

1 Collins, *A Voice from Afar;* Ogle, *Hello, Long Distance!*
2 Babe, *Telecommunications in Canada.*
3 Surtees, *Pa Bell.*
4 Surtees, *Wire Wars.*

CHAPTER ONE

1 Libois, *Genèse et croissance*, 19–23.
2 Prince Edward, duke of Kent, was the younger son of King George III and the future father of Queen Victoria. He was the namesake of Prince Edward Island.
3 Letter from Captain Lyman in Halifax, dated February 1800, to Edward Winslow in Fredericton, quoted in Morrison, "Wave to Whisper," 28.
4 Ibid., 19–36; Collins, *Voice*, 20.
5 Bertho, *Télégraphes et téléphones*, 60.
6 There are four minutes' difference per degree longitude. Since Hamilton is about one-half a degree longitude from Toronto, the difference in solar time would have been about two minutes.
7 *Daily Star*, Windsor, 19 December 1846.
8 Harris had managed to convince a group of businessmen to invest $16,000 in the telegraph line.
9 Green, *Canada's First.*
10 Green, *Telegraphs Statistics.*
11 Warner, "First Fifty Years"; Andrews, "The Heritage of Telegraphy."

12 Warner, "First Fifty Years."

13 This company was not related to the Grand Trunk Railroad Company.

14 Reid, *Telegraph*, 326–31.

15 Reid, *Telegraph*, 338.

16 Easson, "Beginnings," 767.

17 Richeson, *Le télégraphe*, 2.

18 Witteveen, "Telegraph"; Collins, *Voice*, 30–3.

19 *Frederick Newton Gisborne (1824–1892)*, 20 June 1972, Newfoundland Telephone Archives.

20 A type of latex extracted from a tree related to the rubber tree.

21 Field later tried to delete Gisborne's contribution from the official history of the transatlantic cable, going so far as to replace his name with that of his own brother, Dudley Field, in the list of promoters of the venture. See "The Original Projector of the Atlantic Telegraph," *The Times* (London), 1 December 1858.

22 Sharlin, *Making*, 25–32.

23 Reid, *Telegraph*, 508–17; Collins, *Voice*, 37–45; Richeson, *Le télégraphe*, 3–4.

CHAPTER TWO

1 Quoted in Collins, *Voice,* 85.

2 Easson, "Beginnings," 2: 759–69.

3 Collins, *Voice*, 33–5.

4 Forsey, "Telegraphers' Strike," n.p.

5 Bernard, *Long Distance,* 15–16. The figure of 2,500 union members is given in *A Sketch of the Canadian Telegraph System, Its Rise and Development,* CN Archives (NAC), 42.

6 Tillotson, *Canadian Telegraphers*, 15, 73–4, 90–3.

7 Logan, *Trade Unions*, 138–43.

8 *A Sketch of the Canadian Telegraph System*, CN Archives (NAC), 16.

9 Innis, *History*, 21–52.

10 In 1884 Fleming organized the conference in Washington at which the planet was divided into twenty-four time zones based on the Greenwich meridian – located in England, of course.

11 Macdonald, *Dominion Telegraph*.

12 Warner, "First Fifty Years."

13 Macdonald, "Dominion Telegraph." Macdonald dates the decision to create Dominion Telegraph in the summer of 1881 and the beginning of government operation in the summer of 1882. The statistical publication of the federal government (the forerunner of Statistics Canada) sets the date at 1879; see Green, *Telegraph Statistics*, 18. As Macdonald was one of the people active in the saga of Dominion Telegraph, I preferred to follow his chronology.

14 Maconald, *Dominion Telegraph*.

15 In *A Story of the Telegraph*, 123, Murray says that the government lines to the lower St Lawrence and the Gaspé Peninsula were built at the instigation of Pierre-Étienne Fortin, MP for Gaspé. Murray gives Fortin the rank of captain, while the *Canadian Encyclopedia* mentions not that he was a sailor but that he was a doctor. The document on Gisborne in the Newfoundland archives says that the government service was created in 1879.

16 Lalonde, "Riel: Defeated."

17 Macdonald, *Dominion Telegraph*, 35.

18 *La Minerve*, Montreal, 1 April 1885.

19 *La Minerve*, Montreal, 20 April 1885.

20 *La Presse*, Montreal, 24 April 1885.

21 *La Presse*, Montreal, 31 March 1885.

22 *La Minerve*, 17 November 1885. The article was dated the day before in Regina.

23 Innis, *History*, 128n1.

24 Quoted in Nichols, *Story*, 56.

25 Ibid.

26 Nichols, *Story*, 1–79; Collins, *Voice*, 192–4; Record Group 46, Railway Commission, vol. 47, Ottawa, 9 June 1910.

27 Fleming, "Cheap Telegraph Rates," address to the Canadian Press Association, 3 February 1902, cited in Witteveen, *Telegraph*, 21.

28 Quoted in Johnson, *All Red Line*, 53.

29 Erastus Wiman came to a miserable end. He went bankrupt in 1893, was imprisoned in New York, lost the presidency of Great North Western in 1895, and died penniless in 1904. A certain part of Canadian public opinion had always been hostile to him, since he had passed the two largest Canadian telegraph companies, Dominion Telegraph and Montreal Telegraph, to American control.

30 North American populism owed practically nothing to its predecessor, the Russian form of populism called *narodnik*.

31 Murray, *Story*, 124.

32 Postes, télégraphes, et téléphones – government-run communications services.

33 Griset, "1890–1932."

CHAPTER THREE

1 Libois, *Genèse*, 34–5. Harvard University possesses an example of Reis's telephone and has managed to make it work.

2 Bell was usually identified by his middle name.

3 Williams, *Telegraphy*, 110–11.

4 Surtees, *Pa Bell*, 61.

5 In Bell's declaration in the suit brought against Western Union, he stated, "The examiner declined to show me the caveat as it was a confidential document, but he indicated to me the particular claim in my application with which it had

conflicted. I therefore knew it had something to do with the vibration of a wire in a liquid." Quoted in Coon, *American Tel*, 44–55. It must be noted that Bell had deposited a patent application, while Gray had deposited a "caveat." In principle, the patent application is deposited when the apparatus is built and the caveat when it is not built. However, Bell could not deposit a caveat, since the procedure was reserved for American citizens and he was a British subject.

6 Kettle, "The Kettle Text," 10–12. At the beginning of the 1970s, Bell Canada commissioned a history of the company from Toronto journalist and futurologist John Kettle. The book was written but not published. It provides a truly remarkable perspective on the history of Bell and is a valuable source of information.

7 Libois, *Genèse*, 307.

8 In fact, this corporation was first called Bell Telephone Company (1877–79), then National Bell Company (1879–80), then American Bell Telephone Company (1880–89), before becoming AT&T. For simplicity, I will usually refer to this company as AT&T.

9 Datapro Research Corporation, "The Vail Years."

10 Bruce, *Bell*, 283.

11 Thomas Watson, "The Birth and Babyhood of the Telephone," speech given at the third annual convention of the Pioneers of North America, Chicago, 13 Oct. 1913. On Watson's life, see his autobiography, *Exploring Life* (Appleton, 1926).

12 Robert V. Bruce, "Alexander Graham Bell," *National Geographic* 174, no. 3 (Sept. 1988): 358– 84; Williams, *Telegraphy and Telephony*, 120.

CHAPTER FOUR

1 Quoted in Kettle, "Kettle Text," 22.

2 Ibid.

3 Collins, *Voice*, 71; Ogle, *Hello, Long Distance!*, 99.

4 Canada, Department of Agriculture, "Patent Case," 9, 13.

5 Coulombe, "Le téléphone"; Collins, *Voice*, 82; Patten, *Pioneering*, 97–8.

6 This expression is attributed to McCabe, "Regulation," 27.

7 Hereafter, the Bell Telephone Company of Canada will be referred to by the usual shorter name, Bell Telephone, or simply as Bell.

8 Taylor, "Charles F. Sise," 20.

9 Bell Telephone's authorized capital was $500,000. In the first year, only 15,104 shares were issued, worth $377,600. Of this amount, Canadian Telephone received $167,000 in exchange for the right to use its patents on the telephone, while American Bell received $93,900, which corresponded more or less to what it had spent to acquire Dominion Telegraph's telephone installations. Translated into percentages, these were, respectively, 44.6 percent and 24.9 percent. See "Amount and Percent of B.T. Co. Stock Held by American Bell Telephone Company and American Telephone and Telegraph Company, 1880 to Date," Office of the Vice-President, Finances, Bell Canada, Montreal, January 1975 (updated May 1977), BCA,

no. 30763; Fetherstonhaugh, *Charles Fleetford Sise*, 132–4. As well, Canadian Telephone's authorized capital was $300,000. American Bell held $225,000 of that amount (75 percent), the balance going to Western Union through its American Speaking Telephone and Gold & Stock Telegraph subsidiaries. See "The Relationship between the Bell Telephone Company of Canada and the Bell System in the United States in Regard to Patents," document used by Bell Canada in the Northern Electric case during the 1960s, n.d., BCA, no. 24 894–91.

10 This was Duquet's patent, which Sise acquired later.

11 Letter from Sise to Forbes (3 March 1980), BCA; "A History of the Telephone Business," Montreal, Bell (8 October 1952), BCA, no. 9834.

12 In some texts, Bell Telephone Company of Canada is mentioned among the members of the Bell System, but these were just public-relations documents with no legal value.

13 Letter from Forbes to Sise, 13 July 1880 (BCA).

14 Letters from Sise to Forbes, 28 June and 9 July 1880, quoted in "Charles F. Sise, Bell Canada, and the American: A Study of Managerial Autonomy, 1880–1905" (BCA), 21.

15 Muir, "History."

16 Letters from Sise to Forbes, 15, 16, and 18 February 1882 (BCA).

17 Letter from Duquet to Sise, 7 January 1881 (BCA, no. 21732–8).

18 Fetherstonhaugh's figures are different from those mentioned by the head of Bell Canada's history department, G.L. Long, for 1880 in "The Beginning of the Telephone Business in Canada," a document dated February 1963 (BCA):

Company	Edison telephones	Bell telephones
Dominion Telegraph		968
Montreal Telegraph	850	
Toronto Telephone Despatch		400
Hamilton District Telephone		181
Windsor Telephone Exchange		35
London Telephone	72	

19 McCabe, *Regulation*, 37–41.

20 Sise's log no. 3, May 1888 (BCA), cited in Armstrong and Nelles, *Monopoly's Moment*, 126.

21 Even before Bell was created, John Watson Tringham, with the approval of Dominion Telegraph, had founded his own company in Windsor. In January 1880 Tringham signed an agreement with American District Telegraph of Detroit to establish an international link. It was Bell that accomplished this eighteen months later with an undersea cable imported from Great Britain (BCA, no. 31–284).

22 This theory was expounded by J.E. Macpherson, vice-president of Bell, in Bell Canada, *Lectures on the Telephone Business*. See also Pike and Mosco, "Canadian Consumers."

23 Pike and Mosco, "Canadian Consumers," 20.

24 Anonymous, *Telephone History of Nova Scotia* (January 1962) (BCA, no. 21570), 7.

25 Armstrong and Nelles, *Monopoly's Moment*, 72.

26 Anonymous, *The Telephone in British Columbia* (n.d., probably 1963), BCA, no. 27183.

27 Bernard, *Long Distance Feeling*, 8.

28 Brooks, *Edgar Crow Baker*, 115–25.

29 Allen, *Factors*, 23–5.

30 Ibid., 39, 49. The new owners had created a holding company called British Columbia Telephones Ltd in Great Britain to manage all companies they acquired. This was the first time this name appeared. It seems not to have been used in the province.

31 Allen, *Factors*, 51–2; Brooks, *Edgar Crow Baker*, 124.

32 Allen, *Factors*, 56–8.

33 Ibid., 60–1.

34 Letter from Brian M. Longden, head of finance and costs at the Telecommunications Division of the Canadian Transport Commission, to Robert Spencer, Bell Canada historian, 11 February 1874 (BCA, no. D732272); *B.C.'s First Telephones* (n.d., probably 1970; probably published by BC Telephone); Ogle, *Hello, Long Distance*, 150–3; Collins, *Voice*, 126–32; Kettle, "Kettle Text", 22–3. It is to be noted that the merger was made in the name of Vernon and Nelson Telephone – that is, the subsidiary – because it possessed a permanent charter, while the parent company's charter was subject to renewal.

35 Owen, "The Fortunate Isle."

CHAPTER FIVE

1 Canada, Department of Agriculture, "Patent Case," 9–17.

2 Letter from Sise to W.A. Haskell, 28 January 1885 (BCA). Curiously, the letter continued, "All alleged violations occurred before the Am. Bell purchased the [Canadian] Patents," which is not true. Sise was well and truly responsible for what took place between Cowherd's death in January 1881 and the opening of the manufacturing division in July 1882.

3 Auld, *Voices*, 16–24.

4 Nova Scotia, Department of Transport, *The History of Telephone Service of Nova Scotia*, 2, 14.

5 Letter from B.F. Pearson to C.F. Sise, 14 September 1888 (BCA).

6 "Fredericton to Have Modern Dial Exchange," *New Brunswick Telephone News*, January 1947; "The Story of the Saint John Exchange," *New Brunswick Telephone News* (September 1948).

7 Letter from B.F. Pearson to C.F. Sise, 17 September 1888 (BCA); See following section, "Bell Confronts the Competition," for Vail's theory of occupation of the territory by long-distance lines.

8 "The Story of the Saint John Exchange," *New Brunswick Telephone News*, September 1948.

9 Patten, *Pioneering*, 90.

10 Armstrong and Nelles, *Monopoly's Moment*, 108. This conversation has been widely cited in many sources.

11 Dawson, *Proceedings* 1: 219.

12 Lebel, "Le téléphone."

13 Grindlay, *History*, 17–30.

14 This figure is cited almost everywhere but it seems a bit inflated, since many of the companies counted never passed the planning stage. Nevertheless, hundreds of independent companies were in operation at the turn of the century.

15 In 1884 Erastus Wiman, president of Great North Western, had drawn Sise's attention to CP's telephonic ambitions. He tried to forge an alliance between his company and Bell, but Sise bluntly refused. See letter from Wiman to Sise, 20 February 1884 (BCA).

16 Letter from Sise to T.E. Nalsh, regional manager at Winnipeg, 8 November 1888 (BCA).

17 Letters from Sise to J.E. Hudson, 17, 19, and 29 June 1890 (BCA). Emphasis Sise's.

18 Letter from Sise to John E. Hudson, 10 February 1892 (BCA).

19 The venture failed because of the condition regarding burying of cables was attached to the granting of a concession by the City of Toronto.

20 Dawson, *Proceedings*, 1:256–64; letter from Charles Frédéric Beauchemin to the mayor of Toronto, 28 September 1896 (BCA 27636).

21 McCabe, *Regulation*, 48–54; Armstrong and Nelles, *Monopoly's Moment*, 164 n.3.

22 Armstrong and Nelles, *Monopoly's Moment*, 144–62. Having described in great detail the wave of populism that should have led to nationalization of Bell in the Prairies, Armstrong and Nelles conclude, curiously, that anti-Bell opposition had been launched by the politicians and not by discontent subscribers. It was only later, according to Armstrong and Nelles, that public opinion turned against the telephone company. Other sources consulted seemed, on the contrary, to indicate popular resentment towards monopolies in general, including Bell's, with politicians simply channeling this resentment for their own electoral purposes. On the other hand, Armstrong and Nelles are correct when they point out that Bell was a scapegoat because it had no connections in local business circles or legislatures.

23 Collins, *Voice*, 182.

24 McCabe, *Regulation*, 54–9, 60–8.

25 Quoted in Cashman, *Singing Wires*, 116.

26 *Report of Mr Francis Dagger, Employed as Provincial Telephone Expert, with Respect to the Development of the Telephone Service in the Province of Saskatchewan*, Railways Ministry, Regina, 25 March 1908, 6.

27 Ibid., 11.

28 Dawson, *Proceedings*, 1: 19.

29 See the collection of newspaper articles of the time in the Lorimer Brothers' note-books (BCA, no. DF 25098–3).

30 McCabe, *Regulations*, 11–12.

31 Contemporary research proves that between 1880 and 1905, American Bell, AT&T, and Western Electric invested almost $1.2 million more in Bell Telephone than they received in the form of dividends, royalties on patents, and other trans-fers. See Taylor, "Charles F. Sise," 29.

32 Letter from C.F. Sise to F.P. Fish, 24 May 1905, quoted in Armstrong and Nelles, *Monopoly's Moment*, 172.

33 "Telephone rates have just grown," *Montreal Star*, 29 May 1907.

34 Ibid.

35 Armstrong and Nelles, *Monopoly's Moment*, 200–3.

36 "Telephone Tolls," *The Gazette* (Montreal), 14 May 1907.

37 Ville de Montréal v. Bell Telephone Co. (1912), cited in Baldwin, *Échec et renou-veau*, 80. The 1911 request was initiated by Toronto, the 1912 request by Mon-treal, and the 1915 request once again by Toronto.

38 Armstrong and Nelles, *Monopoly's Moment*, 203.

39 McManus, "Federal," 403.

CHAPTER SIX

1 Surtees, *Pa Bell*, 76.

2 Cashman, *Singing Wires*, 23.

3 Collins, *Voice*, 188.

4 *Western Canada: 1880–1909* (BCA, no. 24090), unsigned internal Bell Canada document, n.d.

5 Armstrong, and Nelles, *Monopoly's Moment*, 178.

6 Both the shadow company and the bribery scheme are mentioned in Armstrong and Nelles, *Monopoly's Moment*, 179 and 180.

7 Britnell, *Public Ownership*, 37.

8 According to the MGT annual report for 1908, there were a total of seventeen thousand subscribers in Manitoba, adding those of Bell to those of the govern-ment network. This figure, cited in Ogle, *Hello Long Distance*, 131, includes sub-scribers already served by MGT, which the MGT annual report for 1908 confirms. The brochure *People of Service: A Brief History of the Manitoba Telephone Sys-tem*, published by the Manitoba Telephone System, talks of twenty-five thousand subscribers, of whom fourteen thousand had been Bell's, but this figure seems greatly inflated and appears nowhere else.

9 Quoted in Cashman, *Singing Wires*, 140.

10 Ibid., 146.

11 Integration of Bell's and AGT's installations added up to 1,900 kilometres of long-distance lines, 60 long-distance exchanges, 35 local exchanges, and 2,929 sub-scribers. "History of the Alberta Government Telephones," (February 1921).

12 *Western Canada: 1990–1909* (BCA).

13 Spafford, *Telephone Service*, 5–19.
14 SaskTel, Public Affairs, *Answering the Call.*
15 Innis, *Problems*, 70.
16 Ibid., 76–81.
17 From "The Taxpayer and the Telephone," 2 January 1902 (BCA, no. 12016).
18 Bell Telephone of Canada annual report, 1912, quoted in Fetherstonhaugh, *R.C. Charles Fleetford Sise*, 224–5.
19 Taylor, "Charles F. Sise," 22.
20 Collins, *Voice,* 113; Fetherstonhaugh, *Charles Fleetford Sise*, 223.
21 Collins, *Voice*, 172–5.
22 Armstrong and Nelles, *Monopoly's Moment*, 198–200.
23 Mackay, *Abbreviated History*, binder 1, 70–4.
24 In 1907, Wisconsin had created a Board of Railway Commissioners. See Waverman, *Process*, 59–60.
25 Ogle, *Hello Long Distance!*, 114–16; Newfoundland Telephones *Along These Lines*; Collins, *Voice*, 190.
26 Judson, *Selected Articles*, 193.
27 "Report of the Commissioners of Manitoba Government Telephones for the year 1910," Sessional Paper no. 14, Manitoba, Session 1911, quoted in Britnell, *Public Ownership*, 41.
28 Final report of 20 May 1912 (made public on 14 June), quoted in Britnell, *Public Ownership*, 50.
29 Britnell, *Public Ownership*, 39 and 55; Armstrong and Nelles, *Monopoly's Moment,* 193–5. See also Manitoba Telephone System, *People of Service*, 6.
30 Britnell, *Public Ownership*, 58.
31 Quoted in Ogle, *Hello Long Distance*, 137.
32 Britnell, *Public Ownership*, 126.
33 Annual report of the Provincial Department of Railways and Telephones, 1915, quoted in Britnell, *Public Ownership*, 130.
34 Cashman, *Singing Wires*, 247–8.
35 Stinson, *Wired City*, 46 and 102.
36 Britnell, *Public Ownership*, 165; Waverman, *Process*, 69.
37 Telephone ministry, annual report 1923–24, quoted in Spafford, *Telephone Service*, 38.
38 Innis, *Problems*, 71–2; Spafford, *Telephone Service*, 35–36; Ogle, *Hello Long Distance*, 136.
39 Collins, *Voice*, 197–9.
40 Innis, *Problems*, 70.
41 Allen, *Factors*, 64, 88, 93, 101–8, 118–25, and 141–8.
42 Lebel, "Québec-Téléphone" (BCA).

CHAPTER SEVEN

1 Taylor, "Charles F. Sise," 18; Kettle, "Kettle Text," ch. 4, p. 3.

2 Fagen, *History of Engineering*, 32–3; Smith, *Anatomy* 111–20.

3 Taylor ("Charles F. Sise") and Kettle ("Kettle Text," ch. 4, p. 13) contradict each other: Taylor says 1891, while Kettle says February 1892. It was impossible to find the reason for this divergence.

4 Which was at the time 53 percent for Bell Telephone, 40 percent for Western Electric, and 7 percent for the others (members of the board of directors). See Bell Telephone, Secretariat Department, *History of Northern Electric Company Ownership*, 17 August 1967 (BCA, no. 27494).

5 Letter from Sise to Fish, 30 December 1901, quoted in Taylor,"Charles S. Sise," 27.

6 The other 4.2 percent was held by members of Bell's board of directors. See Bell Telephone, Secretariat, *History of Northern Electric* (BCS).

7 This made 50 percent for Bell Telephone, 43.6 percent for Western Electric, and 6.4 percent for the others (members of the board of directors). See Bell Telephone Secretariat, *History of Northern Electric* (BCS).

8 Stoffels, "In the Beginning," 154–9.

9 BCA document no. 21 7664.

10 Letter from Stirling Ross (Automatic Electric Company, Chicago) to Toronto magazine *Electric Digest*, 26 January 1940 (BCA).

11 *Ernest A. Faller vs. Lorimer & Lorimer*, Department of the Interior, United States Patent Office, 2–213, Washington, DC, 3 February 1902 (BCA, no. 23235), 9–18, 37–8, 82–4, 245–65, 289–97; Smith, *Early History*, 73–86.

12 Stinson, *Wired City*, 47–58; Cashman, *Singing Wires*, 167–73.

13 BCA document no. 25098; "The Lorimer Automatic in France," *Telephony* 12, no. 2 (August 1906); Memorandum on the Automatic Telephone, Telephone Historical Collection, 22 October 1965 (BCA, no. 27309–2).

14 Chapuis, *100 Years*, 167.

15 Dagger, "The Lorimer Machine Telephone"; Chapuis, *100 Years*, 67 n.13.

16 For more details on the principles of the Lorimer switch, see the purely technical work by Aitken, *Automatic Telephone Systems*, 233–43 and illustrations.

17 See collection of newspaper articles in the Lorimer brothers' notebooks, BCA no. D.F. 25098–3, 21797, and 972 (mainly *Expositor*, 7 November 1901; *Hamilton Spectator,* 14 March 1902; *Montreal Herald*, 20 July 1904; *Globe*, 5 August · 1905; *Illustrated London News*, 25 July 1908).

18 Quoted in Chapuis, *100 Years*, 162.

19 Quoted in ibid.

20 Ibid., 165–82.

21 Ibid., 162.

22 Letter from Sise to the Toronto Manager, 24 October 1892 (BCA). The quotation repeats a warning by an expert at Bell. The content of this letter, including the warning from the expert it quotes, were published in a newsletter for managememt dated 22 November 1892.

23 BCA document no. 21 7664.

24 Quoted in Lester, *Evidence*, 9 and 10.

25 Fagen, *History of Engineering*, 252–64.

26 Collins, *Voice*, 191–2; Ogle, *Hello Long Distance*, 42–4.

CHAPTER EIGHT

1 Parsons, *History*, 153.

2 Ibid., 151.

3 For instance, "Lachapelle, one of the men you sent to Toronto, broke his wrist and was sent back last night by Mr Dunstan (Manager in Toronto), who, I understand, paid his doctor's bill and wages up to the time he left. While it is not customary for us to do it, nor is it an obligation on our part, I am glad to say that Mr Sise has approved of our allowing this man his wages until he has recovered." Letter from L.B. McFarlane to D.C. Dewar, dated 12 February 1896 (BCA).

4 Background information, "Chronologie des relations de travail à Bell Canada," 17 November 1971 (BCA).

5 "Who Wouldn't Be a Telephone Girl," *The Watchman* (Montreal; 12 June 1898).

6 Martin, *Hello, Central?*, 97, 102.

7 "Loud talking or laughing must be avoided," says article 6 of the contract operators signed when they were hired. See Bernard, *Long Distance Feeling*, 19–21. Operators had to "sit up straight with no talking or smiling," said the *Toronto Star* (11 February 1907), quoted in Sangster, "1907 Bell Telephone Strike," 139. "Any moral lapse meant instant dismissal," Robert Collins wrote in *Voice*, 144.

8 Maddox, "Women and the Switchboard," 268.

9 Usually, an "auxiliary local" was reserved for wives of union members. This second-class status was, however, progress within the IBEW, where women had been formally excluded in 1893 before being readmitted. See Bernard, *Long Distance Feeling*, 18. The NBEW was founded in 1891 and renamed IBEW in 1897, with the arrival of the first Canadian locals. See Barbash, *Unions and Telephones*, 2.

10 Bernard, *Long Distance Feeling*, 27–39.

11 Ibid., 39–42.

12 Letter from L.B. McFarlane, general manager, to F. Scott, Montreal local manager, 13 August 1896 (BCA).

13 Owen, H.G., *The First Century of Service*, 21.

14 This was attested to by Dunstan himself in a letter to top management in Montreal on 20 December 1906. See King and Winchester, *Report*, 12.

15 Report by James C.T. Baldwin, 30 November 1906, quoted in King and Winchester, *Report*, 11.

16 King and Winchester, *Report*, 19.

17 Ibid., 22.

18 "I beg to state that had this request been received before the operating staff, or rather a portion of it precipitated trouble by striking yesterday about 1 p.m., the company would gladly have accepted." Quoted in ibid., 22.

19 Ontario Hydro had been created in May 1906, and less than a year later a majority in Toronto voted in a referendum in favour of municipalization of the local electricity company.

20 The correspondence between Dunstan and top management shows that Sise, a proponent of strong-arm tactics, wished to rehire "only those who are desirable." It was McFarlane who insisted on rehiring all operators without discrimination. See *Labour Trouble* (BCA, no. 24091).

21 *Toronto Star,* 7 February 1907, cited in Sangster, "1907 Bell Telephone Strike," 144.

22 King and Webster, *Report*, 36.

23 Parsons, *History.*

24 "The way in which King represented the operators referred back to the Victorian image of women." See Sangster, "1907 Bell Telephone Strike," 147.

25 King and Winchester, *Report*, 98.

26 See, in particular, King, *Industry and Humanity,* 156–62.

27 Parsons, *History.*

28 The percentage of women in Bell's labour force in 1880 is not known, but before the job of operator was feminized, it was probably insignificant, as it was in the Canadian labour force as a whole. The first statistics on the proportion of women at Bell date from 1897 (40 percent), and they indicate a rapid rise until 1920 (71 percent), with a slow drop afterwards. See Parsons, *History*, 146.

29 Owen, *The First Century*, 21.

30 "Hilloa! Hilloa!" *Montreal Daily Witness*, end of 1884 or beginning of 1885. Lewis B. McFarlane's personal papers, vol. 2, p. 70 (BCA).

31 "Le français au téléphone," *Le Nationaliste* (Montreal, 14 April 1907).

32 Fortner, "Messiahs and Monopolists," 89–95; Wade, *Les Canadiens français*, 1: 625.

33 *Le Devoir*, 12 July 1911. See also *La Presse* and *La Patrie* of the same date.

34 Omer Héroux, "Encore la Compagnie Bell," *Le Devoir*, 17 February 1912; "Ces Méssieurs du téléphone," *Le Devoir,* 16 March 1912; "Ces Messieurs du téléphone, épilogue," *Le Devoir*, 18 March. 1912.

CHAPTER NINE

1 Bertho, *Télégraphes et téléphones*, 186–7; Holcombe, *Public Ownership*, 418–38; Libois, *Genèse*, 268–72, 275–81.

2 "The Vail Years: Organizing for the Universal Network," Datapro Research Corporation, Industry Briefs (Jan. 1986).

3 Cited in Libois, *Genèse et croissance*, p. 325.

4 Quoted in Conway, "Public Relations Philosophy," p. 39.

5 "The Vail Years."

6 Markus, "Les télécommunications," 63–7; Meyer, *Public Ownership*, 5–7.

7 Codding, Jr, "International Telecommunication Union," 77–9.
8 Holcombe, *Public Ownership*, 433.

CHAPTER TEN

1 At the same time, the Russian naval engineer Aleksander Popov, who taught at the Kronstadt Torpedo School, came up with exactly the same invention but did not develop any commerical applications for it.
2 Sharlin, *Making of the Electrical Age*, 92.
3 Quoted in Collins, *Voice from Afar*, 160.
4 The nine participants were Germany, Austria, Spain, the United States, France, Great Britain, Hungary, Italy, and Russia.
5 The countries represented were Argentina, Austria, Belgium, Brazil, Bulgaria, Chile, Denmark, Egypt, France, Germany, Great Britain, Greece, Hungary, Italy, Japan, Mexico, Monaco, the Netherlands, Norway, Persia, Portugal, Rumania, Russia, Siam, Spain, Sweden, Turkey, the United States, and Uruguay. There were two observer countries: China and Montenegro.
6 Codding, *International Telecommunication Union*, 81 n. 1 and 98–100.
7 The *Canadian Encyclopedia* says East Milton, while Robert Collins and Adrian Waller, in an article in *Equinox*, say East Bolton.
8 Quoted in Waller, "Unsung Genius," 95.
9 Ibid.
10 In 1890 the French physicist Édouard Branley had invented an electromagnetic-wave receiver that improved upon Hertz's oscillator.
11 Sharlin, *Making of the Electrical Age*, 123–4.
12 Waller, "Unsung Hero," 95.
13 Harlow, *Old Wires*, 455–6.
14 See Raby, *Radio's First Voice*, for an anecdotal biography of Fessenden.

CHAPTER ELEVEN

1 Cumulative revenues of all telephone companies have been greater than those of telegraph companies since 1909. The first figures given by Statistics Canada for these two industries date from 1911 and 1912 ($10 million and $5.3 million, respectively). I have assumed that telephone revenues grew by $2.4 million per year (the average figure from 1911 to 1914), while telegraph revenues remained stagnant.
2 Weir, *Struggle*, 86.
3 Bell Telephone annual report, 1923.
4 *Suggested Agreement between the Bell Telephone Co. of Canada and Northern Electric Company, Limited in respect to the use of wire telephony in conjunction with radio broadcasting and the sale and lease of public address systems*, Montreal, 8 January 1924 (BCA, no. 27 321).

5 Weir, *Struggle*, 15, 43, 79–82, 90–2, 137–9.

6 Ibid., 143.

7 Desclouds, "History," 11–13; *Les TCN dans le Nord*, 19.

8 Desclouds, "History," 13; Law, *Development of CN Telecommunications*.

9 Desclouds, "History," 6, 7, 13.

10 Red was the colour of British Empire countries on maps.

11 Request regarding long-distance service, Unitel, Toronto, 16 May 1990, ch. B, p. 9.

12 Weir, *Struggle*, 35–9.

13 All technical details are taken from Bonneville, "Trans-Canada Telephone System."

14 Ogle, *Hello, Long Distance!*, 51–92.

15 Cashman, *Singing Wires*, 347.

16 Ogle, *Hello, Long Distance!*, 174–5.

17 Weir, *Struggle*, 130.

18 Ibid., 161–64.

19 The chain of events and the correspondence between the TCTS and the CRBC are reported in their entirety in the TCTS's submission to the parliamentary special committee on the radio, 12 May 1936 (BCA, no. 15536). See also Weir, *Struggle*, 169–70.

CHAPTER TWELVE

1 Kettle, "Kettle Text," ch. 5, 20–2; *Amount and Percent of BTCo Stock Held by American Bell Telephone Company and American Telephone and Telegraph Company*, 1977 (BCA, no. 30763).

2 Kettle, "Kettle Text", ch. 5, 26.

3 Lester, *Evidence of A.G. Lester*, 17.

4 See Bell Canada, *History of Bell Telephone Company of Canada and Northern Electric Company from Patent, Technical Information, Service Agreement and Ownership Standpoint up to July 1, 1959* (20 April 1961) (BCA, no. 14894–103), 3; *Memorandum in Regard to the Rights of the Bell Telephone Company to Secure Patents from and through the American Telephone & Telegraph Company*, 14 May 1925 (BCA, no. 21741).

5 "Our Relationship with the American Telephone and Telegraph Company," *The Blue Bell* (September 1925; BCA, no. 18776). This article was reprinted separately, which emphasized the importance placed upon it by top management.

6 Memorandum from R.V. Macauley to Frederick Johnson, 23 Aug. 1949. Quoted in BCA #24894-91, 8.

7 Marquez, "Building an Innovative Organization."

8 Bennett, "Une multinationale canadienne."

9 Quoted in Kettle, "Kettle Text," ch. 4, 35.

10 Lester, *Evidence*, 33; Kettle, "Kettle Text," ch. 6, 1–9.

11 Interview with Robert H. Spencer, Bell historian, session 2, 5 September 1968 (BCA, no. 17 992–1, biographical file).

12 C.J. Mackenzie, *Proposals for a Research and Development Organization for Northern Electric Co. Ltd.*, September 1955 (BCA), 4 and 14.

13 Interview with Robert H. Spencer (BCA).

14 The complete list of Bell attendees at the college is A.G. Lester (1949–50), H. Pildington (1963–64), J.V.R. Cyr (1972–73), J. Monty (1979–80), and M. Lisogurski (1987–88). BC Tel has sent one representative: D.M. Ramsay (1988–89).

15 Alex Lester, personal communication.

16 Interview with Alex Lester by Arthur Gosselin, September 1971 (BCA biographical file). According to Lester, the concept of the National Defence College came from Great Britain and was adopted by the United States in 1946.

17 Interview with Alex Lester by Arthur Gosselin, September 1972 (BCA).

18 Lecour, "Cadin Pinetree."

19 Charles R. Terreault, interview with author, 8 August 1990.

20 Lester, *Special Contract*.

21 Ogle, *Hello, Long Distance!*, 189–91; Marcotte, "The North Warning System."

CHAPTER THIRTEEN

1 Collins, *Voice from Afar*, 199.

2 MT&T in fact exchanged its 3,787 shares in the Prince Edward Island telephone company for all of the shares (947) in Eastern Telegraph and Telephone. It was a good deal for MT&T, since it was purchasing a company worth $15,000 for $10,000. It was an even better deal when it resold Eastern, without the Prince Edward Island system, to AT&T for $75,000. See Mackay, *The History of Maritime Tel & Tel*.

3 Previously, under MT&T's direct administration or Eastern's indirect administration, the island system had been called the Telephone Company of Prince Edward Island.

4 Kettle, "Kettle Text," ch. 11, 5; Lester, *Evidence*, 19.

5 Thomas W. Eadie, quoted in Ogle, *Hello, Long Distance!*, 182.

6 Ogle, *Hello, Long Distance!*, 116–17; Owen, "The Newfoundland Project" and "The Fortunate Isle."

7 Hoffman, "History of Telecommunications in Newfoundland," 27–35.

8 Auld, *Voices*, 52–102; *Confiance et service: NB Tel (1888–1988)*; C.A. Kee, *Synopsis of History of the New Brunswick Telephone Company Limited (1888–1957)* (BCA, no. 10 660–1); McKay, *History*.

9 Baldwin, *Échec*, 84; Britnell, "Public Ownership," 61.

10 Britnell, "Public Ownership," 62.

11 Stinson, *Wired City*, 95. On Lowry, see also "Commissioner Lowry Honoured in Retirement," *Telephone Echo* (Winnipeg; MTS internal newsletter), March 1945.

12 MTS annual report, quoted in Britnell, "Public Ownership," 71.

13 Goldenberg, *Manitoba Telephone Commission*.

14 Manitoba Telephone System, *People of Service;* Muir, "History," 69–82.

15 Saskatchewan Ministry of Telephones annual report, 1917, quoted in Britnell, "Public Ownership," 95.

16 Saskatchewan Ministry of Telephones annual report, 1920, quoted in Britnell, "Public Ownership," 102.

17 Britnell, "Public Ownership," 114–16.

18 Spafford, *Telephone Service*, 60–2.

19 One engineer was in charge of transmission, another of switching, and the third of electricity feed. George Spencer, Vice-President, Operations, SaskTel, interview with author, 16 October 1990; SaskTel, *Answering the Call;* Spafford, *Telephone Service*, 70–91.

20 Spafford, *Telephone Service*, 114.

21 Cashman, *Singing Wires*, 263–4.

22 Spafford, *Telephone Service*, 114.

23 Cashman, *Singing Wires,* 330–8; Spafford, *Telephone Service*, 149–53; Stinson, *Wired City*, 101–07.

24 Cashman, *Singing Wires*, 342, 343.

25 Ibid., 336–74, 376–82.

26 AGT annual report, quoted in ibid., 383.

27 Ibid., 396.

28 "British Columbia System Now Controlled by New Company."

29 Allen, "Factors," 62–72.

30 Bernard, *Long Distance Feeling*, 72–3, 125–6; letter from Brian M. Longden, head of finances and costs at the Telecommunications Division of the Canadian Transport Commission, to Robert Spencer, Bell Canada historian, 11 February 1974 (BCA); Jim A. MacInnes, Vice-President, Communications (retired), BC Tel, interview with the author, 21 August 1991; McCarthy, *History of GTE*, 27–47; Williams, *Labour Relations,* 64, 65.

31 BC Tel, *BC Tel's First Telephones*.

32 In fact, Nationale's name changed twice. Right after the acquisition, it was named Corporation de Téléphone et de Valeurs d'Utilités Publiques de Québec. In the following November, it was given its final name.

33 Quoted by Yvon Côté (magistrate), President (retired), Régie des services publics du Québec, interview with author, 26 June 1990.

34 "Jalons chronologiques," Québec-Téléphone (1986; unpublished) Archives Québec-Téléphone; Lebel, "Québec-Téléphone" (BCA); Levesque, *60ᵉ (1927–1987)*; "Une compagnie semi-centenaire"; "Une empire industriel."

35 In fact, there were two reports and two laws on rural telephony. The first report, written by an engineering consultant, Brigadier R.E. Smythe, resulted in a 1951 law that recommended general improvements to the rural network. It was after this that Ontario Hydro received the mandate to specify the means. Brigadier Smythe was also on Ontario Hydro's committee.

36 Grindly, *History of the Independent*, 22–48.
37 Campbell (Major), "Canada Discovers the North."

CHAPTER FOURTEEN

1 Owen, *The First Century of Service*, 1.
2 Collins, *Voice*, 195.
3 Vail, *Views*, 104, quoting from a 1913 AT&T annual report.
4 Ibid.
5 These three motifs are mentioned by Shacht in "Toward Industrial Unionism."
6 Bernard, *Long Distance Feeling*, 46.
7 Cashman, *Singing Wires*, 293.
8 Barbash, *Unions*, 4.
9 Logan, *Trade Unions*, 150–2.
10 *Toronto Globe*, 29 August 1918, quoted in Parsons, *History*, 49.
11 Parsons, *History*, 50.
12 Ibid., 52.
13 Bernard, *Long Distance Feeling*, 231, n.18.
14 This thesis is put forward by Jack Barbash, a historian of unionism in the North American telephone industry, in *Unions and Telephones*, 17.
15 Royal Commission on Industrial Relations, 28 June 1919, quoted in Parsons, *History*, 91.
16 J.H. Brace, General Plant Manager, "Employee Relations in the Telephone Business," lecture delivered at Department of Political Science, University of Toronto, 17 February 1927 (BCA).
17 Quoted in Parsons, *History*, 86.
18 The other technicians left the IBEW to form their own union, the Manitoba Federation of Telephone Workers, which was not a company union. The other non-management employees were registered in a company union.
19 Bernard, *Long Distance Feeling*, 65.
20 Logan, *Trade Unions*, 151, 157.
21 *Labour Relations: 1900–1940*, letter dated 31 March 1956 (BCA, Labour Relations File).
22 In the twelve sectors covered by Labour department statistics, only the laundry and personal services sectors had faster salary increases. See Williams, *Labor Relations*, 401, Table 18.
23 In Bell's territory, the work week in Quebec was aligned with the national average, while the average in Ontario was from 44 to 54 hours. Telephone employees in British Columbia worked from 44 to 48 hours a week, and those on the Prairies were the best off, working 42 to 48 hours a week. See Williams, *Labor Relations*, 424.
24 Ibid., 424–52.
25 Pike and Mosco, "Canadian Consumers."
26 The Wagner Act was adopted in 1935 and contested in the courts by employers; it was declared constitutional by the Supreme Court in 1937. AT&T

waited until that time to apply it, as did other large companies hostile to the legislation.

27 Order in Council CP 1003, 17 February 1944.

28 *ACET, Historique de l'Association*, January 1971 (BCA, Labour Relations file).

29 A referendum had been organized during the autumn of 1946 among the members of the Plant Employees' Association on the issue of accreditation, and a crushing majority had rejected it.

CHAPTER FIFTEEN

1 Quoted in Armstrong and Nelles, *Monopoly's Moment*, 280.

2 McKay, *History of Telephone Service.*

3 Armstrong and Nelles, *Monopoly's Moment*, 275.

4 Pike and Mosco, "Canadian Consumers."

5 Annual report of the Railway Commission, Ottawa, 1922, quoted in Armstrong and Nelles, *Monopoly's Moment*, 279.

6 "Bell inventory details occupy railway board," *Montreal Star*, 22 April 1926.

7 "Les taux du téléphone," editorial in *La Presse*, 27 January 1926; "Auditors to Look into Telephone Company's Books," *Montreal Star*, 12 February 1926.

8 "Commission Declines Request of Toronto and Montreal Men," *Montreal Star*, 19 April 1926.

9 Selection of articles: "Bell Telephone Co. Submits Case and Hearing Adjourns," *The Gazette*, 12 May 1926; "Bell Telephone Co. Hearing Adjourned until July 20th," *The Gazette*, 19 June 1926; "Automatic Phones Have Carried Down Equipment Prices," *The Gazette*, 9 October 1926; "Bell Telephone Claims Immense Savings Under Its Northern Contract, *The Financial Post*, 5 November 1926; "Depreciation Big Discussion Point in Phone Hearing," *The Financial Post*, 19 November 1926; "Les avocats devront présenter des plaidoiries écrites dans la cause de la compagnie de téléphone Bell," *La Presse*, 27 November 1926; "Bell Telephone Case Undecided for Weeks Yet," *The Financial Post*, 17 December 1926.

10 Selection of articles: "Rate Increase for Telephone Co. is Granted," *The Gazette*, 23 February 1927; "Rate Increases are Granted to Bell Telephone Company," *Montreal Star*, 23 February 1927; "Les taux de téléphone," editorial in *La Presse*, 23 February 1927; "Le téléphone," editorial in *La Patrie*, 23 February 1927; "The Telephone Award," editorial in *The Gazette*, 25 February 1927; "Les nouveaux taux du téléphone," *La Patrie*, 3 March 1927.

11 "Commission Declines Request of Toronto and Montreal Men," *Montreal Star*, 19 April 1926.

12 In fact, the provincial company was called British Columbia Telephone Company Limited, and the federal company simply British Columbia Telephone Company.

13 Ross, *Factors*, 156–82.

14 "Telephone Business Unlike Any Other." Curiously the Bell rationale calculates 1:140 for small districts and 1:80 for larger ones.

15 Cayla-Bouchard et al., "Tarification."

16 U.N. Bethell, vice-president of AT&T, quoted in *Report from the Select Committee on Telephone Service*, London, 1922, 22–23 (BCA, no. 29811).

17 Pike and Mosco, "Canadian Consumers."

18 Hay, "Trends."

19 Ibid., 125.

20 Board of Transport Commissioners Decision C.955, 16 August 1957, 10 January 1958, cited in Waverman, *Process*, Appendix 1, 21.

21 McManus, "Federal."

22 Letter from Alexander Graham Bell to his father, 10 March 1876. In Bruce, *Bell and the Conquest of Solitudes*, 181.

23 *The Regulation of Bell Canada.*

24 This was not the first time discrimination was raised as an argument against interfinancing; the populist movement at the beginning of the century had opposed it as well, as this quotation from the Dagger report in Saskatchewan testifies:

Again, if the government owned the local exchanges there would be a strong tendency on the part of the people to expect a uniform rate for service in all towns and villages the population of which most clearly corresponded. These towns would in many cases be several miles apart and their local variations and conditions would be such that what was a profitable rate in one exchange might entail operating at a loss at another. This would make it necessary for the government to adopt a schedule of rates under which the subscribers in one exchange would be paying the profits on their own system and would also be contributing toward a deficit elsewhere. In other words it may be possible for the people in one city to be contributing $5 a year more in cost per telephone in order that the citizens in another part of the province might have their telephones at $5 less than cost. (Dagger, *Report*, 10)

25 Board of Transport Commissioners Decision C.995–170, 12 October 1949.

26 Waverman, *Process*, 99; *The Regulation of Bell Canada.*

27 *The Regulation of Bell Canada.*

28 Quoted in Babe, *Telecommunications in Canada*, 129.

29 Racicot, *Évolution*, 70, 71.

CHAPTER SIXTEEN

1 In 1911 SGT had purchased the Saskatoon independent company, which had been automated in 1907, and inaugurated a systematic policy of automation starting with Regina (1914).

2 Lester, *Evidence*, 12.

3 Letter to G.H. Rogers, 31 March 1936 (BCA, Labour Relations file).

4 Interview with Alex G. Lester by Aurthur Gosselin, Bell historian, September 1972 (BCA, biographical file), 8, 9.

5 Other, smaller automatic switchboards must have been used in the nationalized companies on the Prairies. With 500 extensions and a capacity of 700, the Canadian National switchboard was no doubt the first of its size to be installed in Canada. See articles in the *Gazette, La Presse, Montreal Star*, 20 January 1926; "The evolution of PBXs," 69.

6 In the original plan, there was to be a continental centre situated in St Louis, Missouri. It was never put into service, since the regional centres were always able to manage the flow of telephone traffic. On the other hand, the section centres did not exist in the initial plan and were added later.

7 Fagen, *History of Engineering*, 647–53; Joel, *History of Engineering*, 173–80; Lester, "Telecommunications."

8 Fagen, *History of Engineering*, 391–409.

9 Headrick, *The Invisible Weapon*, 202.

10 Owen, *First Century*, 30.

11 Collins, *A Voice from Afar*, 250.

12 Cashman, *Singing Wires*, 406–16; *From Sea to Sea*, TCTS information pamphlet (BCA MP. 14279); Ogle, *Hello, Long Distance!*, 204–16.

13 Alex G. Lester interviewed by Arthur Gosselin, Bell historian, 58 (BCA).

14 Ogle, *Long Distance Please*, 216.

15 Jack Sutherland (ex-vice-president of RCA and ex-president of CNCP Telecommunications), interview with author, 27 June 1991.

16 Ogle, *Hello, Long Distance!*, 217–24.

CHAPTER SEVENTEEN

1 Rens, *Rencontres avec le siècle*, 144.

2 McFarlane, "Organization in the Telephone Business" lecture in the series *Lectures on the Telephone Business* given at University of Toronto by Bell directors, 1926–27 (BCA).

3 The new multinational was known for the brutality of its marketing methods and its political manipulations, before it withdrew completely from the telecommunications sector in 1986. It was accused, among other things, of participating, in September 1973, in the overthrow of the democratically elected president of Chile, socialist Salvador Allende. See Samson, *Sovereign State*, 24.

4 Brooks, *Telephone*, 171; Libois, *Genèse*, 330, 331.

5 Bellchambers et al., "International Telecommunication Union"; Libois, *Genèse*, 48, 49. At the beginning, the telephone committee was called the International Consultative Committee on Long-distance Telephone Communications, and the radiotelegraphic committee was called the International Consultative Committee on Radio-electric Communications.

6 Quoted in Codding, *International Telecommunication Union*, 140.

7 Ibid., 426–7, 459–61.
8 Codding, *International Telecommunication Union*; Telecommission Study 3a, "International Implications of Telecommunications: The Role of Canada in Intelsat and Other Relevant International Organizations," 28.
9 Headrick, *The Invisible Weapon*, 266.
10 Jean-Claude Delorme, interview with the author, 22 May 1991; St-Arnoud, *Mise en œuvre*, 3–66.
11 Télécommission, Study 3a, 73–7.
12 Federal Law 1949 S.C., chapter 10.

CONCLUSION

1 Alexander Graham Bell, letter to his father, 10 March 1876, quoted in Bruce, *Bell*, 181.
2 Flichy, *Une histoire*, 47.
3 Collins, *A Voice from Afar,* 123.
4 Cherry, "The Telephone System," 120.
5

Type of advertising	Total ads	Local only
Social, sociability	5	2
Business, businessmen	20	20
Household, convenience, etc.	28	28
Public relations, other	30	30
Total	83	80
Approximate ratio of social to others	1/16	1/39

These figures concern Bell's advertising in Canada before the First World War and are drawn from a table titled Counts of Dominant Advertising Themes by Period. The "social/other" ratio changed completely in the 1920s. See Fischer, "Touch Someone," 46.
6 Briggs, "The Pleasure Telephone," 40.
7 Armstrong and Nelles, *Monopoly's Moment*, 141–62.
8 Ferguson, "The American-ness of American Technology," 3–24.
9 Quoted in Sola Pool. *Forecasting the Telephone*, 48–9.
10 McLuhan, in Innis, *The Bias of Communications*, xiii.
11 Guillaume, "Une société de commutation."
12 Katherine M. Schmitt, quoted in Maddox, "Women and the Switchboard," 270. Ms. Schmitt worked at AT&T from 1881 to 1930.

Bibliography

Aitken, William. *Automatic Telephone Systems*. Vol. 1. London: Benn Brothers, 1981.

Allen, Lindsay Ross. "Factors in the Development of the British Columbia Telephone Industry, 1877–1930." Master's thesis, Simon Fraser University, 1990.

American Telephone and Telegraph CAT&T). *Telephone and Telegraph Statistics of the World*. Annual reports. N.p.: AT&T.

Andrews, Frederick T. "The Heritage of Telegraphy." *IEEE Communications Magazine* (August 1989).

Antéby, Élizabeth. *La grande épopée de l'électronique*. Neuilly: Hologramme, 1982.

Armstrong, Christopher, and H.V. Nelles. *Monopoly's Moment (The Organization and Regulation of Canadian Utilities, 1830–1930)*. Philadelphia: Temple University Press, 1986.

Association of Canadian Bankers. *American Long Distance Competition: The Power of Choice*. May 1991.

Auld, Walter C. *Voices of the Island: History of the Telephone on Prince Edward Island*. Halifax: Nimbus Publishing, 1985.

Auw, Alvin Von. *Heritage and Destiny (Reflections on the Bell System in Transition)*. New York: Praeger, 1983.

BC Tel. *BC Tel's First Telephones*. Vancouver: BC Tel, n.d. (1969?).

Babe, Robert. *Telecommunications in Canada*. Toronto: University of Toronto Press, 1990.

Baldwin, John R. *Échec et renouveau (L'évolution de la réglementation des monopoles naturels)*. Ottawa: Economic Council of Canada, 1989.

Ball, Norman R., ed. *Building Canada: A History of Public Works*. Toronto: University of Toronto Press, 1988.

Barbash, Jack. *Unions and Telephones: The Story of the Telecommunications Workers of America.* New York: Harper and Brothers, 1952.

Barnett, William Paul. "The Organizational Ecology of the Early American Telephone Industry: A Study of the Technological Cases of Competition and Mutualism." Doctoral dissertation. University of California at Berkeley, 1988.

Bell Canada. *Lectures on the Telephone Business.* Collections of courses given in the Political Science Department of the University of Toronto by Bell Canada managers, 1926–27.

– Bell Telephone annual report. 1923.

Bellchambers, W.F. Francis, E.J. Hummel, and R.L. Nickelson. "The International Telecommunication Union and Development of Worldwide Telecommunications." *IEEE Communications Magazine* (New York), 22, no. 5 (May 1984).

Bennet Communications. "Bell Canada/Northern Telecom/Bell-Northern Research: Key Dates in the Evolution of R&D." Unpublished document. Pointe-Claire, Quebec, 12 November 1986. Northern Telecom Archives.

Bennett, Gordon. "Une multinationale canadienne à l'assaut des marchés mondiaux." *In Search/En Quête* (Canada, Department of Communications), 4, no. 2 (Spring 1977).

– "L'évolution de la commutation: à pas de géant." *In Search/En Quête* (Canada, Department of Communications) 8, no. 3.

Bernard, Elaine. *The Long Distance Feeling: A History of the Telecommunications Workers Union.* Vancouver: New Star Books, 1982.

Bertho, Catherine. *Télégraphes et téléphones, de Valmy au microprocesseur.* Paris: Livre de Poche, 1981.

– *Histoire des télécommunications en France.* Paris: Érès, 1984.

Bilodeau, Rosario, ed. *Histoire des Canadas.* Montreal: Hurtubise HMH, 1971.

Bolton, Brian. *Les principaux exploitants publics de télécommunications dans le monde (partie II).* Geneva: Internationale du personnel des postes, télégraphes et téléphones, fall 1990.

Bonneville, Sidney. "The Trans-Canada Telephone System." *Bell Telephone Quarterly* 9 (1932): 228–44.

Bosworth, Newton. *Hochelaga Depicta: The Early History and Present State of the City and Island of Montreal.* Montreal: William Craig, 1839.

Briggs, Asa. "The Measure Telephone: A Chapter in the Prehistory of the Media." In *The Social Import of the Telephone*, edited by Ithiel Sola Pool. Cambridge, MA: MIT Press, 1977.

"British Columbia System Now Controlled by New Company." *Automatic Telephone* (January/February 1927).

Britnell, G.E. "Public Ownership of Telephones in the Prairie Provinces." Master's thesis. University of Toronto, 1934.

Brooks, George Waite Stirling. "Edgar Crow Baker: An Entrepreneur in Early British Columbia." Master's thesis. Vancouver, University of British Columbia, 1976.

Brooks, John. *Telephone: The First Hundred Years*. New York: Harper and Row, 1975.

Brothers, James Alexander Roy. "Telesat Canada: Pegasus or Trojan Horse? A Case Study of Mixed, Composite and Crown Enterprises." Master's thesis. Ottawa, Carleton University, 1979.

Bruce, Robert R., Jeffrey P. Cunard, and Mark D. Director. *From Telecommunications to Electronic Services: A Global Spectrum of Definitions, Boundary Lines, and Structures*. Washington: Butterworths, 1986.

Bruce, Robert V. *Bell: Alexander Graham Bell and the Conquest of Solitude*. Boston: Little Brown and Company, 1973.

CN Telecommunications. *Les TCN dans le Nord firent profit de l'expérience des Canadiens*. Montreal: CN Telecommunications, n.d.

Campbell, E.R., Major. "Canada Discovers the North and the Story of the Mad Trapper." *Communications and Electronics Newsletter*. Ottawa: Canadian Forces, 1973/3.

Canada. Department of Agriculture. "Patent Case: The Toronto Telephone Manufacturing Company v. the Bell Telephone Company of Canada." Ottawa, 24 January 1885.

– Department of Industry. *White Paper on a Domestic Satellite Communication System for Canada*. Ottawa, March 1968.

– Ministry of Industry. *Livre blanc sur un système domestique de télécommunications par satellite pour le Canada*. Ottawa, March 1968.

– Statistics Canada. *Historical Statistics of Canada*. 2d ed. Series I 336–341.

Carpentier, Michel, Sylvie Farnoux-Toporkoff, and Christian Garric. *Les télécommunications en liberté surveillée*. Paris: Technique et documentation – Lavoisier 1991.

Carré, Patrice, and Martin Monestier. *Le Télex, 40 ans d'innovation*. Paris: Mangès 1987.

Cashman, Tony. *Singing Wires: The Telephone in Alberta*. Edmonton: Alberta Government Telephones, 1972.

Cayla-Boucharel, Laurette, Marie-France Cicéri, Raymond Heitzmann, and Charles Pautrate. "Tarification, prix de revient et concurrence." *France Télécom (Revue française des télécommunications)*, no. 74 (October 1990).

Chapman, John H., P.A. Forsyth, P.A. Lapp, and G.N. Patterson. *Upper Atmosphere and Space Programs in Canada*. Ottawa: Queen's Printer, 1967.

Chapuis, Robert J. *100 Years of Telephone Switching 1878–1978)*. Vol. 1. Amsterdam: North-Holland Publishing Company, 1982.

Chapuis, Robert J., and Joel, Amos E. *Electronics, Computers and Telephone Switching (A Book of Technological History as Volume 2: 1960–1985 of 100 Years of Telephone Switching)*. Amsterdam: North-Holland Publishing Company, 1990.

Cherry, Colin. "The Telephone System: Creator of Mobility and Social Change." In *The Social Impact of the Telephone*, edited by Ithiel Sola Pool. Cambridge, MA: MIT Press, 1977.

Codding Jr., George Arthur. "The International Telecommunications Union: An Experiment in International Cooperation." Doctoral dissertation. University of Geneva, 1952.

Cole, Barry, ed. *After the Breakup: Assessing the New Post-AT&T Divestiture Era.* New York: Columbia University Press, 1991.

Collins, Robert. *A Voice from Afar: The History of Telecommunications in Canada.* Toronto: McGraw-Hill Ryerson, 1977.

Conway, Connie Jean. "Public Relations Philosophy of Theodore N. Vail." Master's thesis. University of Wisconsin, 1958.

Coon, Horace. *American Tel & Tel: The Story of a Great Monopoly.* New York: Longmans, Green and Co., 1939.

Coopers and Lybrand Consulting Group. *Overview of the Interim Audit.* Vol. 1 of *Management Audit of MTX.* Coopers and Lybrand Consulting Group, 1986.

Cordell, Arthur J. *Sociétés multinationales, investissement direct de l'étranger et politique des sciences du Canada.* Special study no. 22 December. Ottawa: Science Council of Canada, 1971.

Coughlin, Ray. "Si la SCTT m'était contée ..." *Spargo* (COTC internal magazine, twenty-fifth anniversary special issue), 12, no. 1 (January 1975).

Coulombe, J.-T. *Le téléphone à Québec (réalisation canadiennes et américaines).* Unpublished manuscript, n.d. BCA.

Curien, Nicolas, and Michel Gensollen. *Économie des télécommunications (ouverture et réglementation).* Paris: Economica, 1992.

Dagger, Francis. *Report of Mr Francis Dagger, Employed as Provincial Telephone Expert, with Respect to the Development of the Telephone Service in the Province of Saskatchewan.* Regina, 25 March 1908.

Dalfen, Charles M. "The Telesat Canada Domestic Communications Satellite System." *Canadian Communications Law Review (University of Toronto) 1 (December 1968): 182–211.*

Datapro Research Corporation. "The Vail Years: Organizing for the Universal Network." Industry briefs. New Jersey, 1986.

Dawson, S.E. *Proceedings of the Select Committee on Telephone Systems.* 2 vols. Ottawa: King's Printer, 1905.

Desclouds, G.A. "History, Development, Services and Organization of CNCP/Telecommunications." Speech given to Canadian Forces Communications and Electronics School. Kingston, Ontario, 15 October 1970.

Dohoo, Roy M. "The Development of Canadian Satellite Communications." In *Canadian Developments in Telecommunications: An Overview of Significant Contributions,* edited by Thomas L. McPhail and David C. Coll. Calgary: University of Calgary, Department of Communications, 1986.

Easson, Robert F. "The Beginnings of the Telegraph." In *The Municipality of Toronto: A History,* edited by Jesse Edgar Middleton. Vol. 2, 759–69. Toronto: Dominion Publishing Co., 1923.

Easton, Kenneth J. "La télédistribution est née du désir des Canadiens d'avoir directement accès à la télévision américaine." *In Search/EnQuête* 4, no. 4 (Autumn 1988).

"Un empire industriel (les principales étapes de croissance de l'empire industriel de Jules-A. Brillant." *Synergie* (Rimouski) 1, no. 1 (April 1984).

"The Evolution of PBXs." *Telesis* (Ottawa) 4, no. 3 (Autumn 1975).

English, H. Edward, ed. *Telecommunications for Canada: An Interface of Business and Government*. Toronto: Methuen, 1973.

Fagen, M.D. *A History of Engineering and Science in the Bell System: The Early Years (1875–1925)*. Bell Telephone Laboratories, 1975.

Ferguson, Eugene S. "The American-ness of American Technology." *Technology and Culture* 20, no. 1.

Fetherstonhaugh, R.C. *Charles Fleetford Sise, 1834–1918*. Montreal: Gazette Printing Company, 1944.

Fischer, Claude. "Touch Someone: The Telephone Industry Discovers Sociability." *Technology and Culture* 29, no. 1 (1998): 46.

Flichy, Patrice. *Une histoire de la communication moderne (espace public et vie privée)*. Paris: La Découverte, 1991.

Forsey, Eugene. "The Telegraphers' Strike of 1883." *Délibérations et Mémoires de la Société royale du Canada* (Ottawa) 9, 4th series (1971).

Fortner, Robert Steven. "Messiahs and Monopolists: A Cultural History of Canadian Communications Systems, 1846–1914." Doctoral dissertation. University of Illinois, 1978.

Franklin, C.A. "John Herbert Chapman (1921–1979)." *Déliberations et Mémoires de la Société royale du Canada* (Ottawa) 9, 4th series (1980): 68–72.

Gille, Bertrand. *Histoire des techniques*. Paris: La Pléiade, Gallimard, 1978.

Globerman, Steven, and Deborah Carter. *Telecommunications in Canada: An Analysis of Outlook and Trends*. Vancouver: Fraser Institute, 1988.

Goldenberg, Carl H. *Manitoba Telephone Commission*. Report. Winnipeg: Government of Manitoba, 1940.

Goodspeed, D.J. *A History of the Defence Research Board of Canada*. Ottawa: Queen's Printer and Controller of Stationery, 1958.

Grant, Peter S. "Telephone Operation and Development in Canada (1921–1971)." Law thesis. University of Toronto, 1974.

Grant, Peter S., ed. *Telephone Operation and Development in Canada (1921–1971)*. Toronto: University of Toronto Press, 1974.

Green, Ernest. *Telegraph Statistics of the Dominion of Canada for the Year Ended June 30, 1912*. Ottawa, 1913.

– *Canada's First Electric Telegraph*. Paper no. 24. Toronto: Ontario Historical Society's Papers and Records, 1927.

Gregory, John D. *La régie des services publics du Québec et le contrôle des services téléphoniques*. Quebec City: Ministère des Communications du Québec, Éditeur officiel du Québec, 1975.

Grindlay, Thomas. *A History of the Independent Telephone Industry in Ontario.* Toronto: Ontario Telephone Service Commission, 1975.

Griset, Pascal. "1890–1932: la difficile genèse." *France Télécom* (Paris), no. 68 (March 1989).

– "Les fils de Théodore (Genès et dévelopement du Bell System, 2ᵉ partie)." *France Télécom* (Paris), no. 73 (July 1990).

Guillaume, Marc. "Une société de commutation." *Réalités industrielles* (une série des Annals des Mines), April 1993.

Hamelin, Jean, and Yves Roby. *Histoire économique du Québec (1851–1896).* Montreal: Fides, 1971.

Harlow, Alvin G. *Old Wires and New Waves: The History of the Telegraph, Telephone and Wireless.* New York: D. Appleton-Century Company, 1936.

Hary, M., ed. *Disconnecting Bell: The Impact of the AT&T Divestiture.* New York: Pergamon, 1984.

Hay, J.M. "Trends in Telephone Toll Rates." In *Telephone Operation and Development in Canada (1921–1971),* edited by Peter S. Grant, 123–27. Toronto: University of Toronto Press, 1974.

Headrick, Daniel R. *The Invisible Weapon: Telecommunications and International Poitics, 1851–1945.* New York: Oxford University Press, 1991.

Hoffman, K.W. "History of Telecommunications in Newfoundland." Speech to the Newfoundland Historical Society St John's, Newfoundland, 2 November 1978.

Hogue, Clarence, André Bolduc, and Daniel Larouche. *Québec, un siècle d'électricité.* Montreal: Libre Expression, 1979.

Holcombe, A.N. *Public Ownership of Telephones on the Continent of Europe.* Economic Studies, vol. 6. Cambridge: Harvard University Press, 1911.

Innis, Harold A. *Problems of Staple Production in Canada.* Toronto: Ryerson Press, 1933.

– *A History of the Canadian Pacific Railway.* Toronto: University of Toronto Press, 1971.

– *The Bias of Communications.* Toronto: University of Toronto Press, 1984.

Intven, H.G., and L.P. Salzman. *Les télécommunications Canada–Outre-mer dans le contexte mondial (évaluation du milieu d'exploitation de Téléglobe Canada Inc. et de ses répercussions sur la réglementation).* Study for the CRTC. Toronto: McCarthy-Tétreault, March 1991.

Jelly, Doris H. *Canada: 25 Years in Space.* Ottawa: Polyscience Publications with the National Museum of Science and Technology, 1988.

Joel, Amos E., Jr. *A History of Engineering and Science in the Bell System: Switching Technology (1925–1975).* United States: Bell Telephone Laboratories, 1982.

Johnson, George. *The All Red Line: The Annals and Aims of the Pacific Cable Project.* Ottawa: James Hope and Sons, 1903.

Judson, Katharine B, comp. *Selected Articles on Government Ownership of Telegraph and Telephone.* New York: H.W. Wilson, 1914.

Kahaner, Larry. *On the Line: How MCI Took on AT&T – and Won!* New York: Warner Books, 1986.

Keefer, Thomas Coltrin. *Philosophie des chemins de fer.* Montreal: John Lovell, 1853.

Kelly, Tim. *Perspectives des communications.* Paris: Organization for Economic Co-operation and Development, 1990.

Kettle, John. "The Kettle Text." Unpublished text on the history of Bell. 2 vols. N.d. (probably 1975). BCA unclassified.

King, W.L. Mackenzie. *Industry and Humanity.* Toronto: Macmillan Company of Canada, 1918, 1935.

King, W.L. Mackenzie, and John Winchester. *Report of the Royal Commission on a Dispute Respecting Hours of Employment between the Bell Telephone Company of Canada Ltd and the Operators at Toronto, Ont.* Ottawa: Government Printing Bureau, 1907.

Kingsbury, J.E. *The Telephone and Telephone Exchanges: Their Invention and Development.* London: Longmans, Green and Co., 1915.

Klie, Robert H., ed. *Telecommunications Transmission Engineering. Vol. 3: Network and Services.* United States: Bell System Center for Technical Education, 1975.

Lacroix, Jean-Guy, and Robert Pilon. *Câblodistribution et télématique grand public.* Montreal: GRICIS, n.d. (probably 1983).

Lafrance, Jean-Paul. *Le câble ou l'univers médiatique en mutation.* Montreal: Québec/Amérique, 1989.

Lalonde, André. "Riel: Defeated by the Telegraph?" *In Search* (Winter 1977).

Law, D.S. *Development of CN Telecommunications.* CN Public Relations, 1962.

Leacy, F. H. *Statistiques historiques du Canada.* Toronto: Macmillan Canada and Statistics Canada, 1965 (Code T327–341).

Lebel, Monique J. *Québec-Téléphone (de ses origines à nos jours).* Unpublished ms. Rimouski, August 1969. BCA, no. 30381.

Lecour, E.E. "Cadin Pinetree." *Attention Arrow/Le Pointeur* (Ottawa, Department of National Defence) Spring 1990.

Leinwoll, Stanley. *From Spark to Satellite: A History of Radio Communication.* New York: Charles Scribner's Sons, 1979.

Leland Rhodes, Frederick. *Beginnings of Telephony.* New York: Harper and Brothers, 1989.

Lévesque, Sylvie. *60ᵉ (1927–1987).* Historical brochure. Rimouski: Québec-Téléphone, June 1987.

Lester, Alex G. "Telecommunications in Canada." *Engineering Journal* (October 1954).

– "Special Contract: A Story of Defence Communications in Canada." Unpublished ms. Montreal, 1976. BCA Defence File.

– *Evidence of A.G. Lester.* Ottawa: Restrictive Business Practices Commission, January 1970.

Lester, Normand. *L'affaire Gerald Bull (Les canons de l'Apocalypse)*. Montreal: Méridien, 1991.

Libois, Louis-Joseph. *Genèse et croissance des télécommunications*. Paris: Masson, 1983.

"Life Begins at 60." Interview with John C. Lobb. *Forbes* (15 June 1974).

Logan, H.A. *Trade Unions in Canada: Their Development and Functioning*. Toronto: Macmillan Company of Canada, 18948.

Lubberger, F. *Les Installations téléphoniques automatiques*. Paris: Gauthiers-Villars, 1927.

Lussato, Bruno. *Le défi informatique*. Montreal: Sélect, 1981.

Macchi, César, and Jean-François Guilbert. *Téléinformatique (Transport et traitment de l'information dans les réseaux et systèmes téléinformatiques)*. Paris: Dunod, 1979.

Macdonald, J. Stuart. *The Dominion Telegraph*. The Canadian North-West Historical Society Publications, vol. 1, no. 6, Battleford, Saskatchewan, 1930.

Maddox, Barbara. "Women and the Switchboard." In *The Social Impact of the Telephone. The Social Impact of the Telephone*, edited by Ithiel de Sola Pool. Cambridge, MA: MIT Press, 1977.

Maitland, Donald. *Le chaînon manquant*. Report of the Independent Commission on World Development of Telecommunications. Geneva: International Telecommunications Union, December 1984.

Manitoba Telephone Service. *People of Service. A Brief History of the Manitoba Telephone System*. Winnipeg, n.d. [1980].

Marcotte, J.G.M.M. "The North Warning System." *Canadian Defense Quarterly* (Summer 1988): n.p.

Markus, Charles. "Les télécommunications en Grande-Bretagne et leur cadre réglementaire." *Télécommunications 55* (April 1985): 63–7.

Marquez, V.O. "Building an Innovative Organization – Wanted: Small Catastrophes." *The Business Quarterly* (University of Western Ontario) 37, no. 4 (Winter 1972).

Martin, Michèle. "Communications and Social Forms: A Study of the Development of the Telephone System, 1876–1920." Doctoral dissertation. University of Toronto, 1987.

– *Hello, Central? Gender, Technology, and Culture in the Formation of Telephone Systems*. Montreal: McGill-Queen's University Press, 1991.

McCabe, Gerald Michael. "Regulation of the Telephone Industry in Canada: The Formative Years." Master's thesis. Montreal, McGill University, 1985.

McCarthy, Thomas E. *The History of GTE: The Evolution of One of America's Great Corporations*. Stamford, CT: GTE Corporation, 1990.

McKay, A.M. "The History of Telephone Service in Nova Scotia." Summary of five-volume study by the ex-president of MT&T, A.M. McKay, entitled "The History of Maritime Tel and Tel (1877–1964)." Unpublished. Halifax: Transportation Department, July 1983.

– "Abbreviated History." N.p., n.d.

– "The History of Maritime Tel and Tel (1877– 1964)." Unpublished. Halifax: Government & Nova Scotia, Department of Transportation, 1983. MT&T archives.

McLuham, Marshall. Preface. In Harold Innis. *The Bias of Communications* Toronto: University Press, 1984.

McManus, John C. "Federal Regulation of Telecommunications in Canada." In *Telecommunications for Canada: An Interface of Business and Government*, edited by Edward H. English, 419–25. Toronto: Methuen, 1973.

McPhail, Thomas L., and David C. Coll, eds. *Canadian Developments in Telecommunications: An Overview of Significant Contributions*. Calgary: University of Calgary, Department of Communications, 1986.

Messier, Michel. "Intervention étatique et restructuration de l'industrie des services de télécommunications au Canada." Master's thesis. Université du Québec à Montréal, 1991.

Meyer, Hugo Richard. *Public Ownership and the Telephone in Great Britain.* London: Macmillan, 1907.

Michaelis, Anthony R. *Du sémaphore au satellite.* Geneva: Union internationale des télécommunications, 1965.

Middleton, Jesse Edgar, ed. *The Municipality of Toronto: A History.* 2 vols. Toronto: Dominion Publishing Company, 1923.

Moir, John S. *History of the Royal Canadian Corps of Signals.* Ottawa: Corps Committee, Royal Canadian Corps of Signals, 1962.

Mongeau, Jean-Pierre. *Enquête fédérale-provinciale sur la tarification des télécommunications et l'universalité d'un service téléphonique abordable.* Hull: Supply and Services Canada, 1986.

Morin, Claude. *L'art de l'impossible (La diplomatie québécoise depuis 1960).* Montreal: Boréal, 1987.

Morrison, James H. *Wave to Whisper: British Military Communications in Halifax and the Empire, 1780–1880.* History and Archaeology, no. 64. Ottawa: Parks Canada, 1982.

Muir, Gilbert A. "A History of the Telephone in Manitoba." *Historical and Scientific Society of Manitoba*, 1964–65.

Murray, John. *A Story of the Telegraph.* Montreal: John Lowell & Son, 1905.

NB Tel. *Confiance et Service: NB Tel (1888–1988).* Centenary brochure. St John: NB Tel, 1988.

Newfoundland Telephones. *Along These Lines: A History of Newfoundland Telephones.* Brochure. St John's: Newfoundland Telephones, n.d.

Newman, Peter C. *Nortel: Past, Present Future.* Toronto: Nortel, 1995. 2d ed. 1996.

Newman Kuyek, Joan. *The Phone Book: Working at the Bell.* Toronto: Between the Lines, 1979.

Nichols, M.E. *The Story of the Canadian Press.* Toronto: Ryerson Press, 1948.

Nova Scotia. Department of Transport. *The History of Telephone Service of Nova Scotia.* Halifax, 1983.

OECD. *Indicateurs de performance pour les exploitants de télécommunications publiques.* Report no. 22. Paris, 1990.

– *Le service universel et la restructuration des tarifs dans les télécommunications.* Report no. 23. Paris, 1991.

– *Les équipements de télécommunications: transformation des marchés et des structures des échanges.* Report no. 24. Paris, 1991.

Ogle, E.B. *Long-Distance Please.* Toronto: Collins, 1979.

O'Neill, E.F., ed. *A History of Engineering and Science in the Bell System: Transmission Technology (1925–1975).* United States: AT&T Bell Laboratories 1985.

Owen, H.G. *Cent ans déjà, Bell 1980.* Montreal: Bell Canada, 1980.

– *The First Century of Service.* Montreal: Bell Canada, 1980.

– "The Fortunate Isle." *The Blue Bell* 42, no. 9 (October 1983).

– "The Newfoundland Project." *The Blue Bell* (March 1945).

Parsons, George M. "History of Labour Relations in the Bell Telephone Company of Canada, 1880 to 1962." Unpublished ms. Montreal, 1963. CBA #18919.

Patten, William. *Pioneering the Telephone in Canada.* Montreal: Privately published, 1926.

Pelton, Joseph N., and Wu, William W. "The Challenge of 21st Century Satellite Communications: INTELSAT Enters the Second Millennium." *IEEE Journal on Selected Areas in Communications* SAC-5, no. 4 (May 1987).

Pike, Robert, and Mosco, Vincent. "Canadian Consumers and Telephone Pricing: From Luxury to Necessity and Back Again?" *Telecommunications Policy* 10, no. 1 (March 1986): 17–32.

Poitras, Claire. "La construction des réseaux dans la ville: l'exemple de la téléphonie à Montréal, de 1879 à 1930." Ph.D. dissertation. University of Montreal, 1996.

Québec-Téléphone. "Une compagnie semi-centenaire se raconte dans ses livres." *Échange* (Québec-Téléphone ?????? newsletter) 5, no. 2 (n.d.).

– "Jalon chronologique." Unpublished. Québec-Téléphone, 1986. Archives Québec-Téléphone.

Raby, Ormond. *Radio's First Voice: The Story of Reginald Fessenden.* Toronto: Macmillan Co. of Canada, 1970.

Racicot, Michel. "Évolution de la situation de la concurrence dans le domaine des télécommunications au Canada." Unpublished study for Clarkson Tétreault. Montreal, March 1987.

The Regulation of Bell Canada: History; Comparisons with Other Canadian and US Companies. Paper 37, 31 August 1977.

Reid, James D. *The Telegraph in America.* New York: Derby Brothers, 1979.

Rens, Jef. *Rencontres avec le siècle.* Gembloux, Belgium: Duculot, 1987.

Restrictive Trade Practices Commission. *Les Télécommunications au Canada (Partie I – Interconnexion).* Ottawa: Consumer and Corporate Affairs Canada, 1981.

– *Les télécommunications au Canada, Partie II: La proposition de réorganisation de Bell Canada.* Ottawa: Consumer and Corporate Affairs, 1982.

– *Les télécommunications au Canada, Partie III: Répercussion de l'intégration verticale dans l'industrie du matériel de télécommunications.* Ottawa: Consumer and Corporate Affairs Canada, 1983.

Richeson, D.R. *Le télégraphe électrique au Canada (1846–1902).* Histoire du Canada en images, vol. 52. Ottawa: National Museum of Man, National Film Board of Canada, n.d.

Robin, Gérard. *Les Télécommunications.* Paris: PUF, "Qui Sais-je?," 1985.

Sampson, Anthony. *The Sovereign State of ITT.* New York: Stein and Day 1973.

Sangster, Joan. "The 1907 Bell Telephone Strike: Organizing Women Workers." In *Rethinking Canada: The Promise of Women's History,* edited by Veronica Strong-Boag and Anita Clair Fellman. Toronto: Copp Clark Pitman, 1986.

SaskTel, Public Affairs. *Answering the Call (The History of Telecommunications in Saskatchewan).* Regina: Sasktel, 1988.

Shacht, John N. "Toward Industrial Unionism: Bell Telephone Workers and Company Unions, 1919–1937." *Labor History* 16 (Winter 1975): 11–13.

Sharlin, Harold I. *The Making of the Electrical Age, from the Telegraph to Automation.* Toronto: Abelard-Schuman, 1963.

Sherman, L.R. *La concurrence dans le service téléphonique interurbain publique qu Canada.* Report by the Federal-Provincial-Territorial Working Group on Telecommunications. Ottawa, 1988.

Simon, Samuel A. *After Divestiture: What the AT&T Settlement Means for Business and Residential Telephone Service.* White Plains, NY: Knowledge Industry Publications, 1985.

Sinclair, Bruce, Norman R. Ball, and James O. Peterson, eds. *Let Us Be Honest and Modest: Technology and Society in Canadian History.* Toronto: Oxford University Press, 1976.

Smith, Arthur Bessey. *Early History of the Automatic Telephone (1879–1906).* Dublin, CA: Telephone History Press.

Smith, George David. *The Anatomy of a Business Strategy: Bell, Western Electric, and the Origins of the American Telephone Industry.* Baltimore: Johns Hopkins University Press, 1985.

Sola Pool, Ithiel de. *Forecasting the Telephone: A Retrospective Technology Assessment of the Telephone.* Norwood, NJ: Ablex Publishing Corporation, 1983.

– *Technologies of Freedom: On Free Speech in an Electronic Age.* Cambridge, MA: Belknap Press of Harvard University Press, 1983.

Sola Pool, Ithiel de, ed. *The Social Impact of the Telephone.* Cambridge, MA: MIT Press, 1977.

Spafford, Dufferin S. "Telephone Service in Saskatchewan: A Study in Public Policy." Master's thesis, University of Saskatchewan, 1961.

St-Arnaud, Diane. "La mise en œuvre en droit canadien des réglementations et conventions internationales en matière de télécommunications spatiales." Master's thesis. Montreal, McGill University, 1991.

Stinson, Margaret. *The Wired City: A History of the Telephone in Edmonton*. Edmonton: Edmonton Telephones, 1980.

Strong-Boag, Veronica, and Anita Clair Fellman, eds. *Rethinking Canada: The Promise of Women's History*. Toronto: Copp Clark Pitman, 1986.

Sturges, James. *Adam Beck*. Don Mills, ON: Fitzhenry and Whiteside, 1982.

Surtees, Lawrence. *Pa Bell: A. Jean de Grandpré and the Meteoric Rise of Bell Canada Enterprises*. Toronto: Random House, 1992.

– *Wire Wars (The Canadian Fight for Competition in Telecommunications)*. Scarborough: Prentice Hall Canada, 1994.

Taylor, Graham D. "Charles F. Sise, Bell Canada, and the American: A Study of Managerial Autonomy, 1880–1905." *Communications historiques*. Ottawa: La Société historique du Canada, 1982.

Télécommission. *L'univers sans distance*. Communications Canada, 1971.

"Telephone Business Unlike Any Other." *Telephone Gazette* (Bell Telephone Company & Canada, Montreal), no. 6 (October 1909).

Temin, Peter. *The Fall of the Bell System: A Study in Prices and Politics*. New York: Cambridge University Press, 1987.

Tillotson, Shirley Maye. "Canadian Telegraphers, 1900–1930: A Case Study in Gender and Skill Hierarchies." Master's thesis. Kingston, Queen's University, 1988.

Tunstall, Brooke W. *Disconnecting Parties*. New York: McGraw-Hill, 1985.

Union Relations Department, Bell Canada. "Chronologie des négociations avec le STCC." Internal document. Work Relations file, BCA, 1982.

– "Chronologie des négociations avec les téléphonistes." Internal document. Work Relations file, BCA, 1982.

– "Chronologie des négociations avec l'ACET." Internal document. Work Relations file, BCA, 1982.

United States. Department of Commerce. Bureau of the Census. Annual reports, Western Union.

Vail, Theodore Newton. *Views on Public Questions: A Collection of Papers and Addresses*. Privately published, 1917.

Vice, David G. "The Arcom Terminal: Prototype for Future Satlellite Communication Earth Stations." *Telesis* 1, no. 2 (May 1968).

Wade, Mason. *Les Canadiens français de 1760 à nos jours*. 2 vols. Montreal: Le Cercle du Livre de France, 1966.

Waller, Adrian. "Unsung Genius." *Equinox* (March/April 1989).

Warner, Donald G. "The First Fifty Years of the Canadian Telegraph: A Geographical Perspective." Undergraduate thesis. University of Toronto 1975. CN Archives (NAC).

Watson, Thomas. *The Birth and Babyhood of the Telephone*. New York: AT&T, 1951.

Waverman, Leonard. *The Process of Telecommunications Regulation in Canada*. Working Paper no. 28. Ottawa: Economic Council of Canada, January 1982.

Weir, Austin E. *The Struggle for National Broadcasting in Canada*. Toronto and Montreal: McClelland and Stewart, 1965.

Williams, Archibald. *Telegraphy and Telephony.* London: Thomas Nelson and Sons, 1928.

Williams, Jack. *The Story of Unions in Canada.* Toronto: J.M. Dent and Sons, 1975.

Williams, James Earl. "Labor Relations in the Telephone Industry: A Comparison of the Private and the Public Segments." Doctoral dissertation, University of Wisconsin, 1961.

Williams, Trevor, ed. *A History of Technology.* Vol. 7: *The Twentieth Century, c. 1900 to c. 1950, Part II.* Oxford: Clarendon Press, 1978.

Witteveen, Hank. "The Telegraph in Canada." Undergraduate thesis. Ottawa, Carleton University, 1973.

Index